The Practitioner's Guide to POLCA

The Production Control System for High-Mix, Low-Volume and Custom Products

The Practitioner's Guide to POLCA

The Production Control System for High-Mix, Low-Volume and Custom Products

Rajan Suri

CRC Press is an imprint of the
Taylor & Francis Group, an **informa** business
A PRODUCTIVITY PRESS BOOK

CRC Press
Taylor & Francis Group
6000 Broken Sound Parkway NW, Suite 300
Boca Raton, FL 33487-2742

First issued in hardback 2019

ISBN-13: 978-1-138-21064-6 (hbk)

Library of Congress Cataloging-in-Publication Data

Names: Suri, Rajan, editor.
Title: The practitioner's guide to POLCA : the production control system for high-mix, low-volume and custom products / editor, Rajan Suri.
Description: Boca Raton : Taylor & Francis, CRC Press, 2018. | Includes bibliographical references and index.
Identifiers: LCCN 2017045110| ISBN 9781138210646 (hardback : alk. paper) | ISBN 9781351170765 (eBook)
Subjects: LCSH: Production control. | Lean manufacturing. | Card system in business. | New products.
Classification: LCC TS155.8 .P68 2018 | DDC 658.5--dc23
LC record available at https://lccn.loc.gov/2017045110

Visit the Taylor & Francis Web site at
http://www.taylorandfrancis.com

and the CRC Press Web site at
http://www.crcpress.com

To the industry professionals who believed in the potential of POLCA and devoted a lot of their time and effort to prove that it would work—in many cases risking their careers and even their companies in the belief that POLCA would deliver significant results. Without your faith and support, and the resulting successes, this book could not have been written.

Contents

Acknowledgments xv

PART I UNDERSTANDING POLCA

1 **POLCA—Simple Yet Highly Effective: A Tale of Two Companies....3**
Overview of This Book 7
Guide to the Reader 8

2 **POLCA Explained: How It Works 11**
Introducing the MMC Factory Example 12
Authorization Date and Authorization List 15
Paired-Cell Loops of Cards 16
Cell Team's Decision on What to Work on Next 18
Overlapping Loops 21
Differences from MRP and Kanban Systems 23
Putting It All Together with an Example 24

3 **Operational Benefits of POLCA 27**
Tailored to High-Mix, Low-Volume and Custom Production 27
No Complex Software Implementation Required 29
Works With Your ERP System—or Without It 30
Makes Effective Use of Capacity at Each Step 31
Avoids Congestion and Excessive WIP Build-Up 32
System Is Adaptive 33
By Holding Off on Jobs the System Results in Multiple Benefits 33
Provides Formal Framework for Improvement Activities 36
Prevents Inventory Proliferation for Products with Low Demand 37
Allows Complete Flexibility in Product Routings and Works Seamlessly for Custom Products 39
Creates Customer-Supplier Relationships between Cells 40

Installs a Visual System....42
Builds on the Capabilities of People in the System....42
Provides Fast Response through Decentralized Decision-Making....44
Builds on Your Existing Structure and Systems....44

PART II IMPLEMENTING POLCA

4 Pre-POLCA Assessment, Prerequisites, and Preparation for POLCA....49
Clarify the Motivation and Goals for Pursuing POLCA....49
Check for Symptoms That Have Led Companies to Consider POLCA....50
Review Your Company's Operating Environment....51
Set Your Goals....52
Check These Prerequisites Before Deciding to Implement POLCA....53
Review Your Shop Floor Organizational Structure....54
Make Sure That You Have POLCA-Enabling Cells....55
Engage in Effective Planning Before Control....57
Limit the Occurrence of Material and Part Shortages....59
Prepare for the POLCA Implementation....61
Ensure the Availability of Authorization Dates and Lists....61
Implementation Without Using an ERP System....62
Implementation through Your ERP System....64
Ensure That Manufacturing Metrics Will Not Conflict with POLCA Rules....67
Secure the Upfront Understanding and Commitment of Management....68

5 Designing the POLCA System for Your Situation....71
Create an Implementation Team and Designate a POLCA Champion....72
Begin by Determining All the POLCA Loops....72
Implementing POLCA throughout Your Shop Floor....73
Do Not Include Stocking Points in the POLCA Loops....77
Implementing POLCA in a Subset of Your Shop Floor....78
Should You Put in Single or Multiple Loops for a Given Area?....82
Lay Out the POLCA Chains....83
Starting Point for the POLCA Chains....83
Including the Lead Time for the Planning Cell....85
Starting Points for POLCA Chains When Implementing in a Subset of Your Shop Floor....87

Ending Points for the POLCA Chains 89
Define the Procedure for Incorporating Outside Operations into the POLCA System 92
Consider Whether You Will Use the Simpler Release-and-Flow POLCA (RF-POLCA) Version 94
Situations Where Standard POLCA Is Recommended 97
Decide on Details Related to the POLCA Cards 99
Design Your POLCA Cards 99
Decide on the Quantum—The Amount of Work to be Done per POLCA Card 100
Determine the Process for Returning the POLCA Cards 103
Calculate the Initial Number of POLCA Cards in Each Loop 106
Add the Finishing Touches with Some Final Details 109
Identify the Triggers for Decision Time 109
Hedging with Lead Time or with Cards? 113
Adding Safety through Lead Time 113
Adding Safety through Number of Cards 114

6 Accommodating Exceptions within the POLCA Framework ... 119
Dealing with an Unexpected Shortage of a Component Part 120
Introducing the Safety Card Concept 122
Procedure When a Safety Card Is Available 124
Adding a Shortage Tracker 125
Procedure When a Safety Card Is Not Available 126
Dealing with an In-Process Quality Problem 127
Dealing with Machine Downtimes 128
Expediting a Rush Job for an Urgent Customer Need 129
Accommodating Schedule Changes Seamlessly 131
Adjusting for Non-Synchronized Delivery of In-House Manufactured Components to Assembly 133
Preventing Gridlock When Loops Form Cycles 135
Summary: POLCA Accommodates Exceptions Smoothly 140

7 Launching POLCA 143
Continue to Use the POLCA Implementation Team and the POLCA Champion 143
Conduct Training for Everyone Impacted by the System 145
Training for the Core Group 145
Training for the Rest of the Organization 148

Create a Standard Process for Secondary Activities Initiated by Decision Time Rules 150
Schedule Frequent Review Sessions and Management Updates 151
Have a Plan for Rolling Out Adjustments and Corrections 151
Design the Checks That Will Ensure POLCA Discipline Is Followed ... 152
Start Tracking and Debugging Key Metrics 153

8 How We Designed and Launched POLCA in Three Days 157
GUEST AUTHOR: ANANTH KRISHNAMURTHY
The Setting: A Factory in Canada 157
Day One: POLCA Overview and Factory Tour 158
Day Two: Checking the Prerequisites and Some Initial Design 159
Checking the Remaining Prerequisites 160
Initial Design: The POLCA Loops 161
Deciding on the Quantum 161
Designing the POLCA Card 162
Day Three: Final Design Decisions 163
Defining the Decision-Time Rules and Exceptions 164
Decision to Launch 165
Postscript: Keys to Success and a Challenge to Others 166
About the Author 167

9 Sustaining POLCA: Post-Implementation Activities 169
Create a Steering Committee and Stay the Course 169
Track the Key Metrics and Celebrate Successes 172
Use Surveys and Feedback Forms to Assess the Qualitative Benefits ... 173
Conduct Regular POLCA Audits 175
Periodic Audits 175
Random Audits 178
Use POLCA to Drive Long-Term Improvements 179

PART III INDUSTRY CASE STUDIES

10 From the United States: Using POLCA to Eliminate Material Flow Chaos in an Aluminum Extrusion Operation 183
GUEST AUTHORS: JEFF CYPHER AND TODD CARLSON
The Start of Our POLCA Journey 185
Experimenting with a Small-Scale POLCA Implementation 187
System Design and Training 189
System Startup and Early Results 192
Extending POLCA to the Whole Factory 193

What We Learned, and Advice to Other Practitioners......194
Impact of POLCA on Our Organization......196
About the Authors......197

11 From Canada: Applying POLCA in a Pharmaceutical Environment......199
GUEST AUTHORS: JUSTIN BOS AND SUSAN FERRIS
Why Did We Pursue POLCA?......200
The POLCA Kaizen Workshop......201
Deciding on All the Fundamentals of the POLCA System......202
Rules to Break Rules......203
Senior Leadership Plays the POLCA Game......205
Training, Training, Training......206
The POLCA Boards and POLCA Dashboard......206
Life with POLCA......209
About the Authors......210

12 From the Netherlands: Creating Flow and Improving Delivery Performance at a Custom Hinge Manufacturer through QRM and POLCA......211
GUEST AUTHORS: GODFRIED KAANEN AND ROBERT PETERS
SECTION I: THE JOURNEY TO POLCA, by Godfried Kaanen......211
I Get a Rude Wake-Up Call......212
Our Initial Improvement Efforts......212
The Pastry Shop Needs a Breakthrough......213
Our First Step: The Rhineland Model......215
We Are Introduced to POLCA......216
Implementing the Color-Coded Visual System......219
Initial Results from POLCA......222
SECTION II: FROM PHYSICAL POLCA TO DIGITAL POLCA, by Robert Peters......223
Implementing the Quantum through Load-Based POLCA......227
Impact of PROPOS and POLCA on the Company's Operations......229
About the Authors......230

13 From Belgium: Metalworking Subcontractor Reduces Lead Times by 85% Using QRM and POLCA......233
GUEST AUTHORS: PASCAL POLLET AND BEN PROESMANS
Motivation for Rethinking the Shop Floor System......233
Implementing the Stove Cell......234
Introducing Color-Coded Visual Work Management......235

Expanding to the Rest of the Factory with POLCA 238
Customizing the System for Better Use of Scarce Resources 240
Results: The People Side 241
Results: The Commercial Side 242
About the Authors 243

14 From Poland: Dancing the POLCA in a Glass Factory 245
GUEST AUTHOR: KAROL BĄK
The Factory and Process Flows 246
The Cutting Operations and Resulting Inventory 247
The Machining Department and Sequencing Problems 248
Finishing Processes 249
Initial Design of the POLCA System 249
Initial POLCA Loops, Quantum, and Number of Cards 249
Authorization Method 251
Simplifying the Design: Moving to POLCA "Lite" 251
Results of the POLCA Implementation 253
About the Author 254

15 From Germany: Yes, Even Small Companies Can Benefit from Implementing POLCA 255
GUEST AUTHOR: MARKUS MENNER
Evaluating the Company's Situation 256
Opportunities and Challenges for Implementing POLCA 257
Cell Formation and POLCA Launch 258
Results from the Implementation 260
About the Author 261

Introduction to the Appendices 263

Appendix A: Introduction to MCT: A Unified Metric for Lead Time 265

Appendix B: Overview of Quick Response Manufacturing (QRM) 285

Appendix C: Perspectives on Applying MRP, Kanban, or Some Other Card-Based Systems to HMLVC Production Environments 301

Appendix D: Capacity Clusters Make High-Level MRP Feasible 323
GUEST AUTHORS: IGNACE A.C. VERMAELEN AND ANTOON VAN NUFFEL

Appendix E: The Sociotechnical Success Factors of POLCA 353
GUEST AUTHOR: JANNES SLOMP

Appendix F: Experiences with Using a POLCA Simulation Game ... 365
GUEST AUTHOR: HANS GERRESE

Appendix G: Explanation of Formula for Number of POLCA Cards ... 375

Appendix H: POLCA: A State-of-the-Art Overview of Research Contributions ... 381
GUEST AUTHOR: JAN RIEZEBOS

Appendix I: Additional Resources for POLCA Training and Games ... 423

Index ... 425

About the Author ... 435

Acknowledgments

I am deeply indebted to many people who helped transform POLCA from a theoretical concept to a practical working system with proven results. In 1998, when I first described the complete concept of POLCA in a chapter in my book, *Quick Response Manufacturing*, it was an idea based purely on my intuition about manufacturing systems. I had devised the details of the system using thought experiments based on my knowledge of the dynamics of interconnected systems combined with my experiences about the behavior of people in production environments. Although I managed to put together a lot of details on the design of POLCA—the chapter consisted of over 40 pages—there had been no practical trials of this system at that point. Thus, it required a leap of faith and considerable courage for people from industry to implement the system and try it out. A few pioneers from industry, assisted in their efforts by researchers from academia, implemented POLCA in their factories and proved that the system not only worked but produced results even beyond our combined expectations.

I am grateful to the following individuals from both industry and academia who were instrumental in installing the earliest implementations of POLCA and/or exploring its potential through research activities. Also included in this list are several people who reviewed a first draft of this book and provided extensive comments that greatly helped to improve the content. In alphabetical order (by first name), I would like to thank Aije Klaren, Al Drifka, Al Hetchler, Ananth Krishnamurthy, Ankur Jain, Antoon van Nuffel, Arthur Vandebosch, Ashesh Sinha, Ashish Chaudhari, Atul Tripathi, Azmi Issa, Bashar Eljawhari, Ben Bomstad, Ben Proesmans, Bill Molek, Bill Ritchie, Bob Dempsey, Bob Mueller, Brad Miller, Brian Korrison, Brian Larson, Brian Lenderink, Brian Rosenboom, Carl Treankler, Chris Wood, Christy Nunes, Craig Freedman, Cynthia Bruns, Dale Sunstrom, Dan Tate, Dave Holmgren, David Groom, Debjit Roy, Deng Ge, Derek

Bailey, Devika Suri, Diederik Claerhout, Doug Chadwick, Duygu Unlu, Eric Hengst, Ferry Senten, Francisco Tubino, Frank Rath, Gary Lofquist, Gerardo Hernandez, Godfried Kaanen, Greg Dettman, Greg Diehl, Greg Sandeno, Gregg Jacob, Hans Gerrese, Haoxun Chen, Hendri Kortman, Ignace Vermaelen, Inneke Vannieuwenhuyse, Jacob Pieffers, Jan Riezebos, Jannes Slomp, Jason Bachman, Jason Peto, Jeff Amundson, Jeff Matti, Jeff Cypher, Jennifer Anderson, Jessica Nguyen, Jens Johansen, Jerry Neilson, Jitesh Mehta, John Grosskopf, Jon Denzin, Justin Bos, Karissa Kletzien, Karol Bąk, Kathy Pelto, Keith Newman, Ken Blodgett, Kristof Souwens, Kurt Norling, Larry Cronce, Lynn Van Dyke, Maciej Malewicz, Malini Suri, Mark Dawson, Mark Haslam, Markus Menner, Marty Grell, Matthias Preter, Merve Ozen, Mick Petzold, Mike Kiens, Mike Nasiopulos, Mike Plotecher, Mike Schneider, Neha Patil, Nico van der Dussen, Nico Vandaele, Nicole Suri, Okan Gürbüz, Pascal Pollet, Paul Diedrick, Paul Slater, Paul van Veen, Peter Jahn, Peter Luh, Ragini Saxena Peters, Rahul Shinde, Ravi Suman, Rob Herman, Robert Peters, Robert Tomastik, Robert Yoerger, Rony Cremmery, Rusi Debu, Ryan Mijal, Sanket Bhat, Scott Gilson, Shyam Bhaskar, Sriramkumar Thangavel, Steve Fladwood, Steve Honerlaw, Sue Klingaman, Susan Ferris, Sushanta Sahu, T.J. Doppler, Teri Blumenthal, Terri Walter, Terry Hanstedt, Thomas Luiten, Todd Carlson, Todd Hansel, Tom Groose, Tom Schabel, Travis Kalous, Tugce Martagan, Vinay Kulkarni, Vince Caira, Xiaohui Zhou, and Yasemin Limon.

The folks at C&M Printing—Mike Londerville, Jeanette Londerville, and Natalie Reilly—provided much-needed support in the printing, binding, and distribution of several early drafts of my manuscript.

In closing, I would like to thank Michael Sinocchi of Productivity Press for encouraging me to write this book, and also thank the team consisting of Alexandria Gryder and Jay Margolis at Taylor & Francis, Jonathan Achorn of the Manila Typesetting Company, and freelance graphic artist John Gandour for their help during the production and printing of the final bound copy.

UNDERSTANDING POLCA

Chapter 1

POLCA—Simple Yet Highly Effective: A Tale of Two Companies

Although they were separated by half a continent and a major ocean, Alexandria Industries in Minnesota and Bosch Hinges in the Netherlands were facing exactly the same problems. Both companies produced a high variety of products in low volumes—Alexandria made aluminum extrusions, and Bosch manufactured specialty metal hinges—and both had long lead times of six to eight weeks. Despite having this large window of time, both organizations were struggling with late deliveries, which required daily expediting efforts in attempts to get jobs out on time. For several years the two companies had tried a plethora of manufacturing and control strategies but none had alleviated this ongoing problem. Then, although separated by some 5,000 miles, management at each company independently decided to implement POLCA, which was a relatively novel idea at that time. Within a few months both companies had reduced their quoted lead times by more than 50%, and, even with this smaller time window, they were able to achieve near-perfect on-time delivery. And thus it was that these two companies joined the ranks of the pioneers of POLCA in both the new world and the old one.

The benefits of POLCA implementation were not confined to the shop floor; at each company, it had larger implications for the business as a whole. With customers taking note of their shorter lead times and excellent delivery records, both Alexandria Industries and Bosch Hinges experienced dramatic increases in sales. In addition, the reduction in overhead activities of rescheduling and expediting jobs as well as the reduction in expenses such as rush shipments to customers meant that the two companies also saw significant increases in profitability. Hours of management time—previously spent in "hot job" meetings and related expediting efforts—were liberated, allowing senior managers to focus on more strategic efforts that helped improve and grow the business. Finally, the reduction in work-in-process (WIP) by more than 50%, combined with getting jobs to customers quickly, meant big improvements in cash flow for both companies. More details on the POLCA implementations and results achieved at these two companies will follow in Chapters 10 and 12.

So why was POLCA able to achieve, in a relatively short time, what these two companies had been unable to accomplish for many years? The answer to this question provides an excellent executive summary of the strengths of POLCA. First, let's understand what POLCA does for a company. POLCA is a card-based visual control system that manages the flow of jobs through the shop floor: at each of the main operations it controls which job should be worked on next in order to meet delivery targets. Specifically, it ensures that upstream operations use their capacity effectively by working on jobs that are needed downstream, while at the same time preventing excessive WIP build-ups when bottlenecks appear unexpectedly. POLCA is particularly effective in high-variability environments where the variability arises for external reasons such as unpredictable demand, many different types of orders, and dynamic changes in customer requirements, as well as internal reasons such as changes in equipment and labor status and other day-to-day conditions on the factory floor.

Next, to get an overview of why POLCA is so effective, we need to understand the acronym: POLCA stands for *Paired-cell Overlapping Loops of Cards with Authorization*. Each of these terms signifies a key feature of POLCA. As background, POLCA is aimed at companies that have already organized their shop floor into several manufacturing cells; however, it can also be used with companies that still have individual workcenters, or even a combination of cells and standalone workcenters. The three key features of POLCA that we mention at this stage are:

1. POLCA connects pairs of cells (or workcenters) with loops that contain circulating cards, which provide capacity signals between each pair of cells. These signals help with the decision regarding which job should be worked on next at each cell.

2. The operating rules for the loops are intentionally designed so that for a job going through several cells the loops for this job actually overlap, like interlocking links in a chain.
3. In addition to the card signals, in deciding which job to work on next, the cells also need to check the "Authorization Dates" of jobs—these are dates calculated for each cell that the job visits, based on the required completion date for a job.

You can see that these three features are directly related to the terms in the full name for POLCA, i.e., *Paired-cell Overlapping Loops of Cards with Authorization*. More details on these features and how they work will be provided in the next chapter. The purpose here is to explain the acronym and provide enough initial information for you to understand the points that follow.

We can now summarize the main factors that contribute to the success of POLCA.

- **POLCA is designed for high-mix, low-volume and custom (HMLVC) production environments.** In today's world, companies are increasingly seeing demand for smaller batches, higher variety of products, and even customized products that are tailored to each order. Because of the application of computer-aided design and manufacturing, this trend—commonly referred to as "mass customization"—is only going to get more pronounced. Hence, a system aimed at this environment is critical for companies aiming to succeed in tomorrow's production situation. Henceforth we will use the acronym HMLVC (defined above) to refer to this production environment.
- **POLCA is simple and easy to understand.** In the next chapter, you will see that at each cell the team of workers just needs to check three simple conditions to determine which job to run next at this cell.
- **It is a visual system.** In recent years there has been a recognition that visual systems have many advantages. Application of this approach is commonly called "visual management." In keeping with this idea, POLCA uses cards that are attached to jobs or posted on notice boards to give clear visual signals that can be understood by both shop floor workers and management, and which provide instantaneous feedback and evaluation of the current status in any area.
- **It does not require complex software implementation.** There are many sophisticated scheduling software packages available today,

but they suffer from two major drawbacks: (i) because the real world changes frequently, plans that are calculated—even if based on state-of-the-art optimization algorithms—are often obsolete soon after (or even before!) they are released to production; and (ii) complex software systems can take months, or even years, to implement, customize, and work out all the bugs for a given environment.

- **POLCA works with your ERP system—or without it!** In the next chapter, you will see that if you are already using an Enterprise Resource Planning (ERP) system, POLCA can seamlessly work with this system. On the other hand, if you don't have an ERP system, you can still use POLCA just by performing simple calculations for each job—calculations that are easily implemented in a spreadsheet.
- **POLCA builds on the capabilities of humans in the system.** Many so-called "optimum" scheduling systems lose sight of the fact that such systems are deployed by people, and these people need to buy in to the system and support it. This oversight has resulted in ineffective or even failed implementations. In contrast, experience with many companies has shown that people on the front line like using POLCA, and they help to ensure that it works well.
- **It builds on cells, teams, and ownership.** In keeping with the latest manufacturing strategies, many companies are implementing cells and creating self-directed shop floor teams that have ownership of their areas. You will see in the next two chapters how POLCA not only builds on this structure but also takes advantage of it to get improved performance from these teams. In addition, it supports better coordination and cooperation between teams.
- **It uses decentralized decision-making, resulting in agility and responsiveness.** Instead of implementing a centralized planning system that attempts to optimize the whole operation, POLCA uses a simpler central system and pushes real-time decision-making to the cells and teams. This enables fast reaction times to real-world events.
- **POLCA is an adaptive system.** The combination of card signals, decision rules, and decentralized structure used in POLCA is designed so that the system continuously adapts to unforeseen real-world events such as changes in priorities or unanticipated bottlenecks. This adaptation includes outcomes such as: ensuring that upstream operations use their capacity more effectively by working on jobs that are needed downstream; at the same time avoiding building up too much work when bottlenecks appear unexpectedly; adjusting the job sequence at

every operation to ensure that jobs are being processed in the right order to meet their due dates; and signaling to one team that another team might need assistance in order to keep jobs on track for their due dates. In particular, for HMLVC companies with a high variety of jobs that are made infrequently, or custom orders that have not been made before, estimated operation times may not be very accurate, resulting in unplanned bottlenecks. POLCA helps to alleviate such situations in real-time.

All these points will be explained in detail in later chapters and illustrated in practice through case studies from various industries. However, you can see from the above that POLCA offers many benefits and is able to achieve, in a relatively simple way, what many sophisticated scheduling systems have been unable to achieve for decades.

Overview of This Book

This book is intended as a comprehensive guide to both understanding and implementing POLCA. Although the book is primarily aimed at practitioners, academics and researchers will find substantial areas of the book that are relevant to teaching and research.

The remainder of Part I provides an understanding of POLCA. Chapter 2 explains how the system works and, in particular, the key rules that determine the sequence in which jobs will be tackled and how jobs and cards flow through the loops. Then in Chapter 3 we get into more depth on the benefits of POLCA; with a better understanding of how the system works, we can go into more detail than was provided in this chapter and understand how the benefits derive in practice.

Part II is a systematic guide to implementing POLCA. It is intended to serve as a detailed roadmap for teams at companies that wish to implement POLCA, as well as for consultants and academics that might be engaged in advising industrial clients. We begin in Chapter 4 by laying down the prerequisites for POLCA. This is not only to ensure that you are implementing POLCA in the right environment and for the right reasons, but also that some fundamentals are in place before proceeding with POLCA. Chapters 5 and 6 get into the detailed design of the system, including dealing with various special circumstances that might apply in different production environments, as well as practical techniques for dealing with unexpected situations that arise, particularly in HMLVC companies. Next, it is time for the rubber to hit the road. Providing you with the keys to a successful launch of the system is the subject of Chapter 7. Since managers often worry about the length of time that it will take to implement a new system, Chapter 8 presents an inspiring case study of a company that designed and launched its POLCA system *in three days*! Practitioners and consultants are aware of the fact that improvement projects can start out with much enthusiasm, but then lose momentum and even get dropped after some time; therefore, we end Part II of the book with Chapter 9, which provides advice on how to sustain POLCA and ensure that it continues to operate successfully in the long run.

A major goal of this book is to demonstrate that POLCA is not just a theoretical concept; it has been implemented in factories around the world with impressive results. Part III of this book documents six such implementations with case studies from the U.S., Canada, the Netherlands, Belgium, Poland, and Germany. The companies involved range in size from a large multinational to a factory with only a dozen employees! Industrial applications span from the more common metalworking operations such as extrusion, metal-cutting, and fabrication, to more unusual factories such as glass and even pharmaceuticals. It was important to us that you, the reader, should experience these case studies from the eyes of the people who were involved. Thus, the six case studies in Part III are all written by guest authors—senior managers, employees, and consultants at the companies that implemented POLCA—who share their personal journeys, insights, and, of course, results. This also lends credibility to the results presented, as they are directly supported by the writings of these guest authors.

The last part of this book consists of numerous appendices. These will be of particular interest for academics and researchers looking for more details on some aspects, as well as for practitioners or consultants who have an interest in such details. Some of these appendices have also been contributed by guest authors who are experts in their field. You can find an overview of all the appendices at the beginning of that portion of the book.

Guide to the Reader

If you are new to POLCA and wondering if you should invest time in learning more about it, don't be daunted by the size of this book! As a manager or consultant looking to see if POLCA has potential for your business or your clients, you can get sufficient insight by reading this chapter along with

Chapters 2 and 3. You may also find it helpful to read one or two of the industry case studies in Part III, choosing a context that seems closest to yours. This small set of readings, which you could probably accomplish in one or two evenings, should give you sufficient understanding to make the call as to whether you should dive deeper into the details of POLCA in the rest of the book. So, even as a novice or someone who is not yet convinced about the need for POLCA, you can get considerable value from this book in just a few hours of your time.

On the other hand, if you have already decided that you are moving forward with implementing POLCA, or at least seriously exploring the possibility of it, then you should start with Part I and follow it with a sequential reading of Part II, reading the chapters in strict order. Even if you already know a lot about POLCA, it is important to read through Part I first to remind yourself of the basics and also to become familiar with the terminology that will be used in this book. The case studies in Part III and the appendices are all optional; you may read any or all of them based on your needs and interest.

For readers from the academic or research community, again it is important to start with reading all of Part I. Next, you should read Chapters 4 and 5. If you are interested in more practical details, you can also read Chapter 6. The remaining chapters in Part II are less critical for you, unless you will be involved in a practical implementation with a company. However, as an academic or researcher, you will probably find all the appendices to be of value. In particular, for researchers, Appendix H will be an invaluable guide to the related literature.

In summary, regardless of your background or goals, we recommend that you launch into reading the next two chapters of Part I. You will find these chapters easy and quick to read; hopefully they will keep you entertained and at the same time pique your interest into learning more about POLCA and how companies have benefited from using this system.

Chapter 2

POLCA Explained: How It Works

This chapter gives an introduction to the key features of POLCA. The aim is to provide an overview of how POLCA works, so that you can understand the operational benefits of POLCA discussed in the next chapter. Additional details on the POLCA system and on implementing POLCA will follow in Part II of this book.

POLCA is designed to address two significant trends in the manufacturing world today. The first is that, in keeping with the latest thinking on effective organizational structures, companies are reorganizing their shop floor operations into manufacturing cells, which we will henceforth refer to simply as cells. A more precise discussion on cells in the context of POLCA will follow in Chapter 4, but for now, briefly, a cell is a collection of equipment (e.g., machines, workstations, benches, and tools) along with a team of people that is responsible for the operation of this collection of equipment. The aim of the cell is to complete a sequence of operations for a predefined class of jobs.

The second trend addressed by POLCA is that companies are experiencing demand for smaller and smaller order quantities of products, and beyond that, for individually customized products with an order quantity of one. This was already referred to in Chapter 1 as the high-mix, low-volume and custom (HMLVC) environment.

Despite these two major trends, there have been no significant changes in the way that companies use their materials requirements planning (MRP), scheduling, and shop floor control systems. Specifically, traditional planning, scheduling, and control systems do not explicitly take advantage of the cell structure; we will see in this chapter and the next that POLCA uses this structure to simplify product routings and it also takes advantage of the ownership of the cell teams. (Note: companies that do not have cells can still benefit from POLCA as discussed later in this chapter and illustrated with some of the case studies in Part III.) A second point about traditional MRP systems is that they have difficulty in coping with the high degree of variability in an HMLVC company. As explained in more detail in Appendix C, in the HMLVC environment MRP systems can result in lengthening lead times, late deliveries, and an increasing cycle of rush jobs needing to be expedited. In contrast, POLCA is particularly effective in environments with unpredictable demand, a high variety of orders, changes in customer requirements, and unplanned or shifting bottlenecks due to the variability in products being made and day-to-day changes in equipment and labor status. In this dynamic environment, POLCA ensures that upstream operations use their capacity effectively by working on jobs that are needed downstream, while at the same time preventing excessive WIP build-ups when bottlenecks appear unexpectedly. The industry case studies in Part III will demonstrate that after implementing POLCA companies have reduced their lead times as well as achieved high levels of on-time delivery. POLCA achieves these goals via a card-based control system that manages the flow of jobs through the shop floor. We now explain how this system works via a detailed example.

Introducing the MMC Factory Example

POLCA is based on the fundamental recognition that a company operating in the HMLVC environment that has decided to reorganize its shop floor into cells will typically have a structure as depicted in Figure 2.1. This example describes a hypothetical company, Madison Machining Corporation (MMC) that is a contract manufacturer supplying machined parts to several Original Equipment Manufacturers (OEMs). There are three points to be observed in Figure 2.1. First, we see that there are several cells focused on various types of operations and products—for example, machining cells for different types of parts, cells for some finishing operations, and packing. In keeping with the concept of a cell as outlined at the beginning of this chapter, each of these cells consists of different machines that typically complete several operations on a given job. Larger companies could have multiple cells of each of these types, each focused (say) on different materials and

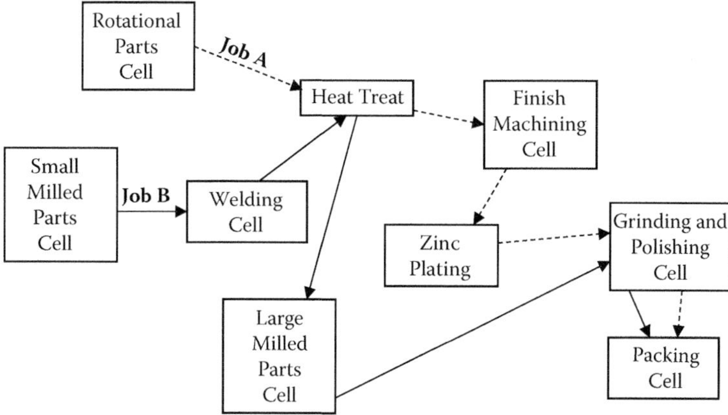

Figure 2.1 Company that supplies machined parts, organized into multiple cells. Examples of two different routings are Job A (with the dashed lines) and Job B (with solid lines).

technologies. Second, we see in the figure that there are some standalone workcenters, like a heat treat furnace and a zinc plating line. And third, we note that a given customer order will be routed to cells that have the appropriate capabilities for that order. In the figure, Job A and Job B are examples of orders with different routings. Extrapolating from this example, we see that even after organizing into cells, a typical HMLVC company could still have a network of many possible routings through its shop floor, along with all the variabilities, such as unpredictable demand and unplanned or shifting bottlenecks, as described above.

In this context, we need to address two important questions related to planning, scheduling, and control of jobs:

1. Should we rethink the MRP system structure if the company is organized into cells?

   Comment: Traditionally, an MRP system deals with operations, but now multiple operations have been combined into new units called cells.
2. In the real-world environment, there are always day-to-day and even minute-to-minute schedule changes—for example, due to quality issues, equipment problems, or customer requirements—and therefore, the original plan that resulted in the current schedule may no longer be the best way to proceed. So how do we ensure that upstream cells are working on jobs that are truly needed by downstream cells?

   Comment: The schedule that has been released may result in upstream cells working on jobs that are not needed till much later, thus creating

> *excess work-in-process (WIP) and congestion. At the same time, jobs that are now needed sooner may not receive the attention that they should from these upstream cells.*

We will see that the POLCA system helps us address both these questions in a simple yet effective way.

Note that, although POLCA is designed to manage the flow of jobs through multiple cells, it can also be used to manage routings that involve standalone workcenters, where jobs go between cells and workcenters, as in Figure 2.1. In fact, as seen from the case studies in Part III, some companies have also used POLCA to route jobs entirely between individual workcenters and they have seen significant benefits in such situations as well. In order to keep the language in our explanations simple, we will use the term *cell* throughout the rest of this chapter; however, *a "cell" in the explanations below can be interpreted as being either a manufacturing cell or a standalone workcenter.* Similarly, we will use the term *cell team* for the operators responsible for the equipment in the cell; again, this term can be interpreted as being a single operator if there is only one person responsible for a standalone workcenter.

Also, thus far we have used the word "job" without defining it formally. For now, you can think of a job as either an actual customer order that is flowing through various cells on the shop floor (such as Job A and Job B in Figure 2.1), or a work order to make some component parts that will be used later—in general, such a work order would also have a routing that takes it through various cells. We will provide a more thorough definition of "job" in Chapter 4.

The approach in the rest of this chapter will be to appeal to your intuitive understanding of a manufacturing system. Rather than burden you with many more such definitions and details, we will provide a quick overview of POLCA so you can get a holistic view of how the key elements come together to make it work; this will also allow you to appreciate the simplicity of the core system. Then, in subsequent chapters we will provide insights on why the system works well in practice, as well as details on many specifics that need to be considered in order to make it usable and practical for different manufacturing scenarios.

Before proceeding, let's remind ourselves of the acronym: POLCA stands for *Paired-cell Overlapping Loops of Cards with Authorization*. Each of these terms signifies one of the key features of POLCA. Despite the long acronym, the operation of POLCA is actually quite simple. We will see below that at each cell, in order to decide which job to work on next, the cell team just needs to answer three easy questions: (i) What's the next job on the list for our cell? (ii) Has the material for this job arrived at our cell? (iii) Do we have the right POLCA card? Details on items such as "the list for our cell" and "the right POLCA card" will follow in this chapter, but even without these details you can see that these questions are straightforward and easy to answer. Surprisingly, with each cell team just following this simple process, the resulting performance of the whole production system is remarkably good, as verified by the cases studies in Part III. This speaks to the underlying design of the POLCA system, which makes it both simple and yet highly effective.

Now let us go into more detail on the key features of POLCA.

Authorization Date and Authorization List

The concept of *Authorization Date* is the starting point of decision-making in the POLCA system. (We will use initial capitals for this concept throughout the book, to remind the reader that this is a term with specific meaning.) Based on the due date of a job (the date by which it must be completed), the system calculates Authorization Dates for each of the cells in the job's routing. This calculation is not complex at all: the dates are obtained using simple backward scheduling from the due-date, taking into account the planned lead times for each cell in the job's routing. We will explain the calculation of Authorization Dates in Chapter 4, but for now, the following understanding suffices. The Authorization Date represents the ideal date for when a job should be started in a given cell: starting earlier than this date could lead to unnecessary use of capacity as well as an excessive build-up of WIP; on the other hand, starting the job later risks missing the deadline for its completion at the end of the system.

Through a simple sorting of the data for all the jobs, the system produces, at a given point in time, a list of all the jobs that are yet to be processed in each cell. This is called an *Authorization List*. (For similar reasons as above, we use initial capitals for this concept as well.) Figure 2.2 shows an example of an Authorization List for Cell A, printed out at the beginning of the shift on January 15. If you are familiar with the "Dispatch List" that is published in a typical MRP system for each workcenter, you will see the resemblance between that and an Authorization List; however, there are some important differences, which will become apparent soon.

The Authorization List only contains jobs that have *not yet* been launched into this cell. Once a job is launched into the cell, it is removed from this cell's list. Referring to Figure 2.2, you can see that the Authorization List is further sorted by Authorization Date, and the job with the earliest date is at the top of

Madison Machinery Corp – Authorization List			
Cell Name: **Cell A**		Date: **January 15**	
Job ID	**Authorization Date**	**Next Cell**	**Additional Job Data...**
R2D2	January 13	D	...
C3P0	January 15	B	...
NR07	January 15	D	...
AK08	January 16	F	...
SS28	January 17	B	...
...	...	...	...
...	...	...	...

Figure 2.2 Example of Authorization List. The yellow-highlighted cells display Authorized jobs and the darker-shaded cells show jobs that are Not Authorized.

the list. The first rule in POLCA is that jobs with a date that is today or earlier are termed *Authorized*, and jobs with a date of tomorrow or later are said to be *Not Authorized*. In Figure 2.2, you can see that all the Authorized jobs are highlighted in yellow so that they are easily visible for the cell team members.

Another key aspect of the Authorization List is that, for each job, the list also displays the "Next Cell," which is the cell that this job will go to next, based on its routing. This leads us to the next feature of POLCA.

Paired-Cell Loops of Cards

In POLCA, if jobs flow between any two cells—let's say from Cell A to Cell B—then these cells are connected by a POLCA loop (Figure 2.3). This loop contains a number of cards, called POLCA cards, that circulate in the loop. These cards are specific to this loop and are labeled based on the origin and destination cell; in this case they would be called A/B cards, and are labeled with these two

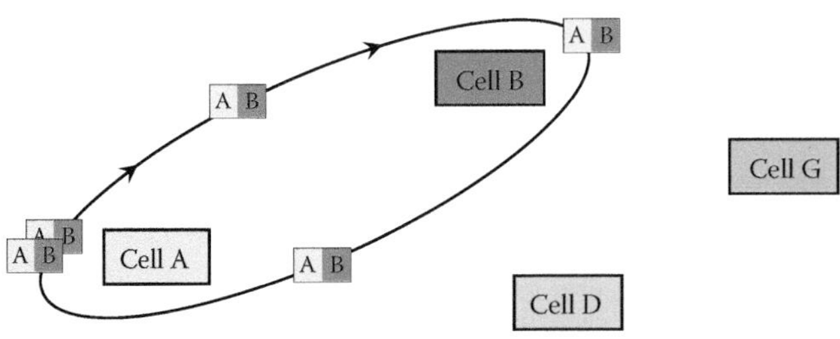

Figure 2.3 POLCA Loop from Cell A to Cell B with five circulating POLCA cards.

cell names. In addition, to enhance the visual nature of the system, each cell is assigned a color and the cards are color-coded as shown in Figure 2.4; this allows people to quickly associate a card with the origin and destination cells.

In essence, POLCA cards signal the availability of capacity at destination cells. This is how the cards work. When a job at Cell A is Authorized, and if this job is destined for Cell B next, then the Cell A team needs to have an A/B card available in order to launch the job into their cell. Typically, each team will have a bulletin board where it organizes all the POLCA cards currently available to that cell. This is called a *POLCA Board*: Figure 2.5 shows an example from one of the case studies in Part III. If an A/B card is available on Cell A's POLCA Board, the job can be launched into the cell (subject to a couple of other conditions discussed below). Once the job is launched, the A/B card is kept with this job or with its paperwork to signify that the card is associated with that job. When Cell A completes the job, it sends the job along with the A/B card to Cell B.

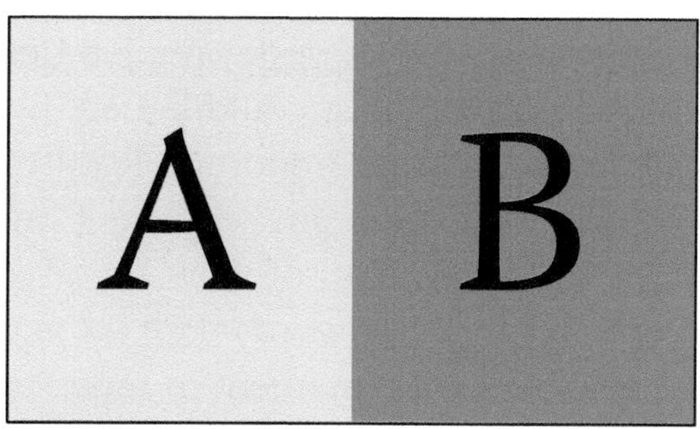

Figure 2.4 A/B POLCA card.

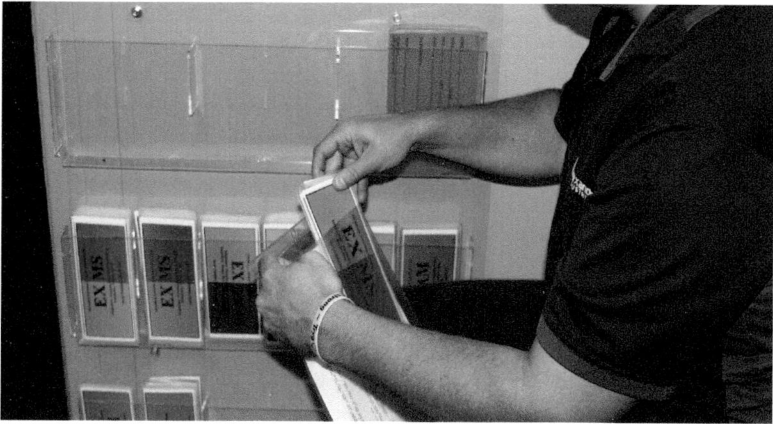

Figure 2.5 POLCA Board in use at Alexandria Industries. This is described in more detail in Chapter 10.

We will get to Cell B's decision-making below, regarding when it can start this job. For now, assume that Cell B does start this job. It also keeps the A/B card with the job. Only when Cell B *finishes* working on this job does it send the A/B card back to Cell A. Thus, you see the next key feature of POLCA—the card coming back from Cell B essentially conveys the message to Cell A: "We finished one of the jobs you sent us; you can send us another." In other words, a returning POLCA card signifies the *availability of capacity* in a given downstream cell.

Here is a quick preview of why this signal might be beneficial. Suppose there are five cards in the A/B loop. (In a later chapter, we will see how to calculate the ideal number of cards in a given loop.) Now, if no A/B card is available to Cell A, this means that there are five jobs already at Cell B, or else being processed and on their way to Cell B. For whatever reason, not one of those five has been completed by Cell B. Working on yet another job would just result in more jobs piling up for Cell B, which seems to be experiencing some kind of bottleneck at this time. Instead, as we will see next, the POLCA rules will redirect Cell A to work on a job destined for a different cell, and, in fact, it will be a cell that could benefit by receiving this job.

This preview has illustrated another important feature of the POLCA system. POLCA cards ensure that upstream cells work on jobs that will continue to flow through the factory instead of working on jobs that will end up just sitting at a bottleneck. In other words, POLCA attempts to ensure the most effective use of capacity at each moment in time.

We briefly remark on a convention that facilitates interpretation of POLCA system diagrams. You will see in Figure 2.3 that the POLCA loop has arrows going in the direction from Cell A to Cell B (the direction of job flow), but there are no arrows placed on the part of the loop where cards return from Cell B to Cell A (without jobs). More details on this convention and the need for it will be provided in Chapter 5, but we mention it here so you can understand the diagrams that follow prior to Chapter 5.

Now we can put the above-described features together to get a complete picture of the POLCA logic that is executed at each cell.

Cell Team's Decision on What to Work on Next

During the operation of POLCA there are certain events that trigger a decision-making process by the team at a cell, which then enables the team to decide which job to work on next. These triggers will be specified in Chapter 5. When such an event occurs, we will say that it is *Decision Time*. Once a Decision Time occurs, the team just needs to answer the three questions that were previewed earlier and are now shown in detail in the steps in Figure 2.6. These questions are: (1) What's the next job on the Authorization List? (2) Has the material for this job arrived from the upstream cell? (3) Do we have the right POLCA card for this job? If there is an Authorized job on the list and the answers to questions (2) and (3) are "Yes" then this job is launched into the cell and the team can start working on it. If the answer is "No" to the questions in Steps 2 or 3, then the team goes back to the question in Step 1 and looks for the next job on the list. In the event that all the Authorized jobs receive a "No" in Step 2 or Step 3, and there are no more Authorized jobs, the team will not launch any other production jobs into the cell (even if the material has arrived and these jobs are in the schedule for the near future); instead, it will engage in secondary activities such as continuous improvement projects, or assisting other cell teams, or other tasks that are on a pre-decided list. In Chapter 7, we will discuss the various types of secondary tasks that should be considered.

For the current chapter, we will leave the description at this level without additional details. The purpose of this high-level description is to quickly

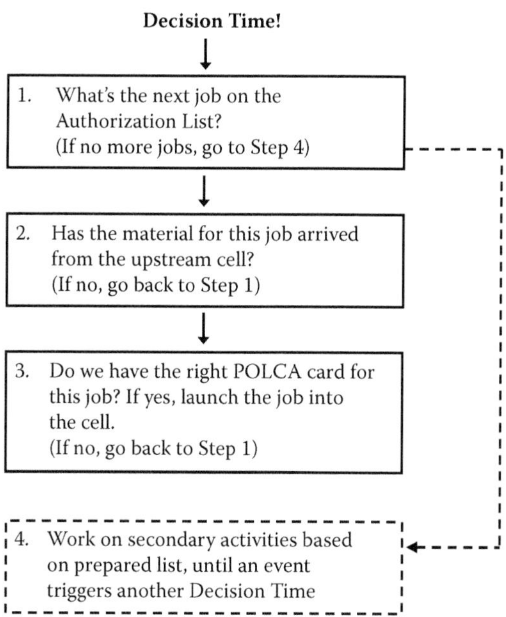

Figure 2.6 Decision Time Flowchart that specifies the cell team's decision on what to work on next.

show you the simplicity of POLCA: the few steps in Figure 2.6 essentially define the core of the whole system!

To illustrate the team's decision-making process, let's go through the steps with an example. Let's use the Authorization List for Cell A in Figure 2.2. At the start of the decision process, in Step 1, the "next" job would be the first Authorized job on the list. Since this job has the earliest Authorization Date it is the most important one to work on, so this logic matches with common sense. In fact, in Figure 2.2 the Authorization List is printed for January 15 but it still contains a job (R2D2) with an Authorization Date of January 13. This means that this job should have been started two days ago and so it needs to be worked on as soon as possible. Indeed, R2D2 will be the first job that the team will pick from this list. Next, in Step 2 the team checks to see if the material for R2D2 has arrived at the cell. If it has, then in Step 3 it checks to see if the right POLCA card is available. The Authorization List shows that for job R2D2, the next cell is Cell D. Since the team is in Cell A, and the destination is Cell D, the team needs to have an A/D POLCA card available. If indeed this card is available, then the job is launched into the cell, along with the A/D card, as explained previously.

If in Step 2 the material has not arrived, or in Step 3 an A/D card is not available, then the team will go back to Step 1 and look at the next job on the Authorization List. From Figure 2.2 we see that this is job C3P0. So now the team will repeat the decision logic for this job. This sequence is repeated until either the team is able to launch a job, or else no more jobs are Authorized. If the team is not able to launch any job at this time, then it is directed to work on secondary activities until another Decision Time occurs.

At this point you may already have several questions about the above logic:

- "Yes, it's simple," you may be thinking, "but does it make sense?"
- You could well ask, "If the R2D2 job is already two days late, but the right POLCA card isn't there, won't it hurt R2D2 to delay it further? Why should we hold off on launching it?"
- Or you might also have thought, "What if the team doesn't have POLCA cards for any of the authorized jobs, but it does have an A/F card for the job AK08 that would be authorized by tomorrow (see Figure 2.2)? Since the cell's capacity is going to waste, doesn't it make sense to start this job right now, since there may be other urgent jobs tomorrow?"

It's good if the descriptions thus far got you pondering such issues, because you are getting involved in thinking through the details of how and why POLCA works. In the next chapter, we will explain why there are manifold benefits to holding off on launching jobs such as R2D2 that don't have the right POLCA card available, and also to not starting jobs like AK08 earlier than their Authorization Date, even if means that the current cell will be idled in each of these situations. These and many related issues will be discussed in considerable detail in the next chapter. For now, we will focus on completing the description of how POLCA works.

Overlapping Loops

The remaining portion of the POLCA acronym has to do with the term "overlapping." Let's return to the first example of a job going from Cell A to Cell B and consider what happens when A completes the job. At this point the job, along with the A/B POLCA card, moves on to Cell B. Suppose also that the next cell for this job after Cell B is Cell G, which implies that there is a B/G POLCA loop as well (Figure 2.7). Now, note that when the job arrives at Cell B, it does not necessarily mean that the team can start working on it. Every cell must follow the Decision Time rules in Figure 2.6. Referring to Step 1 in the figure, we see that first Cell B must get to the point where this job is the next job on the Authorization List. Since the job has already arrived, Step 2 is satisfied, and the team goes on to Step 3. Next, the team needs to check if a B/G POLCA card is available in Cell B; if yes, then the B/G POLCA card is allotted to this job and the job is launched into Cell B. However, remember that the job also arrived with an A/B POLCA card, and that card will

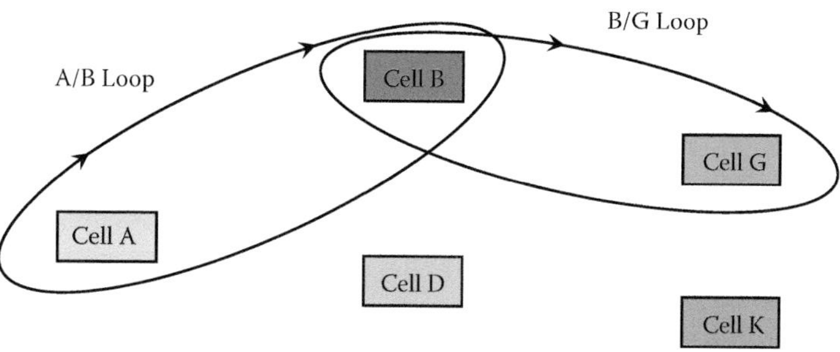

Figure 2.7 Illustration of overlapping loops in POLCA.

not be sent back to Cell A until the job has been finished by Cell B, as we explained in earlier sections.

Our job is already carrying the A/B card with it, and in addition it now has the B/G card allotted to it. So while it is being worked on in Cell B, this job will have *two* POLCA cards with it. Thinking this through, you can see that except in the first and last cells in a routing, a job will always have two POLCA cards with it when it is being actively worked on in any of the intermediate cells in its routing (Figure 2.8). So the POLCA loops overlap throughout the routing except at the first and last cell.

Although this overlapping effect is simply a consequence of applying the POLCA Decision Time rules at each cell, it actually has several positive effects on the interactions between cell teams; we will elaborate on these in the next chapter.

This completes the description of the basic operation of POLCA. We now take a moment to make a few comments that highlight the differences between POLCA and two other commonly used systems, MRP and Kanban, since this may answer some questions that may have also occurred to you while reading the preceding description.

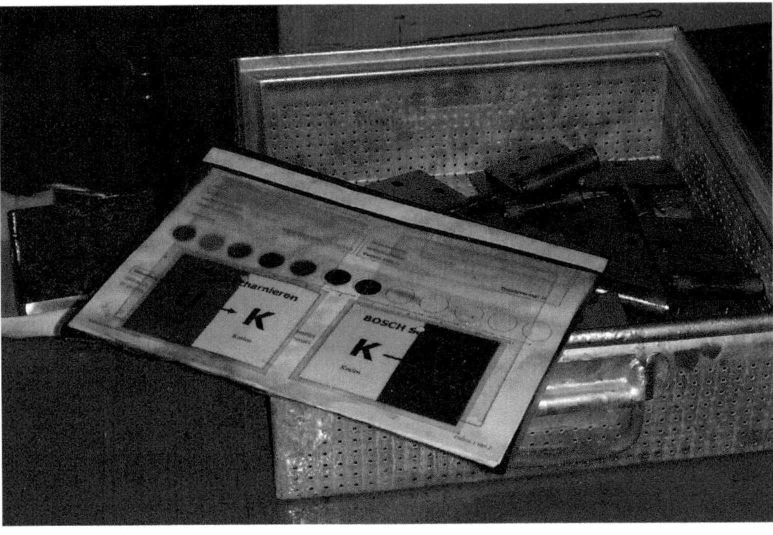

Figure 2.8 Shop packet at Bosch Hinges (the Netherlands) showing two POLCA cards associated with a job while it is in Cell K. The job came from Cell U carrying the U/K card on the left, and since it is going to Cell V next, it needed the K/V card on the right in order to be launched into Cell K. More details on this example are in Chapter 12.

Differences from MRP and Kanban Systems

Readers familiar with MRP systems will have noticed that the Authorization List resembles a Dispatch List created by an MRP system. Indeed, there is a strong resemblance, but there are also some notable differences: (i) a typical Dispatch List uses the term "Start Date" for jobs, but in POLCA we use "Authorization Date" because jobs may or may not be started on this date, based on other rules in the POLCA system; (ii) Dispatch Lists are typically created for each operation, while Authorization Lists are created at the cell level—the details of moving jobs through various operations in the cell are managed by the cell team; and (iii) in most manufacturing operations, if a work center is idle, jobs on the Dispatch List can be launched before their Start Date to keep the work center occupied, while in POLCA the Decision Time rules in Figure 2.6 are firm, and a job *cannot* be started before its Authorization Date. The benefits of sticking with this rule will be discussed in the next chapter.

Also, readers familiar with the Kanban card system will have noticed an important difference between that system and POLCA. Let's revisit the job flow from Cell A to Cell B, as described in the previous section on "Paired-cell Loops of Cards." In Kanban, when Cell B picks the next job to work on, it immediately returns the attached Kanban card to Cell A. However, in POLCA, Cell B will return the card only after it has *completed* the job and sent it on to the next step. You may well ask, "Why not send the card back as soon as the job is launched, so that Cell A can be working on the next job?" The answer is based on a key difference between POLCA and Kanban. Kanban is an *inventory signal*: the signal gets triggered when a certain quantity of parts is used up, and the signal tells the previous operation to make up the inventory by supplying that quantity of parts. On the other hand, POLCA is a *capacity signal*: the signal is triggered when a job is completed, and the signal tells the previous cell that it is okay to send another job to this cell. This difference between an inventory signal and a capacity signal is very significant; it underlies why POLCA works for low-volume and custom parts while Kanban is not suited to these environments (see the next chapter for details). In fact, in POLCA it is critical that the card be sent back only when the destination cell has actually completed the job, for the following reason. Since POLCA is particularly targeted at HMLVC environments that have higher variability and jobs with differing work content and requirements, we don't want a cell to send any signal until it is completely finished with a particular job.

Let's dive deeper into why a capacity signal is needed at all, and why it might be beneficial. We will expand on the brief discussion in an earlier section. Consider the example of the Decision Time logic described in the previous section for the Cell A team. In Step 1 of the flowchart in Figure 2.6, suppose that R2D2 is the next job on the Authorization List. For the Step 2 question, let's say the team confirms that the material for this job has arrived from the upstream cell. So now the Cell A team is at Step 3, and it sees from the Authorization List that this job is destined for Cell D next, so it needs an A/D POLCA card. Suppose there are seven POLCA cards in the A/D loop. (The number of cards in each loop is calculated to keep the system working effectively; this calculation is provided in Chapter 5.) However, when the team looks at its bulletin board, it sees that there are no A/D cards available. What does this tell us? This means that all seven cards are in use with other jobs, so these jobs are either still being worked on in Cell D, or are being worked on in Cell A and are on their way to Cell D. Essentially, either Cell D is backed up with work, or there is enough work already on its way to Cell D. In either case, sending another job to Cell D will not be productive since Cell D won't be able to work on it in the near future and the job will just sit, adding to the WIP. Like sending more cars into a traffic jam, sending more jobs into a bottleneck resource is not a good idea! On the other hand, there may be other cells that are waiting for work from Cell A. For example, we see from Figure 2.2 that the next job on the Authorization List, namely C3P0, is destined for Cell B. If there are A/B POLCA cards available at Cell A, this means that if Cell A works on this job there is a good chance that B will also be ready to work on it and the job will keep moving. So skipping the job destined for Cell D and working on the job for Cell B instead is in fact a good idea from the point of view of the overall system flow.

From this description, you can see once again the point mentioned earlier: one of the major things that POLCA accomplishes is to make sure that upstream cells work on jobs that will be useful to downstream cells, as opposed to working on jobs that will end up sitting in a logjam. In other words, since real-time events can interfere with the original plan, POLCA helps to ensure that your capacity is being used effectively in light of the most recent changes in the actual operations. We will discuss this and related points in more detail in the next chapter.

Putting It All Together with an Example

We end the chapter by returning to the MMC factory in Figure 2.1 and using it to provide a summary of how POLCA works. We will illustrate this via the routing for Job A shown in that Figure. Let's consider all the POLCA loops that will be used to manage the flow of this job through MMC's factory. As explained further in Chapter 5, the starting cell for a routing using POLCA is actually not on the shop floor but in the office, with the people involved in planning and releasing jobs to the factory; we will call this the Planning Cell. For this example, let's say the functions performed by the Planning Cell are restricted to preparing the shop packet with paperwork and routing details for a job, gathering the material needed for the job, and delivering the shop packet and material to the first cell in the job's routing. With the Planning Cell as a starting point, and referring to the steps for Job A in Figure 2.1, all the steps for Job A are shown in sequence in Figure 2.9, along with the POLCA loops that connect each step. We call this the *POLCA Chain* for Job A since the diagram clearly resembles the interlocking links in a chain. In descriptions that follow in later chapters we will often refer to the full set of POLCA loops for a job, so the term "POLCA Chain" is a valuable construct and we will use it throughout the book.

For ease of reference we have also given two-letter acronyms for each cell in Figure 2.9. Now here is how POLCA will manage the flow of Job A. The first step is that, based on the due date (let's say for shipping) of the job, the system will assign Authorization Dates for each cell—this was explained briefly earlier, and more on how to do this is in Chapter 4. Then the job will simply flow through the factory based on the Decision Time rules. The first event for Job A will be when the Planning Cell (PL) decides (based on the standard Decision Time rules) that it can launch the job into Planning. This means that Job A has to be Authorized at PL, any material required for this job should be available, and a PL/RP card must be available at the

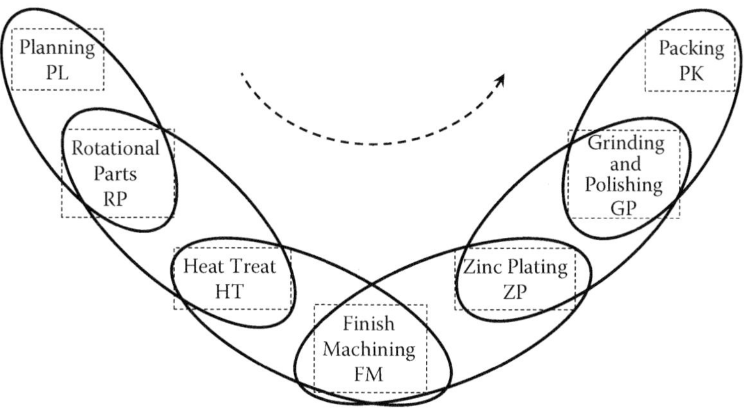

Figure 2.9 POLCA Chain for Job A at Madison Machining Corp.

Planning Cell. Then PL will process any paperwork needed for Job A and deliver the shop packet and material, along with a PL/RP card, to Cell RP.

Next, the team at RP will follow the Decision Time rules. Since we are following Job A, we need to get to the point where it is the next job on the Authorization List for Cell RP. Then the team will verify that the material for Job A has arrived. Since the job is going to HT next, the next step is for the team to check if an RP/HT card is available. Once these conditions are satisfied, the team will launch Job A into the RP Cell along with the RP/HT card attached to the job or its paperwork. Note that Job A arrived with a PL/RP card, which cannot be returned until Cell RP has finished the job, so now Job A will have two cards attached to it: the PL/RP card that came from the upstream cell, and the RP/HT card that is needed for the downstream cell. This is, in fact, the "overlapping" feature of the loops that is part of the POLCA acronym.

When the RP team completes its work on Job A, it will send the job along with the RP/HT card to the next cell (HT), and it will also return the PL/RP card to PL. After the job arrives at HT, the team at HT will go through a similar set of steps as the RP team, and so on for all the remaining routing steps for Job A, until it reaches the Packing Cell (PK). The only small change for the rules at PK is that since there is no following cell, the team does not need to look for a downstream POLCA card. So Step 3 in the Decision Time logic can be eliminated for this team (such details are part of the POLCA system design and training discussed in Part II of this book).

This completes the description of how POLCA would work for Job A at MMC. You can see that the rules are simple and operation of the system is straightforward. Once everyone understands and follows the rules, it is clear to each team what to work on next, and it is not complicated to manage the flow of jobs through the factory floor. Now that you understand how POLCA works, the next chapter will explain the numerous operational benefits that derive from this relatively simple system.

Chapter 3

Operational Benefits of POLCA

After implementing POLCA companies have experienced numerous benefits. In Chapter 1 we gave a brief overview of these benefits, but at that point we had not explained the operation of the system. Now that you have seen how POLCA works, we are able to describe these benefits in more detail; knowing how the system operates enables us to provide in-depth explanations of why the benefits occur. In Part III of this book you will find case studies from various industries, which document that these benefits have indeed been realized in practice.

We begin by noting that a strong point of POLCA is that it is *simple and easy to understand*, as seen in the previous chapter. While this is not exactly an operational benefit, nevertheless it is an important attribute of POLCA because it leads to several of the benefits described below, such as ease of implementation and buy-in from the people involved. We now proceed to explain the main benefits. Each section below will be devoted to discussion of one significant benefit.

Tailored to High-Mix, Low-Volume and Custom Production

The success of mass production over the last century was based on the fact that companies could spread the cost of processes, equipment and tooling over high volumes and thus lower their product costs. However, with the spread of CAD (computer-aided design), CAM (computer-aided manufacturing), and N/C (numerically controlled) machines, it is now possible to make products in low volumes and at reasonable costs, stretching to the point that many manufacturers also offer to customize their product for each order. Since customers can now get products that are more suited to their specific needs, companies are seeing orders for a higher variety of products, to be made in low volumes, and beyond that, for customized products with batch sizes going down to single digits or even one. As stated in Chapter 1, we refer to this production environment using the acronym HMLVC (high-mix, low-volume and custom). This trend to lower volumes and customized products is only going to get more pronounced, and so companies that want to be successful going into the future need to figure out how to operate effectively in an HMLVC environment—without a lot of chaos and constant expediting of "hot" jobs.

As one way of dealing with this shift from mass production to HMLVC, companies have been reorganizing their shop floors from functional departments into cells. (We gave a brief definition of a "cell" in Chapter 2, and that suffices for the context of this chapter.) In such a cell-based organization, simpler products can be completed from start to finish in a single cell. However, more complex products might require a large number of processing operations and/or component parts and/or subassemblies, and trying to accommodate all the needed machines and capabilities in one cell would make the cell too large and unmanageable. In such cases, complex products are produced by combining capabilities from multiple cells. For example, upfront cells could focus on making different families of components, followed by several subassembly cells with varying capabilities, and final assembly cells for different groups of products. In the environment of high variety and possible customization of products, orders would typically require routings through many different sequences of cells.

Even though the use of cells is growing, most scheduling systems have not changed their structure for this environment. In following sections, we briefly explain why two of the most common approaches today—Material Requirements Planning (MRP) and Kanban—do not work in the HMLVC environment. (Additional details on this are in Appendix C.) On the other hand, POLCA specifically takes advantage of the cell structure and team ownership and builds on these two aspects. In addition, POLCA is specifically designed for the HMLVC environment in that first, it allows completely flexible routings with connections between any sets of cells, and second, as you will see from the sections below, it helps to manage capacity and flow in a high-variability, dynamic environment.

No Complex Software Implementation Required

From the description in the previous chapter, we have seen that POLCA can be implemented using simple physical cards along with a set of rules that dictate which job is to be started next at any given cell. The only software needed to support the POLCA system is the software that calculates the Authorization Dates. If a company has an ERP system, then this is already built into the MRP logic in the system. Otherwise, this calculation can be done easily with a spreadsheet or database package. All that is needed from this software are some elementary arithmetic calculations along with sorting operations—both of which can be easily implemented in today's spreadsheet and database packages by a knowledgeable user and without even requiring an expert programmer. (See Chapter 4 for a description of these calculations.)

This ease of implementation is a significant advantage of POLCA over other sophisticated scheduling systems, such as Finite Capacity Scheduling (FCS) software and Manufacturing Execution Systems (MES). Industry experiences with many scheduling software packages have shown that such systems can take months to implement, including working out the details of all the special situations that occur in a given business. During this implementation phase, the day-to-day operations can be significantly degraded due to both the time spent by people in getting the system installed and working properly, as well as errors which need to be corrected. Thus, companies can experience reduced productivity during the several months of implementation.

Added to this is the actual cost of purchasing or leasing the scheduling software, and the extensive training that is needed for company personnel involved in using the system. For some systems, these costs can go into the hundreds of thousands of dollars. The main costs for POLCA are almost trivial in comparison: they involve the time for planning the system, training the relevant people, and creating the Authorization Date calculations and the POLCA cards.

The third and possibly most important point about sophisticated scheduling systems is that in the real world of manufacturing, there are frequent unanticipated changes that impact the flow of products. For example, an unexpected order could arrive from an important customer that needs to be served soon, customers could change their requested delivery dates, a supplier could be late with raw material delivery, a quality problem could require rework on a product, an employee could call in sick, a machine could break down, and so on—anyone with experience in a real production environment could add many more items to this list! Note that many of these changes are even more likely in the world of HMLVC production. Thus, a schedule that is supposedly optimum, derived by running a state-of-the-art algorithm in the scheduling system, can be obsolete soon after it is released to the shop floor! Therefore, it could be argued that this schedule is no longer optimum, and was it really worth all the computer power to calculate it in the first place? Also, fixing the schedule to accommodate the unexpected change then requires human intervention, and after this has been done numerous times, people in the company no longer trust the published schedule in the first place, and start to work around it. This leads to a downward spiral where the scheduling system is used less and less by the people on the shop floor. Eventually the whole company reverts to a culture of working on hot jobs, where "hot" is determined by a network of informal communications between various colleagues around the shop floor.

In contrast to the above sequence of events, industry applications have proven that POLCA is both simple and robust, and in addition, it gets strong buy-in from the personnel in the company; several of the following sections will reinforce these points, as will the industry case studies in Part III.

Works With Your ERP System—or Without It

From Chapter 2 we see that the main data that POLCA requires for its operation are: job routings, Authorization Dates, and the Authorization List for each cell. If you already have an ERP system, then these three items are readily available as follows: (i) job routings are part of the standard data in an ERP system; (ii) Authorization Dates can be derived via the normal backward-scheduling logic used in the MRP portion of the ERP system and so are also easily obtained; and (iii) the Authorization List resembles the Dispatch List, which is also standard in the MRP part of an ERP system—only the "Next Cell" field needs to be added to this list, and the heading of "Start Date" needs to be reworded to "Authorization Date"—both of which are almost trivial additions requiring very minor changes to the query that generates the usual Dispatch List. Since most manufacturing companies have already invested millions of dollars in installing their ERP systems, this is good news. For such companies, POLCA is a very simple "plug-in" to enhance their system.

In fact, for HMLVC companies that have installed ERP systems it is good news for another reason. When such companies discover that their ERP system, despite its cost and complexity, is not doing a good job of managing their HMLVC production, they often resort to buying and installing separate software to do their planning/scheduling, such as the FCS or MES software mentioned earlier. This software replaces some of the modules in the ERP system, which means these companies have essentially wasted some of their original investment in the ERP system. In contrast, POLCA leverages off of the system and modules that are already installed.

Finally, for companies—particularly smaller manufacturers—that do not have an ERP system, there is no major issue either! Let's revisit the three data items above. We can reasonably expect that any manufacturing company would have job routings. For the other two data items, as explained in the previous section, these are easily generated using simple arithmetic and sort operations available in a spreadsheet or database software. Hence POLCA is easily implemented in such companies as well.

Makes Effective Use of Capacity at Each Step

As just explained, in the dynamic HMLVC environment, even the best plans can become obsolete very soon. Let's consider two different scenarios:

1. A cell has developed a bottleneck due to a temporary equipment failure or a quality issue, and is now backed up more than expected; so, it is no longer the best idea for upstream cells to work on jobs destined for this cell. Or a customer may have requested a delay in delivery date. Again, it is now not critical for upstream cells to work on the job for that customer.
2. An urgent job has been accepted by management to help out an important customer that has an unanticipated need. Or there are jobs that have been delayed for some reason but now they should receive attention. Or simply due to the dynamic imbalance in the HMLVC environment there are downstream cells waiting for work. In each of these situations, it is now important for upstream cells to (potentially) allocate their capacity towards resolving these issues.

POLCA helps to manage both these types of situations effectively. If a cell develops an unexpected bottleneck, it will not return POLCA cards to upstream cells, and so those cells will not send it more jobs. At the same time, the POLCA rules will direct those upstream cells to work on jobs that are actually needed by downstream cells (as signaled by the presence of POLCA cards in the upstream cells), hence reducing the waiting time of those cells compared to if the upstream cell had worked on the originally planned job. If a due date has been pushed back by the customer, or if a rush job has been accepted, these events will be reflected in the Authorization Dates for jobs as soon as the system is refreshed. Hence, at the next Decision Time, the system will ensure that the cells adjust to these changes. For example, instead of working on a job that would just go to a bottlenecked cell and wait, the decision rules would ensure that the cell works on the recently accepted rush job.

As seen from these examples and many other similar situations, the POLCA cards and rules help to ensure that upstream cells work on jobs that are truly needed by downstream cells, and also adjust their work sequence to meet the times that jobs are actually needed by customers.

Avoids Congestion and Excessive WIP Build-Up

Building on the preceding point, POLCA also helps companies avoid excessive build-up of Work-in-Process (WIP). As an example, consider a schedule released by a conventional MRP system. Suppose this schedule has upstream cells A, B, and C, all working on numerous jobs that are destined for Cell D. Now let's say Cell D has a problem right away and is immediately bottlenecked. In the conventional system, the upstream cells would continue working on jobs based on the Dispatch List released to them. These jobs would all pile up outside Cell D, resulting in excessive WIP as well as congestion on the shop floor. This has other ramifications too, such as creating safety issues; making it difficult for fork-lift operators transporting other jobs; and potential damage to the parts sitting around the shop floor.

You could argue that in the situation just described, a supervisor would eventually notice that there was a problem, and would stop some of the jobs from being started and make other changes in the schedule. True, this could happen, but there are three issues with this scenario. First, if the upstream operations are separated by quite a distance from the bottlenecked cell, it may take a while for this problem to be noticed and then communicated across the shop floor. Second, now a supervisor is responsible for deciding which jobs will be stopped and which ones will be moved forward. As this type of event could occur often, a lot of the supervisor's time will be consumed in this constant rescheduling. And third, if there are multiple supervisors involved in different areas of the shop floor, then there could be many meetings (and arguments!) that must take place to agree on the schedule changes. All of this simply takes us back to the chaotic environment of hot jobs, machine operators frustrated by the constant changes and lack of focus, and supervisors and managers having to baby-sit jobs to get them out the door.

In contrast, with POLCA the cards and rules will simply take care of the rescheduling that is required. As each team executes its Decision Time rules, appropriate jobs will be held back or moved forward as needed. No management intervention is needed, the system just runs itself! As proof of this, consider a comment made at a company making customized metal parts. Three months after implementing POLCA we conducted a survey of the staff, and one of the comments we received was: "Hours of management time has been liberated each week, allowing managers to engage in more strategic tasks, instead of in constantly expediting jobs on the shop floor."

System Is Adaptive

As seen from the preceding sections, a POLCA system is adaptive in many ways: jobs that are late will "bubble up" to the top of the Authorization List as they get to downstream cells and receive priority; jobs that are finished earlier than expected will not be Authorized and will therefore be held back, thus reducing unnecessary use of capacity and materials as well as minimizing WIP build-ups; when unanticipated bottlenecks develop, POLCA not only controls the WIP build-up but also stops upstream cells from wasting their capacity on jobs that will just sit, instead making them work on jobs that are needed in other places in the factory; and so on. Thus, we see that the POLCA system automatically adapts to the changing HMLVC environment by making adjustments as needed in real-time.

By Holding Off on Jobs the System Results in Multiple Benefits

In traditional manufacturing approaches, it is considered important to ensure that machines are busy and employees are not standing around idle; they are working on jobs. However, as seen from POLCA's Decision Time rules, if the right POLCA cards are not available for Authorized jobs, and no other jobs are Authorized, then the cell team is not allowed to start another job just to keep its machines or people busy. In this section, we elaborate on why this is in fact beneficial in the long run.

Let us consider an example using the Authorization List for Cell A that was explained in Chapter 2 and is shown again here for your convenience in Figure 3.1. As shown in the figure, this list is printed on January 15. Suppose there is no A/D POLCA card available for the team in Cell A. Then it cannot launch the first job (R2D2), and must skip to the next job (C3P0) on the Authorization List. From Figure 3.1 we see that this job is destined for Cell B, so the team needs an A/B POLCA card to start the job. Suppose also that no A/B POLCA card is available at this time. Then the team will have to skip this job as well and go to the third job. In fact, looking at Figure 3.1, we see that the team will have to skip a total of three jobs on the Authorization List because the third job is going to Cell D and we already know that no A/D POLCA card is currently available. Observe that the next job on the Authorization List has an Authorization Date of January 16 and is destined for Cell F. Now suppose an A/F POLCA card is indeed present on the team's POLCA Board. Remember that, in this example, the current date is January 15. The rule in POLCA is that even though the right card is available, this job cannot be started until January 16.

In fact, since the three jobs with earlier dates had to be skipped due to lack of POLCA cards, and this job cannot be started yet, this will mean that the team cannot start any jobs at this moment in time. This is important to note: in a typical shop floor operation, if people don't have work to do, a

Madison Machinery Corp – Authorization List			
Cell Name: **Cell A**		Date: **January 15**	
Job ID	**Authorization Date**	**Next Cell**	**Additional Job Data...**
R2D2	January 13	D	...
C3P0	January 15	B	...
NR07	January 15	D	...
AK08	January 16	F	...
SS28	January 17	B	...
...	...	...	...
...	...	...	...

Figure 3.1 Authorization list for Cell A on January 15.

supervisor will look ahead in the schedule and try to start some jobs even if they get made much earlier than needed. In POLCA the rule is clear: all jobs with Authorization Dates of today's date or earlier are Authorized, while all other jobs with dates in the future are Not Authorized and cannot be started.

Of course, the preceding situation in Cell A would make traditional management or supervisors very uneasy. "Just because of this darn POLCA system we implemented, we now have five people in Cell A standing around doing nothing—how can that possibly be good for our company?" While at first glance it would seem that this complaint has merit, in actual fact there are several reasons why these rules in POLCA have resulted in better performance for many companies even though there are times when the preceding situation occurs on the shop floor.

Let us elaborate on the reasons why it might actually be *better* for the cell team to *not* start work on a job that is Not Authorized, even if it means the whole team does not have any jobs to work on at this moment:

- Every time you put capacity into a job that is not needed, you potentially steal capacity from another job that might have needed it. But, you could argue, if people are standing around, how could you be stealing capacity from another job? To see the answer, remember that on January 15, the team skipped three jobs that were authorized but did not have POLCA cards available. Observe that the first of these jobs (R2D2) had an Authorization Date of January 13. That means it should have been started two days ago. Let's say that since the team is idle and it sees that the A/F POLCA card is available for the job AK08 with the Authorization Date of January 16, it decides to start this job anyway. The first machine for this job has a two-hour setup and just as the team finishes the setup and starts machining the first piece of the job, the A/D POLCA card for the January 13 job shows up and this job also needs to be started on that same machine. Now what will the team do? Having spent two hours on a setup, it is unlikely that the team will stop the job on that machine and tear down the setup and redo a new two-hour setup for the January 13 job. More likely is the scenario that the team will finish the January 16 job and then start the January 13 job. If the January 16 job takes most of the shift to finish and the January 13 job cannot be started until the next day, what has just happened? A job that was two days late is now going to be at least three days late, while a job that was not needed is going to sit on the shop floor for an extra day.

- Working on a job that is Not Authorized means that it will be ahead of schedule and will just add to the WIP. As more and more cells break this rule it will result in a significant increase in congestion on the shop floor.
- When you start a job earlier than planned, you could potentially use material that was earmarked for another job, and when that job arrives the needed material may not be available, causing further delays.
- When the cell team is idled as a result of the POLCA rules, this provides an opportunity to engage in continuous improvement or other useful activities; we elaborate on this point in the next section.
- Another possible good outcome would be for the cell team to offer to help another team that is backed up with work. This point is also discussed in a later section.

In summary, companies implementing POLCA have witnessed that the long-term benefits of sticking with the POLCA rules outweigh the short-term benefits of just trying to keep machines and people busy. This will be further underscored in the following sections as well as through the case studies in Part III.

Provides Formal Framework for Improvement Activities

As mentioned above, when a team doesn't have specific production jobs to work on, there are many other useful things the team can do—spend time in cross-training, study a setup in order to come up with ideas for setup reduction, engage in preventive maintenance tasks, and other improvement activities. Today's leading manufacturing strategies place great importance on continuous improvement activities, but the truth is, if workers never have any non-production time, when can a team engage in any of these continuous improvement activities? When implementing POLCA, teams can be coached that whenever their production activity is stopped due to POLCA's Decision Time rules they can consider using these periods of time productively for improvement.

Chapter 7 recommends a systematic approach to utilizing this available time for improvement activities. The benefit of having such a methodology in place is that when production is stopped the team immediately has a direction to pursue, and time is not wasted on deciding what the team should do.

Prevents Inventory Proliferation for Products with Low Demand

As already mentioned, Kanban is an alternative card system that is popular for controlling production, so let's understand why POLCA might be better at controlling inventory for low-volume products. The main feature of a Kanban system—as described by its common name of "Pull"—is that parts are "pulled" through the factory as they are used in downstream operations. We can understand the way this system works by starting from the finished product. When a container of finished goods is shipped out of the warehouse, a signal is sent to the previous operation to restock that container. That operation has the partially completed material for this product waiting in its stock area, and when it draws a container of this material to work on, it then sends a signal to its previous operation to resupply that material. These signals are often in the form of physical cards called Kanban cards. Each such signal therefore needs to specify both a part number and a quantity that needs to be made. Thus, for the Kanban system to work, you need to have these partly completed products stocked in containers at each stage of the manufacturing system. If you make a small variety of parts and all with relatively high demand, then this system keeps the material moving; it is simple to implement, and it works very well.

However, consider what happens to the Pull approach when you have a large variety of products and many of them have very low annual demand. Suppose a company makes axles for non-automotive applications, such as for construction and mining equipment. Let's say this company stocks in its finished goods a container with six axles of a certain type: an axle that is used in only a few specialized machines, and so typically the company gets an order from a distribution center only about once a year for one of these containers. Let's see what happens when this order arrives. The container is shipped and a Kanban signal is sent to the previous operation. Within a couple of days, the previous operation completes the production of the six axles and restocks the warehouse. Now these axles are going to sit in the warehouse for about a year before another order is received. In other words, you have an inventory turnover rate of only once a year for this product! If most of your products have low demand, then your overall inventory turnover is also going to be very low, perhaps two or three times a year at best. In an era when management expects to achieve inventory turns of twenty or more a year, this would be atrocious.

But it is actually worse than this, because not only are the finished axles sitting in the warehouse, but along the whole flow path of these particular axles there are partly completed products sitting at various stocking points throughout the shop floor, waiting for a pull signal from an upstream operation so that they can be worked on and sent to that operation. The same is true for all the other axles being made at the company. Figure 3.2 illustrates this for a company that starts out with three types of bar stock that undergo various machining operations and end up as more than a thousand different types of axles. You can see the proliferation of inventory that is caused by the Kanban containers needed for this production situation. This is not an unknown phenomenon; in fact, people that design Kanban systems know that these inventories will be needed throughout the shop floor and they even have a name for these intermediate stocks: they are called "supermarkets." When the production volume is high, there is no need for concern about supermarkets because the items move through the supermarket quickly and the inventory turnover rate is high enough. Thus, Kanban works well in these environments; but in low-volume environments, instead of controlling inventory, it actually adds unnecessary inventory.

So why would POLCA work better in this situation? The key is that, unlike Kanban, in POLCA jobs are not launched just because a card is received. There has to be an Authorization as well. More precisely, if you recall the Decision Time flowchart in Chapter 2, the check for Authorization is the first step, and the check for the right POLCA card occurs only after a job is Authorized.

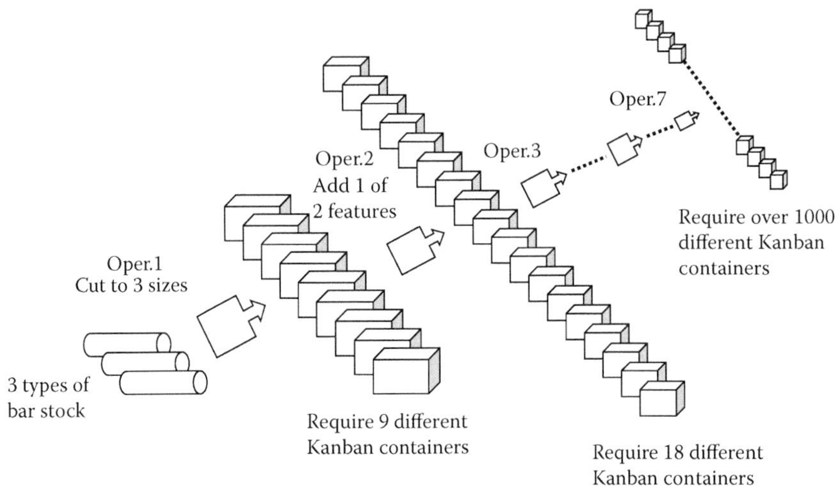

Figure 3.2 Inventory needed at each operation to support the functioning of a Kanban system for a company making over 1,000 types of axles.

Let's see how this would impact the example of the axle with very low demand. When the container with these six axles is shipped, this will trigger a review of this part. The planner responsible for these particular axles knows that it will probably be another year before more are needed. So, this planner will periodically review the stock of these axles at the distribution center. Based on the consumption of this stock, and the lead time for manufacturing the axles, the planner will create a production job for another batch of axles, with a due date at which it needs to arrive in finished goods. This due date will then be used by the planning system and the job will show up on the Authorization List in due course, and then the POLCA rules will govern the flow of this job. You can see that the axles will be restocked closer to when they are needed, as opposed to being restocked right away. It could be, for example, that the replenishment order is not even started for nine months. In the case where the demand for these axles is falling, it could even be that no more axles are made for 18 months.

If it seems that this is a lot of fuss about carrying extra inventory of just six axles, remember that this is just one part. For an HMLVC company, there might be thousands of such parts. In that case six extra pieces of each part would add up to tens of thousands of extra pieces of inventory. This is where POLCA will show large benefits.

Allows Complete Flexibility in Product Routings and Works Seamlessly for Custom Products

In POLCA, a product can be routed from any cell to any other cell; there are no restrictions on this. If there is already a POLCA loop between these cells, then the existing loop will take care of the product's flow. Even if a custom product is being made, most of the time it will likely flow between cells that are already connected, just because of the type of manufacturing sequences being used to make products at the company. However, if a custom product requires a connection between two cells that were previously not connected with POLCA, it is easy to add a new POLCA loop, as explained in Chapter 7. In summary, products can have any possible routing through the factory without creating issues for the POLCA system or rules.

To compare this aspect of POLCA with the Kanban system, let's return to the example of the axle manufacturer in the earlier section, and suppose it receives an order for a custom-engineered axle for a large mining machine. How would the Kanban system work in this situation? It won't work at all.

You can understand this by realizing that Kanban is a replenishment system: you begin by shipping finished goods and then sending a signal to have them replenished—but you can't have something in finished goods if it has never been made before; in fact, it hasn't even been engineered yet. Indeed, at each step of the whole Kanban shop floor operation, you pick partially completed material from your stocking point and then ask for that material to be replenished. So, this flow cannot operate for a custom-engineered product from the point at which customized operations begin and onwards.

In contrast, with POLCA, there is really no difference between the way in which regular products and custom products are treated. A custom product has a routing, and as just explained, POLCA allows any routing through the shop floor. A custom product also gets assigned Authorization Dates just like a regular product does. Hence, the POLCA rules work in the normal fashion for a custom product as it moves from cell to cell, and custom products flow seamlessly through the shop floor.

Creates Customer-Supplier Relationships between Cells

In each POLCA loop, the upstream cell has the role of a supplier and the downstream cell has the role of a customer. This joining of cells into customer-supplier pairs creates productive interactions over time. Because the cell structure gives teams ownership and motivates them to perform well, the supplier cells are keen to keep their customers happy. In reverse as well, the customer cells want their suppliers to know their needs in order that they are better served. Thus, these teams start interacting and sharing information that can help each customer-supplier pair improve its mutual performance.

As one example of such beneficial interaction, suppose Cells B and G are connected by a B/G POLCA loop containing four POLCA cards. At a given Decision Time, let's say the Cell B team finds that the only Authorized jobs are those going to Cell G, but there are no B/G cards available in Cell B, and also no B/G jobs still in Cell B. This means that all the B/G cards are currently with jobs at Cell G, which must be backed up since none of the four cards has been returned (Figure 3.3). Since the Cell B team has no other jobs that it can launch, it could decide to go to Cell G and offer to help with some of the work and alleviate the temporary backlog—or at least offer that some of the Cell B team can help out in Cell G. This will help Cell G and at

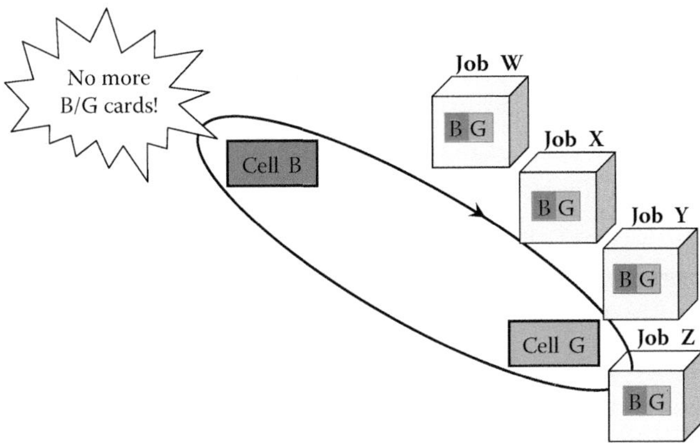

Figure 3.3 Situation where all the B/G cards are with jobs at Cell G.

the same time free up some B/G cards to return to Cell B, and then the Cell B team can launch some of the Authorized B/G jobs. This will also help to prevent those B/G jobs being delayed even more. Note that this interaction has numerous positive benefits: (i) Cell B's offer to help creates goodwill with the Cell G team; (ii) the backlog at Cell G gets alleviated, potentially helping several other jobs too; and (iii) Cell B is able to start working on jobs that would otherwise have been further delayed.

You could argue that such inter-team interactions could occur in any system, so what is special here? The fact is, in a large factory with many teams and hundreds of jobs, it is not often clear to one team that another team is backlogged. You don't need to schedule a meeting of supervisors to figure out that Cell G needs help and that Cell B can help out—the POLCA cards clearly indicate that there is a backlog and also which team is causing it. Further, the supplier team (Cell B above) has an incentive to help the customer team (Cell G) because in so doing it will also help its cell (i.e., Cell B) achieve better performance with upcoming jobs. In contrast, if in a traditional shop floor setting, a supervisor tells some operators to go and work in another area, the workers may not want to work on those processes and could be resentful of the directive to do so. In our example, the Cell B team (or part of the team) volunteers to help out; the workers are not being forced, and so there is no resentment at all.

This is but one example of such interactions; companies have experienced other examples of benefits to do with tooling, quality, and process improvements, through the customer-supplier communications that arise between pairs of cells in POLCA.

Installs a Visual System

Even though there is widespread use of computers in scheduling and shop floor operations, there is strong belief in management circles that clearly visible physical objects can be very effective in keeping day-to-day operations on track. This trend is known as "visual management," and POLCA follows the ideas of visual management in many ways.

First, the large and visible POLCA cards attached to material-handling containers such as bins or skids indicate at a glance where that job is coming from and where it is headed. Second, the POLCA Boards in each cell provide an instant status summary for both the cell team as well as for management. For instance, if a manager walks past Cell A and notices that all the slots for A/D cards are empty, that would indicate that there might be a capacity issue in Cell D. This could just be a temporary situation based on the dynamics of jobs for that day. However, if this same manager notices several times during the week that there are no A/D cards on the board in Cell A, then he or she could check further to see if indeed corrective action is needed for the capacity in Cell D.

As a third instance, let's revisit the earlier example of a cell team deciding to help another team. By looking at its POLCA Board and seeing which cards are missing, a team can get a quick idea of which other teams might need help, and decide if some of its members should help out with those cells.

Builds on the Capabilities of People in the System

Even the most sophisticated technology can be ineffective if people don't want to work with it. As explained in several places earlier in this book, many advanced shop floor control systems have been rendered ineffective by the fact that personnel in the company don't trust the system's decisions, or prefer to work around the system rather than to use it.

A strong point of POLCA is that it doesn't attempt to replace the humans in the system, but rather it builds on their capabilities. POLCA is specifically designed for an organization composed of self-steering teams that have ownership within their areas. As long as teams follow the core POLCA rules, they have a lot of latitude as to how to execute their operations within their span of control. So, the first step in POLCA is to get employees to buy in to the POLCA rules, specifically, the Decision Time rules in Chapter 2. Employees should be shown the benefits that result for the whole organization, as well as for their own jobs, from following the POLCA rules. This should be done through initial training sessions. (We will discuss these and other training sessions in more detail in Chapter 7.) Once this buy-in has been obtained, people will make an initial commitment to following the system.

Next, teams experience that within the framework of these rules, they still have a lot of latitude as to how to execute their jobs. For example, after a job is launched, the team has full ownership of the timing and sequence of operations on that job (within its cell) as well as the order in which this job is done at a given machine relative to all the other open jobs that are currently in the cell. As another example, if no jobs are currently Authorized, the team has the ability to decide which other tasks it will spend time on.

The next reinforcement to the buy-in from employees comes from the visual system. When a computerized scheduling system suddenly resequences upcoming jobs, this can be frustrating; a team might have been gearing up for a particular job and planned the usage of machines busy with other jobs to make way for it, only to find that this job has been replaced by another one. On the other hand, when the team looks at the next few jobs in the Authorization List, it can also view the POLCA cards on its board. If it can see that a particular type of POLCA card is missing, it can anticipate that those jobs will not be launched in the near future. Instead of being blindsided by a computer decision, the employees know that the missing card indicates that the downstream cell is overworked, and so they understand why the job will be skipped. Simply understanding the reasons behind a change helps with buy-in and trust in the system.

Finally, the full buy-in and trust in the system comes when a team experiences a personal incident where POLCA helps them. For instance, if the team is short one member due to sickness, and starts falling behind in its work, in a traditional operation upstream cells would continue sending work and there would be a large buildup of WIP and jobs at the cell. This is both inconvenient, as people have to work their way around all these jobs, as well as demoralizing since this pile of jobs reminds them that they are behind and makes them look bad to people walking by. However, with POLCA in place, the upstream cells would soon stop sending more jobs and provide the team with some breathing room, while it works to catch up over the next couple of days. This will be noticed by the team members, who would be relieved rather than stressed. Even beyond this, it would be a nice surprise to have an upstream team offer to send one or two people to this cell to help out (as per the example in an earlier section). Since this offer is unsolicited, it doesn't come with any strings attached, and is particularly welcome. These and other similar experiences will help to cement the buy-in from employees. This was underscored by feedback received from one of the companies that implemented POLCA, which told us that it wasn't just management that was pleased with the results, but "It's the shop floor employees who are singing the praises of POLCA!"

We should not underestimate the importance of the points in this section. Getting employees to support a system is huge. It is often taught in courses in systems optimization and analytics that it is better to have a system that is reasonably good (suboptimal) as long as it is being used properly, than to have a system that is optimal that no one understands or trusts and eventually ends up not being followed.

Readers that are interested in a deeper discussion on the people-oriented issues in POLCA system design and implementation will benefit from reviewing Appendix E. This appendix uses classic sociotechnical theory to explain in more formal terms why POLCA has been successful in practice.

Provides Fast Response through Decentralized Decision-Making

The POLCA system creates direct communication and feedback between cells involved in the routing of a job. As previously mentioned, the system provides signals so that cells adapt their work to accommodate the changing conditions. This is in contrast to traditional systems where a central planning department needs to periodically review and update priorities and dates, and/or numerous planners and supervisors have to meet frequently to negotiate changes in schedules. In addition, as also mentioned, in POLCA the teams have a good deal of latitude in how to execute the operations within their own cells, as long as they follow the core POLCA rules. The end result of this decentralized communication and decision-making is to create an organization structure that provides fast response to customers, even in the complex and difficult HMLVC environment.

Builds on Your Existing Structure and Systems

This chapter has shown that rather than requiring companies to radically change their operation and to implement whole new software modules, POLCA builds on your existing structure and systems. Specifically, companies that have implemented cells already have in place a good structure of self-directed teams that have ownership of the resources within their cells. POLCA takes advantage of this structure and builds on it. In addition, most companies have some form of ERP or MRP system in place. POLCA also builds on the capabilities of such a system, with some minor modifications required for the Decision Time rules.

These points are valuable to managers who are intent on implementing POLCA. Employees are always wary of new systems and any change proposed by management. People jokingly talk about "The flavor of the month," implying that management will introduce a new idea only to have it replaced by a different idea in the near future. Because of the features of POLCA just discussed, management can explain to employees that POLCA is not a change in direction, but rather, an enhancement of the current strategy and direction that will make the company more effective in the increasingly HMLVC conditions that it is facing.

In summary, this chapter has explained why implementing POLCA provides numerous benefits. In particular, the fact that it fits easily into the existing structure and systems, does not need complex software, and is accepted by and even liked by the shop floor employees, makes it a strong candidate for HMLVC companies to consider using in their operations. All these claims will be reinforced by the case studies in Part III of this book.

This concludes the first part of the book, which has given you an overview of the POLCA system, how it works, and its benefits. In the next part of the book we will go into detail on how companies can implement POLCA and also tailor it to their operations. For readers that are moving forward with implementing POLCA, or at least seriously exploring the possibility of it, we recommend that you work your way through Part II, reading the chapters in strict order, as each chapter uses and builds on concepts from preceding chapters. For readers that are new to POLCA, we hope Part I has piqued your interest in the system, and perhaps motivated you to consider implementing it. At this point, you may wish to skip Part II, and instead look through one or more of the case studies in Part III based on your industry and your needs, returning to Part II when you are ready to move forward with POLCA.

IMPLEMENTING POLCA

Introduction

Now that you have a basic understanding of how POLCA works and its operational benefits, the second part of this book provides a systematic guide to implementing POLCA. It is intended to serve as a roadmap for teams at companies that wish to implement POLCA, as well as for consultants and academics that might be engaged in advising industrial clients. Based on experiences from numerous industry implementations, we have found it effective to proceed with the implementation of POLCA in four phases:

1. *Conducting a pre-POLCA assessment.* This includes clarifying the motivation and goals for implementing POLCA, checking that certain fundamentals are in place before proceeding with POLCA, and engaging in activities to prepare for POLCA. These topics are covered in Chapter 4.
2. *Designing the POLCA system.* This is addressed in Chapters 5 and 6, which get into the detailed design of the system, including dealing with various special circumstances that might apply in different production environments, as well as practical techniques for dealing with unexpected situations that arise particularly in HMLVC companies.
3. *Launching the POLCA implementation.* Once the design is completed, Chapter 7 provides you with the keys to a successful launch of the system. Since managers worry about the length of time it takes to

implement a new system, Chapter 8 presents an inspiring case study of a company that designed and launched its POLCA system *in three days*!

4. *Engaging in post-implementation activities.* These activities aim to evaluate, fine-tune, and sustain the POLCA system. Chapter 9 concludes Part II of the book with advice on how to sustain POLCA and ensure that it continues to operate successfully in the long run.

Chapter 4

Pre-POLCA Assessment, Prerequisites, and Preparation for POLCA

As explained in the introduction to Part II, this chapter covers the first phase of implementing POLCA. We will divide this phase into three main topics:

- Clarifying the motivation and goals for pursuing POLCA.
- Checking some prerequisites before implementing POLCA.
- Preparing for the POLCA implementation.

The three following sections of this chapter will cover each of these topics in turn.

Clarify the Motivation and Goals for Pursuing POLCA

The starting point for your POLCA journey is to arrive at a consensus for the reasons why your company is pursuing POLCA as a solution, and also, what are the organizational goals that you intend to achieve through the implementation. The following sections will help you with these points.

Check for Symptoms That Have Led Companies to Consider POLCA

This section lists frequently cited symptoms experienced by companies that decided to implement POLCA. You may find it helpful to review these symptoms for your own situation, and, at the same time, this list may help stimulate people at your company to think of additional reasons. Symptoms that have led companies to consider implementing POLCA include:

- Long quoted lead times, but still a highly unreliable delivery record as indicated by low on-time delivery performance.
- Daily efforts of expediting that involve classifying jobs as "hot" jobs that need to be rushed through several workcenters. This results in frequent changes in priorities, creating frustration for supervisors and shop floor workers.
- Shifting bottlenecks, so that there is not one clear constraint that needs to be managed. Instead, various workcenters experience "feast or famine"; in other words, being heavily backlogged on some days, and on other days having no jobs at all.
- Work-in-process (WIP) pile-ups around the shop floor, but the location of the pile-ups changes from day to day or week to week.
- Complaints from downstream workcenters that upstream workcenters are not in sync with their needs. This is further indicated by the fact that upstream workcenters try to work far ahead in the schedule and yet their deliveries to downstream workcenters are inconsistent and/or are not the right jobs that are needed at the present time.
- Related to this is the frequent occurrence of the "hurry up and wait" problem: workers are directed to rush jobs through their workcenters, only to find that these jobs end up sitting in a queue at another workcenter.
- Supervisors and shop floor workers spending a lot of time in meetings instead of engaging in production activities. If any person is spending, on average, more than half-an-hour during a shift in such meetings, this is to be considered excessive and a significant opportunity for improvement.
- You have tried implementing the Lean approach along with Kanban but it has resulted in significant supermarket inventory in terms of both monetary value as well as space occupied. (See Appendix C for more explanations on this point.)

- In the case that the shop floor has been mostly reorganized into manufacturing cells in an attempt to improve performance, many of the above-listed issues are still occurring.
- Finally, and worst of all, you are getting frequent complaints from your salesforce about unhappy customers and even lost sales as a result of these performance problems.

Review Your Company's Operating Environment

Checking off the preceding symptoms is a good starting point, but this list alone does not mean that POLCA is necessarily the solution. A second major topic to consider is the environment in which your company operates. The following points reinforce that POLCA would be helpful in resolving the symptoms:

- You manufacture a high variety of products, most of them ordered in small quantities, or even custom-engineered products, ordered only once. As a result, you manufacture few or no products to stock; instead your operation uses one or more of these strategies: make-to-order, assemble-to-order, and engineer-to-order.
- You experience unpredictable demand, difficult to forecast accurately.
- Demand is lumpy—some periods are very busy while during others you experience a lull in orders—but you can't predict ahead of time which of these situations will occur.
- There is a constantly changing part mix, with many currently manufactured products being dropped and new products being added on a weekly basis.
- Jobs have long, multistep routings through various workcenters. For the purposes here, any routing with four or more steps can be considered long.
- There are many different routings in the company for various products; for example, dozens or even hundreds of different routings.
- Even if your shop floor has been organized into cells that are working well within themselves, jobs typically need to visit multiple cells as in the example in Chapter 2 (Figure 2.1).
- There are several one-to-many connections (one cell feeding several downstream cells) and/or several many-to-one connections (several upstream cells sending jobs to one downstream cell) and or

many-to-many connections of cells. The case studies of the hinge manufacturer (Chapter 12) and the glass factory (Chapter 14) provide examples of such complex connections.

Set Your Goals

Using the above lists to help you identify your particular motivation for POLCA should then help you to set the goals for your implementation. Below are typical goals set by companies implementing POLCA, and again, you can use these to help you hone-in on your specific goals.

- Lead time reduction targets. Companies implementing POLCA have seen significant reductions in lead time, often 50% or more—see the case studies in Part III. It is significant to note that lead time reduction itself creates manifold benefits including improved on-time delivery, fewer schedule changes, less expediting, and so on—this is explained further in Appendix A and again verified in the case studies. Hence, reducing lead time should be a key goal of the POLCA implementation. (Appendix A also provides a precise metric for lead time.)
- WIP and inventory reduction targets. Both of these can be significantly impacted by the use of POLCA along with the resulting lead time reduction.
- Elimination of hot jobs. This may sound idealistic, but indeed, companies that implemented POLCA have seen an almost complete elimination of hot jobs.
- Reduction of time spent in planning and expediting activities. As mentioned in a preceding section, you should target specific numbers for the time spent by any person on such activities, such as less than half-an-hour per shift.
- On-time delivery targets. This is where the improvements start to cumulate into results that customers can see, and you want to be sure to target goals that will ensure happy customers.
- And finally, putting it all together, you want to target the elimination of complaints from sales people and loss of customers due to delivery performance.

In order to measure your progress toward these goals, it is important that you find ways to quantify them as much as possible by creating appropriate metrics, and also gather baseline data for where you are today on each of these metrics.

The above are key goals that companies have used, but you will need to tailor them to your circumstances. Reviewing the case studies in Part III might also help you with setting your goals since you can look at both, the motivations that drove companies toward POLCA as well as the specific performance improvements that they were able to achieve. In particular, the latter may help you in setting and justifying your quantitative goals.

Check These Prerequisites Before Deciding to Implement POLCA

Even if the list of symptoms in the preceding section seems all too familiar, and your company's environment matches many of the characteristics listed above, this doesn't mean you should rush into implementing POLCA. You need to check that some fundamentals are in place first. These prerequisites will be covered in the following subsections.

In order to be clearer in the descriptions below, it is time for us to define more precisely what we mean by a "job" in the contexts that follow. For our descriptions of the POLCA system, a job is essentially the same as what is often called a work order (or shop order) along with the material needed to complete that work order. Such a work order defines a routing, which is the flow of this job through the necessary cells. We illustrate this definition with four situations:

1. *A job is a customer order.* Job A and Job B in Chapter 2 (Figure 2.1) are examples of such a situation.
2. *A job is released to make components that will be assembled into a customer order.* For instance, for a company making custom gearboxes, if the gears and shafts are manufactured as needed for each customer order, then specific jobs will need to be released to make these components. As an example, one job could be the work order to make a particular gear; this work order would start with the raw material (typically a gear blank) and end at the point where the gear was used in the gearbox.
3. *A job is used to make components that will be needed later.* For example, a company making many types of axles could cut and machine bar stock to various diameters and lengths, to be used later for specific

customer orders. In this case the work order for a given job would route the material to the necessary cells and end at the point where the material was delivered to a stock room.

4. *A job is used to replenish finished goods.* As an instance of this, in Chapter 3 we had the example of the company that stocks axles in a warehouse. When the warehouse needs to be restocked with a particular axle, a work order would be released to the shop floor for a job to make a batch of those axles, and the routing for this work order would end at the point where the completed axles are delivered to the warehouse.

In each of these examples, a job (or its associated work order) consists of a routing that takes the material through a number of cells. With this concept in place, we can now proceed to consider the first—and possibly most important—prerequisite.

Review Your Shop Floor Organizational Structure

For companies that have dozens of workcenters and complex job routings, and which suffer from many of the issues listed in the previous section, POLCA is typically not the first step. In such companies, a job can travel back and forth across the shop floor in what is popularly called "spaghetti flow." For more than 20 years, manufacturing strategy literature has pointed out that the first step is to simplify this flow by grouping jobs into families with similar routings, and to create manufacturing cells with dedicated equipment and workers for each family of jobs. The subject of cellular manufacturing has been the focus of many books and other publications for over two decades, and there are many resources available on how to create cells, so we will not go into detail on this point. However, below we will identify some key requirements of cells in order to enable POLCA to work.

For smaller companies, say with less than 20 workcenters, there may not be enough of a critical mass of jobs and/or equipment for cells to be economically feasible, and in such cases POLCA can indeed be considered as a first step without cells. Chapter 15 describes the case study of just such a small contract machining company that benefited from implementing POLCA between individual machines. At the other end of the spectrum, some companies might have very large specialized machines that cannot be moved and placed in cells. Chapter 11 describes a pharmaceutical company with this situation that also benefited from implementing POLCA from workcenter to workcenter. Thus, we see that, indeed, POLCA can be successfully implemented between workcenters, without creating cells.

However, in the majority of situations, we have found that cell creation is a necessary first step and provides the right environment for POLCA to succeed. In the rest of this chapter, our descriptions will assume that cells are in place.

NOTE: If you are in the situation of the companies just described and will not have cells on your shop floor, then in the discussions below you can simply replace the word "cell" with "workcenter" and evaluate the points in that context.

Even after reorganizing into cells, companies with more complex products find that jobs need to go through multiple cells in order to be completed. For example, there could be cells for various groups of operations related to sheet metal work, separate cells for welding and grinding operations, additional cells for metal-cutting operations, cells for subassembly and eventually cells for final assembly—and a job may need to visit several of these cells as illustrated by the example in Chapter 2. As you will see from the case studies in Part III of this book, companies have found POLCA to be beneficial if jobs typically need to visit multiple cells. In particular, in situations where there were multiple connections between upstream and downstream cells throughout the shop floor, the case studies show that companies found POLCA to be effective in coordinating the needs of the cells and smoothing out the job flow.

Make Sure That You Have POLCA-Enabling Cells

Even if you already have cells, before proceeding further it is important to verify that these cells have certain characteristics that make them suitable for POLCA. Without these basics in place at the cells, the POLCA implementation will not achieve its full potential, or may fail altogether. Hence, we will refer to cells that satisfy these criteria as "POLCA-enabling cells."

The following are the requirements for a manufacturing cell to qualify as a POLCA-enabling cell:

- The cell consists of a collection of equipment (e.g., machines, workstations, benches, and tools) along with a team of people that is responsible for the operation of this equipment.
- The equipment and the people are dedicated to the cell.

- The equipment is collocated. In other words, the resources that form the cell must be located in close proximity to each other, in an area that is clearly demarcated as belonging to the cell.
- The aim of the cell is to complete a collection of operations for a predefined set of jobs. (Note that the sequence of operations need not be the same for each job. In fact, in the HMLVC context, it is more than likely that individual jobs will have different sequences of operations through the cell.)
- The cell team has complete ownership of the cell's operation. When jobs arrive at the cell, the team first decides (based on the POLCA Decision Time rules) if the job can be launched. For jobs that have been launched into the cell, the team decides which operation on which job should be done next, who will perform each operation, who will assist whom, and so on. Note that this is in contrast to the traditional organization where a supervisor decides on priorities of jobs and assigns tasks to workers.
- There is some degree of cross-training among the people in the cell. It is not necessary for each person to know how to perform every possible operation needed by jobs in the cell, but, ideally, there should be at least two people that are trained for each operation. The aim is that the cell should have some amount of flexible capacity.
- A key goal for the cell team should be to minimize the total time that jobs spend in the cell. Thus, a key metric for the team must be related to minimizing lead time of jobs within the cell. (See Appendix A for an example of such a metric.) To support this lead time goal, and to enable success of POLCA, it is also necessary that traditional metrics based on resource utilization, operational efficiency, and standard cost be deemphasized. This is a critical issue and will be discussed in more detail later in this chapter.

In the rest of this book we will assume that when the term "cell" is used, it always refers to a POLCA-enabling cell. We also remind the reader that as mentioned above and depending on your context, you can interpret "cell" to mean a single workcenter. Finally, as explained in Chapter 2, companies can have a combination of cells and individual workcenters such as Heat Treat or Plating, and POLCA can be used between cells and workcenters. Again, in such cases you should simply interpret "cell" to mean a single workcenter when applicable.

Returning to the example of standalone and specialized resources such as Heat Treat or Plating, if the main issue you are trying to resolve is getting jobs to and from the rest of the shop to one such resource, you should look into *time-slicing* as a potential solution, rather than POLCA. We will not go into the details of time-slicing here; you can find a description in Appendix C of the book *It's About Time* by Rajan Suri, Productivity Press, 2010. Broadly speaking, if your scheduling issues mainly revolve around jobs getting to and from one or two such specialized resources, then you should consider time-slicing, while if the shop issues involve managing jobs through multiple workcenters or cells, along with one or several standalone resources, then time-slicing will not be sufficient and POLCA would be the recommended way to go.

Engage in Effective Planning Before Control

It is important to understand that POLCA is a *control* system, not a *planning* system. The function of a control system is to help execute a plan effectively. However, if you begin with a poor plan, then even the best control system can do little to make it succeed. Let's use a simple example to illustrate this point.

> Example: A 300-passenger jet airplane is at Amsterdam's Schiphol airport, preparing for a nonstop flight to Chicago's O'Hare airport. When it leaves the gate and heads for the runway, it only has 100 liters of jet fuel in the tank. Now this is a clear example of a bad plan! Even the best airplane control system will not enable the plane to fly over the Atlantic and reach Chicago. On the other hand, suppose the plane leaves the gate with 100,000 liters of fuel. The flight plan shows that this allows for two extra hours of flying time if needed. Now this is a reasonable plan. Next, while the aircraft is flying over the Atlantic, it encounters some turbulence. This is where the control system kicks in, and makes sure that the plane remains stable and stays on its flight path.

Although this is an extreme example, it does demonstrate the difference between planning and control, as well as the importance of starting with a good plan.

You may well ask, then: if you have a good production plan, why do you need a production control system at all? To answer this, let's use a company that relies on Material Requirements Planning (MRP) as an example. An MRP system by its very name indicates that it is a planning system. However, when the plan is executed, then that plan, no matter how good, still needs to be managed and fine-tuned. First there is the issue of time scale. If jobs have lead times of several days or even weeks, then a lot of time elapses from the release of a job until it is completed. During this time, as the real world intervenes, there will be unanticipated customer orders or customer-requested schedule changes, as well as unexpected problems and other events. You need a control system to cope with these changes and to help you navigate the "turbulence" by making real-time modifications to execute the plan as best as possible. POLCA is designed to be just such a system, but—just like the transatlantic flight—it needs to work in the context of an effective planning system.

So, what are the characteristics of an "effective" planning system that would enable POLCA to work well? Following are the main prerequisites related to planning that you should have in place before embarking on the use of POLCA.

- You should have the ability to perform rough-cut capacity planning over your typical planning horizon. The length of the planning horizon depends on the lead times for your products as well as the volatility of your demand, but, for example, in many companies, a capacity planning horizon of one to three months is normal. If you have a high variety of products and/or customized products, you won't be able to predict actual demand, but this requirement states that you should be able to roughly forecast average workload on your main workcenters using some aggregate planning models. *If your available capacities are significantly below the actually required levels during the period, then POLCA will be limited in its ability to help you achieve your plan.*
- As part of the rough-cut capacity planning, there should also be a provision for sufficient spare capacity in the system. While the precise amount of spare capacity would differ based on the variability of demand and job characteristics, as a simple guideline, companies have found that planning for an average of 15% spare capacity on critical workcenters over the planning horizon is a good rule of thumb. Note that this requires an awareness from top management to support this goal, instead of trying to maximize utilization based on traditional performance measures. Once again, this is an important management issue discussed later in this chapter.

- Based on your rough-cut capacity planning, you should also be able to put in place reasonable lead time estimates for jobs. In particular, if your lead time estimates are unrealistically short, POLCA will not be able to help you achieve your delivery targets. This may result in people in your organization concluding that POLCA doesn't work, whereas the root cause of the problem is infeasible lead times. Thus, it is important to start with reasonable lead times. But don't be intimidated by this requirement—if you are unsure about setting reasonable lead times, Chapter 5 shows how you can allow for some safety margins initially and then fine-tune the system easily.
- You should review your batch-sizing policies to ensure that they are not creating excessively lumpy workloads. For instance, occasional large batches with long run times will have this effect. If this is the case, work on cutting down batch sizes, engaging in setup reduction programs if necessary.

Limit the Occurrence of Material and Part Shortages

If jobs at your factory are frequently delayed due to missing raw material or shortages of component parts due from *external* suppliers that are needed for an assembly operation, such problems will not be solved by POLCA and they need to be resolved first. It is important to understand the specific issue that such shortages create for the POLCA system. Let's say Cell N is a fabrication cell and it completes job MT5819 and sends it to Cell S, which is an assembly cell. With POLCA in place, these cells will be connected by an N/S POLCA loop, so job MT5819 will have an N/S card attached to it (Figure 4.1). Test Cell T is next after Cell S in this job's routing. Let's say Cell S goes through its Decision Time logic, MT5819 is the next Authorized job, and an S/T card is available, so job MT5819 is launched into Cell S. Now, suppose that during the assembly operation in the cell, the team finds it is short of a component that was supposed to be delivered by an external supplier. At the time the production plan was created by the MRP system, it was expected that this component would arrive prior to the commencement of this assembly operation, but the supplier is late in its delivery. The team cannot continue with assembly of this job and job MT5819 has to wait in Cell S until this part arrives.

Let's understand how this impacts the operation of POLCA. Job MT5819 has two POLCA cards attached to it, namely the N/S card and the S/T card

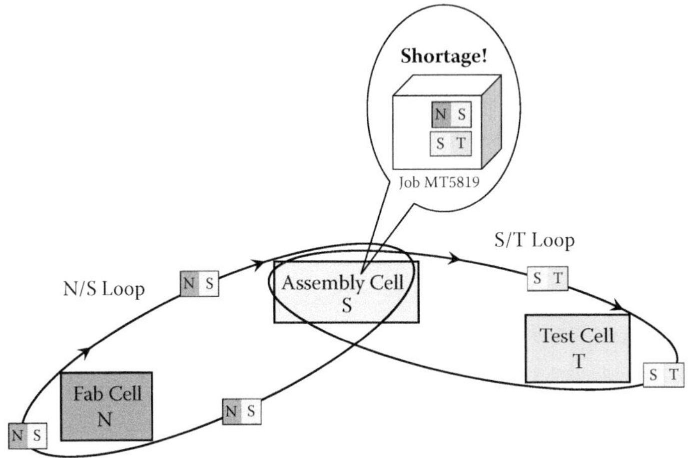

Figure 4.1 Missing component from external supplier detected while job is in process in Assembly Cell S.

(see Figure 4.1). While this job is waiting in Cell S, both these cards will be stuck with the job. In particular, the N/S card will not be returned to Cell N. However, the Cell S team isn't working on job MT5819, so it should signal to Cell N that it has capacity to start another job. But in the absence of such a signal, Cell N will not start another job for Cell S. Looking in the other direction, this job similarly is holding up an S/T card. There might be other jobs that are Authorized in Cell S and going to Cell T, and they could use this card. This means that Cell T is also going to receive fewer jobs than it should. If this situation only happens occasionally, then we don't need to worry about it. But what if there are a lot of component parts from external suppliers that are delayed in their arrivals to Cell S? It could be that eventually all the N/S POLCA cards are stuck with jobs in Cell S, and Cell N cannot work on any more jobs for Cell S. On the other hand, Cell S has a lot of available capacity (since it can't work on any of the jobs that are missing parts), but it won't get any jobs to work on. Similarly, several downstream cards will be stuck in Cell S and those downstream cells could also be starved for work.

The good news is that if this type of situation occurs only occasionally, then POLCA does have a way of dealing with it: see Chapter 6. However, the key point here is that before implementing POLCA, you need to review the frequency of occurrences of such material or part shortages. As a rule of thumb, if more than 10% of the time the reason that jobs are delayed is due to material/part shortages, then those issues need to be resolved first. If these occurrences are less than 10%, then the POLCA system can be used along with the special rules for shortages described further in Chapter 6.

In summary, if missing raw material or missing component parts account for a significant portion of the reasons for your jobs being delayed, you need to resolve those issues first, before proceeding with a POLCA implementation.

This completes the section on prerequisites that should be checked before proceeding to POLCA. You may find that addressing these issues helps to reduce or resolve some of the symptoms listed at the beginning of this chapter. However, even after putting in place all the prerequisites, if many of the symptoms still persist, it means that you do need a better control strategy and you can make the decision to proceed with implementing POLCA.

Prepare for the POLCA Implementation

Now that you have ascertained that you will implement POLCA, you need to engage in a few activities to prepare for implementation. These activities precede the detailed design of the POLCA system, which will be covered in Chapters 5 and 6. The remainder of this chapter will cover these activities.

Ensure the Availability of Authorization Dates and Lists

The POLCA system requires for its operation Authorization Dates and Authorization Lists. These concepts were explained in Chapter 2, but a brief review follows. For a particular job, for each cell in its POLCA Chain the Authorization Date represents the ideal date for when that job should be started in that cell. For a particular cell, at a given point in time, the Authorization List is a list of all the jobs that are yet to be processed in that cell. Note that the Authorization List only contains jobs that have *not yet* been launched into this cell. These two items (Authorization Dates and Authorization Lists) are then used in the Decision Time rules at each cell for when a job is to be launched into the cell, as explained in Chapter 2.

Hence, an important step in the preparation for POLCA is that these two items should be available. As discussed in Chapter 3, POLCA works both with an ERP system and without it. If you have an ERP system, the MRP logic is already in place within this system for producing the Authorization Dates and the Authorization Lists, with only minor adjustments as explained below. If you do not have an ERP system, these two items are easily generated using elementary operations available in standard spreadsheet or database packages. Both these options are now explained further. We will begin with the situation where these items will be obtained without using an ERP system, since this will explain how these two items are generated. If you have an ERP system then this first explanation will also form the basis to illustrate how you can obtain these items using your existing system.

Implementation Without Using an ERP System

If you do not have an ERP system, you can easily create the Authorization Dates and Authorization Lists with some simple calculations.

Let's begin with the Authorization Dates. We will illustrate the calculations via the example presented in Chapter 2, for a job that must start its processing in Cell A, then be processed in Cell B, and finally go to Cell G, and when it is completed in Cell G it is ready to ship. The POLCA chain for this job is shown again in Figure 4.2. Let's say the job has a due date of December 18 (for this company's operation, let's assume that the due date means the date by which the job must arrive in the shipping area). Next, suppose the planned lead times for the cells are as follows: five days for Cell A, six days for Cell B, and three days for Cell G. *We will assume here that move times (the times to move a job from one cell to the next) are included in the planned lead times; if not, you can easily adjust the calculations below in an obvious way to add these move times.* Finally, to keep the example simple and clear, we will assume the factory operates seven days a week.

Now we can perform the simple calculation for Authorization Dates for each cell. Essentially, the Authorization Date specifies the date on which the job should be started, in order to reach the next step on time. Refer to Figure 4.3, as you follow this explanation. We begin from the due date (December 18) and subtract the lead time for Cell G, three days, to get December 15 as the Authorization Date for Cell G. Similarly, subtracting

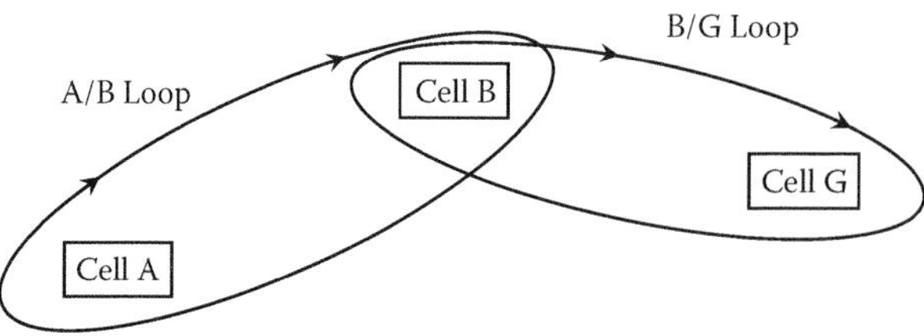

Figure 4.2 POLCA chain for Authorization Date example.

another six days from this, we get December 9 as the Authorization Date for Cell B. Finally, subtracting another five days gives us December 4 as the Authorization Date for Cell A. As you can see from the above and from Figure 4.3, the calculation of Authorization Dates is very simple, and can be easily programmed into a spreadsheet, or a database, or other easy-to-use software.

The next step in this calculation is to create an Authorization List for each cell. This is a list of all the jobs that need to be processed by a given cell in the foreseeable future (see Figure 2.2 in Chapter 2 for an example). This list can be created daily, or before each shift, or even more often, as may be required for a given company's specific situation. For this explanation, let's say the Authorization List is created daily, before the start of the day's operations. To create this list, we take all the jobs that have been accepted and have assigned due dates, but have not yet been completed. For each job, we determine the Authorization Dates for each cell in the job's routing, as already explained. Next, we look at this data by cell. For example, for Cell A, we look at all the jobs destined to visit Cell A and which have not yet been launched into Cell A. Next, we sort these jobs by their Authorization Date at Cell A (earliest date first). We also look up the next cell (the cell that follows Cell A) in the routing for each job, and include it in the data being displayed for that job. This sequence of calculations results in the Authorization List for Cell A. Again, you can see that these calculations only involve simple sorting and look-up operations, which are easily available in any commonly used spreadsheet or database software.

From the preceding paragraphs, this section has shown that your information systems (or IT) personnel can create the Authorization Dates and Authorization Lists with only a small amount of effort, which does not involve any complex programming.

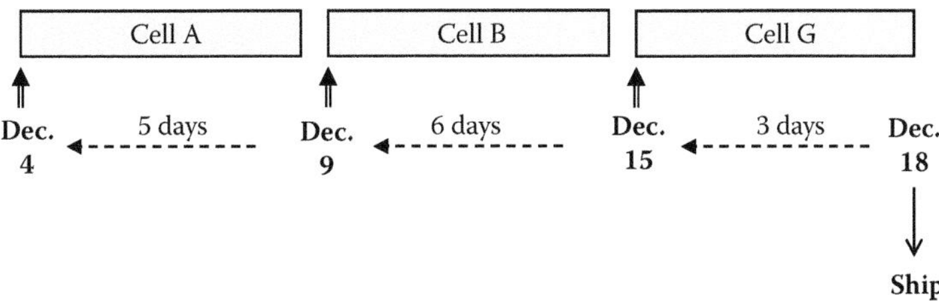

Figure 4.3 Calculation of Authorization Dates for the example. (Start with the ship date on the right and work your way back to Cell A to derive each of the dates.)

Implementation through Your ERP System

If you already have an ERP system, now that you have understood how the Authorization Dates and Authorization Lists are generated, it is easy to see how you can use this system to create these two items. An ERP system already contains standard MRP logic to calculate the "start dates" for each operation on a given job. These dates are obtained using simple backward scheduling from the due-date, taking into account the planned lead times for each operation in the job's routing, in a similar way to the procedure shown in Figure 4.3. The main adjustment to be made for POLCA is that *the dates need to be calculated at the cell level, not the operation level.*

Let's explain this adjustment with an example. Figure 4.4 shows a job that goes through six operations (Op1, Op2,..., Op6). In the standard MRP system this would be represented by a routing with six steps, with six separate lead times and six resulting start dates: SD1, SD2,..., SD6, as also shown in the figure. For POLCA, let's say four of the operations are performed in the first cell in the POLCA Chain for this job, and two are performed in the second cell, as shown in Figure 4.5. Then the first four

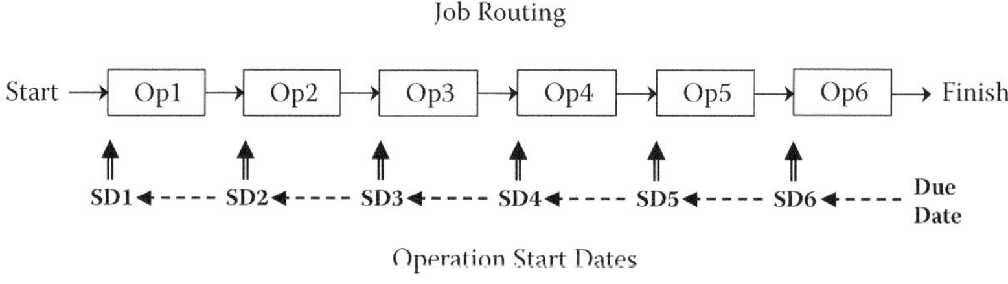

Figure 4.4 Start Dates calculated by standard MRP system for a job with six routing steps.

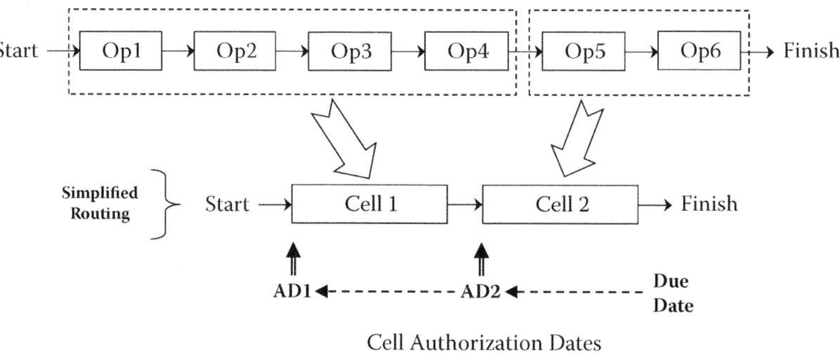

Figure 4.5 Authorization Dates calculated by standard MRP system for the same job now going through only two cells.

operations would be represented by one step in the routing, with one lead time, which would be the planned lead time for the first cell. Similarly, the next two operations would be represented by one more step in the routing, with lead time that is the planned lead time for the second cell. The minor restructuring that is required for your MRP system results in what is called High-Level MRP, or HL/MRP. However, the start date logic in the standard MRP system can still be used in the normal way. If a job goes through two cells, each with a given planned lead time, then as far as the MRP system is concerned, this just looks like a two-step routing. So, the MRP system can calculate its start dates using the above-mentioned backward-scheduling logic. As also shown in Figure 4.5, these start dates, now for each cell, will serve as the Authorization Dates (AD1 and AD2) required by POLCA! (The reason for using a different name for these dates, as explained in Chapter 2, is because a job may not actually be started on this date; additional POLCA rules also come into play before the job can be launched into the cell.)

If you have not already created the HL/MRP with cell-level routings then there is an important issue for you to review when setting the cell lead times. It is not a good idea to simply add together the existing operation-level lead times to get the lead time for a cell. Since each lead time in MRP includes a queue time, this would mean that you were adding together multiple queue times into the cell lead time, and this is likely to be much more slack time than is needed by the cell. To understand this issue better, consider the first four operations in Figure 4.5, which are combined into Cell 1. In the original MRP routing, each of the operations (Op1,..., Op4) would have a queue time included in its lead time. So, if you add these lead times together, you have allowed for four queue times. However, if Cell 1 is a POLCA-enabling cell, then as explained earlier in this chapter, the cell team with its ownership, cross-training, and focus on lead time is likely to keep jobs moving along through the cell once they are launched. So, one queue time for the cell might be sufficient, perhaps with a bit of an additional buffer to allow for minor queueing within the cell. At any rate, four queue times would create too much padding in the lead time (which would then result in too many jobs being released to the shop floor and excessive WIP). But don't worry too much about getting this lead time exactly right, because the POLCA system operation will help you fine-tune it, as explained in later chapters. A pragmatic approach to creating your initial HL/MRP lead times is as follows:

- If the cells have already been operating for some time, check the historical performance data for the cells and use that to set an initial lead time for each cell. (For readers familiar with statistics, the average lead time plus one standard deviation might be a good starting point containing a little padding but not too much.) Also, although individual jobs might have varying lead times depending on the number and complexity of their operations in a given cell, it is okay to start with one common lead time for a cell for two main reasons: first, this is simpler and requires less data calculation and far fewer inputs to your HL/MRP system; and second, the cell team has one target that it needs to meet for all jobs, so its goals are clear and not varying by job. Even though jobs will have different requirements, the POLCA system will make minor adjustments for each job as it flows through the cells.
- If you don't have sufficient historical data for the cells and have to rely on the existing operation-based lead times in your system, then the following will give a good starting point for the cell lead times. For a given cell, first estimate the average working time for all the operations performed in that cell (setup plus run times). Next add a reduced amount of queue time, for example, half the total queue time for the current operations. In other words, if a typical job has four operations performed in the cell, then add the amount equal to two queue times in your current operations. Based on discussions with your planning staff you can decide if you would like to be more aggressive (e.g., cutting the queue times by 75%) or more conservative (say, reducing the queue times by only 25% at first). In any case, note that it is important to review the total of these queue times and take at least some action—the case study of Alexandria Industries in Chapter 10 illustrates why doing so is important.

Next, let's discuss the availability of Authorization Lists. Again, an ERP system already has built-in MRP logic to produce what are called Dispatch Lists. These lists contain, for a given workcenter, details on all the jobs that are going through the workcenter, or are expected to go through the workcenter over a given time horizon (e.g., the next four weeks). In the HL/MRP context, the list just needs to be generated at the cell level, not the workcenter level. In addition, for the Authorization List, as part of the data for each job on the list, we need to include the next cell in the routing. Also, the list should only contain jobs that have not yet been launched into the cell. And finally, the list should be sorted by Authorization Date, with the earliest date first (in other words, at the top of the list). All these requirements are easily met by formulating an appropriate query in any modern ERP system. The result of this computation will produce the Authorization List as needed for a given cell at a given point in time, as shown in Figure 2.2 in Chapter 2.

As a reminder about one of the rules in POLCA, jobs with a date of today or earlier are termed Authorized, and jobs with a date of tomorrow or later are said to be Not Authorized. Therefore, as an additional option, companies can decide if they want all the Authorized jobs to be highlighted when the system produces the Authorization List, so that they are easily visible for the cell team members, as also shown in Figure 2.2.

Ensure That Manufacturing Metrics Will Not Conflict with POLCA Rules

It is important that you review your shop floor metrics prior to implementing POLCA. If your cell teams, supervisors, or factory managers are measured via metrics based on traditional notions of efficiency, this can be frustrating to the shop floor employees and can also result in the POLCA rules not being followed. Let's understand why. As seen from the Decision Time rules, there will be instances when no jobs can be launched into a cell. If machine utilization or worker utilization are key metrics, these will give teams or managers an incentive to launch a job even when it is not allowed by the POLCA rules. Similarly, if efficiency is an important metric, and there are two jobs waiting to be launched that have similar setups or tooling, a team may want to work on both these jobs. But what if only one is Authorized or only one has the right POLCA card, and before the second one can be launched, a third job gets the green light based on the POLCA rules? Again, the efficiency metric might tempt the team to ignore the POLCA decision and launch the second job instead.

So, what is the answer? The key is for management to shift its emphasis to metrics that are based on lead time—more specifically, the metric should encourage behavior that reduces the lead time for jobs to go through a cell. You can brainstorm ideas and construct metrics appropriate for your situation, but there are existing metrics that might be applicable to your situation and this will save you considerable effort. One such metric, becoming increasingly popular in the last few years, is Manufacturing Critical-path Time (MCT), explained in Appendix A. In that appendix, you will also see how to use MCT to construct another metric that helps motivate a team to reduce lead times in its cell.

If you are nervous about eliminating long-standing metrics, you can still monitor your existing metrics—such as utilization and efficiency—from a management level. However, these traditional metrics should be deemphasized for evaluating the shop floor personnel, and metrics based on lead time should be elevated in importance for these personnel. It may be difficult for management to let go of using the traditional metrics on the shop floor, but as company performance improves—such as reduced lead times and improved on-time delivery—management will be reassured that backing off those local utilization and efficiency metrics is justified based on the system-wide results that have been achieved.

Clearly, for metrics and incentives to be changed in a major way, you need top management to be on board and supporting the move to using POLCA. This leads us to the final, and arguably the most important, prerequisite for successful implementation of POLCA at your company.

Secure the Upfront Understanding and Commitment of Management

Several of the preceding sections have highlighted the importance of having top management on board prior to implementing POLCA. The first step to this is ensuring that a cross-functional team of senior managers understands the basics of POLCA, the motives for implementing it, and the changes that will be needed to support the POLCA system and ensure that the implementation is successful.

There are several areas for which this management commitment needs to be in place. Most of these items will be discussed in more detail in the following chapters; however, they are listed here to serve as a checklist for your reference. Your management must commit to:

- Signing off on a clear motivation and goals document for the POLCA implementation.
- Reviewing the structure of cells involved in the POLCA implementation, and if needed, ensuring that necessary changes are made so that these cells and teams qualify as POLCA-enabling cells.
- Defining a policy for spare capacity on critical resources.
- Accepting that the POLCA rules may occasionally result in some cell teams not being able to launch jobs, and being okay with the policy

that during such times teams can engage in activities such as continuous improvement instead of their production work.

- Rethinking shop floor metrics to support the points in the preceding two bullets.
- Investing in sufficient training for all those involved in the POLCA system and implementation. (Chapter 7 will provide more details on the types and extent of training that is recommended.)
- Assigning a POLCA champion, described in the next chapter.
- Engaging in change management methods to help ensure a successful transition to the POLCA system.

If the totality of these items seems too onerous for management, there is the possibility of beginning with a lower-risk option. Some companies have started their POLCA journey by first implementing the system in a portion of their factory, and then extending the implementation when it was clear to management and employees that the system was working and beneficial. For success of this approach, this chosen subset should be relatively self-contained with clearly defined boundaries, and interaction with the rest of the factory only at these boundaries. The way to properly implement POLCA in a portion of your factory is discussed in more detail in Chapter 5. An example of a company that started with POLCA in a portion of its factory is Alexandria Industries, whose case study is described in Chapter 10.

On the other hand, if your company is relatively small, or if it is larger but cells are generally connected to many other cells and it is hard to find a subset, it might be best to "bite the bullet" and proceed with implementing POLCA on the entire shop floor, so that there is no confusion and it is a level playing field for all jobs and all teams. As examples of companies that chose this path, see the case studies in Chapters 11, 12, 14, and 15.

This completes our description of activities that should be performed prior to designing your POLCA system. The remainder of Part II will take you through the details of designing your POLCA system, launching the implementation, and sustaining its operation.

Chapter 5

Designing the POLCA System for Your Situation

Once you are convinced that the prerequisites for POLCA are in place at your company—or at least that they will be sufficiently addressed prior to the POLCA launch—you can proceed with designing the system. This chapter will cover the major points that will help you design and customize the details of the POLCA system for your particular situation.

There are several details and possible special situations to consider, so this chapter is divided into the following major topics to help you navigate through all the issues:

- Creating a POLCA Implementation Team and designating a POLCA Champion.
- Determining all the POLCA loops.
- Laying out the POLCA Chains.
- Deciding on details related to the POLCA cards.
- Adding the finishing touches with some final details.

Each of these topics will be covered in one of the following sections in this chapter.

Create an Implementation Team and Designate a POLCA Champion

POLCA implementation needs to be carried out by a cross-functional team composed of manufacturing managers, planners, schedulers, selected operators from the cells involved, material handlers, and other shop floor personnel who would be influenced by the implementation. We also recommend including someone from Human Resources: the importance of this will become clear from chapters that follow. So, the first step in implementation is for management to put in place a *POLCA Implementation Team*. As part of this, at the outset, management should designate one of the team members as the *POLCA Champion*. This person should attend POLCA training sessions, do the background reading, and learn enough about POLCA in order to serve as the in-house expert on the system. Thus, the POLCA Champion can be the central point of contact to whom questions regarding the design or operation of the POLCA system can be directed. The POLCA Champion also serves as the liaison between the Implementation Team and upper management.

During the implementation, management should not expect the POLCA Champion to take on these tasks in addition to his or her regular duties. Depending on the size and complexity of the proposed implementation, management should reassign some of this person's duties. Based on successful implementations we can state that, typically, POLCA Champions have had 50% or more of their existing workload reassigned so they can devote sufficient time to the POLCA implementation.

Begin by Determining All the POLCA Loops

The first step in designing your POLCA system is to decide on the set of POLCA loops. This can be done using the procedures described in this section, which will cover these points in following subsections:

- How to identify all the POLCA loops if you want to implement POLCA throughout your shop floor.
- An important note about not including stocking points in the POLCA loops.
- The process for determining the loops if you will implement POLCA in only a subset of your shop floor.
- Deciding on whether to put in single or multiple loops for a given area.

Based on the goals that you laid out during the Pre-POLCA Assessment (Chapter 4), decide first whether you want to implement POLCA throughout your shop floor, or if you want to begin with a test implementation as a proof of concept for a limited portion of your operation. Reviewing the case studies in Part III of this book will help you with this decision. In general, smaller companies have decided to launch POLCA at one time throughout their shop floor, while larger companies have started with a limited area as a test case. Examples of smaller companies that launched POLCA throughout their shop floor are BOSCH Hinges (Chapter 12), Szklo (Chapter 14), and Preter (Chapter 15); a larger company that started with a partial implementation and then extended it to the rest of the factory is Alexandria Industries (Chapter 10). A different factor to be considered is provided by the example of Patheon (Chapter 11)—although this is a large company, due to the interlocking of routings throughout their factory they decided that it would not be easy to demarcate a subset for a POLCA trial, and so they launched POLCA all at once for the whole plant.

Even if you intend to implement POLCA only in a portion of your shop floor, you should still read the section that follows as there are several situations that you should learn about which apply in both cases.

Implementing POLCA throughout Your Shop Floor

If you intend to launch POLCA throughout your operations, then the POLCA loops are determined using this procedure. The starting point is the process by which jobs are released to the shop floor. (As a reminder, we defined what we mean by "job" in Chapter 4.) Smaller manufacturing companies usually have a person typically called a planner or scheduler that releases jobs. In larger companies, there may be a group of people, and in a similar way they may be called the planning or scheduling group. We will refer to this person or group of people as the *Planning Cell*. (For now, we will assume that the main function of the Planning Cell is to review the set of jobs that should be released on a given day, and for each such job, checking that the needed material is in stock, and then preparing the work order and shop packet for the job. Later in this chapter we will consider the situation where the Planning Cell may be involved in some more tasks too.) If you have more than one planning team that can release jobs, then you will have multiple planning cells, and you can repeat the logic here for each of the planning cells. This could be the case, for example, when a company has two teams of planners, each one serving a particular subset of the company's customers.

We will continue our description for the case where there is one Planning Cell; let's call it PC1. You should actually start your POLCA loops at the Planning Cell; this was mentioned briefly in Chapter 2, and we will provide more reasons behind this recommendation later in this chapter. Therefore, you will need a POLCA loop from PC1 to each cell to which PC1 could release a job as the first step in the routing. Asking the Planning Cell team about this will identify this first set of loops. There are some additional considerations with regard to the starting point in a job's routing; these will be discussed in the section "Lay Out the POLCA Chains" later in this chapter.

You have now identified the first set of POLCA loops from PC1 to some destination cells. Next, take each cell on the shop floor in turn, and for that cell ask the Planning Cell team: When jobs are completed by this cell, what other cells might those jobs go to next? For example, if jobs from Cell A can potentially go to Cell D, G, and K, then you need to plan for three POLCA loops: A/D, A/G and A/K. You perform the same analysis for every cell, and this generates the complete set of POLCA loops.

If you are in doubt whether jobs flow between certain cells, analyze some historical data on routings of jobs, for instance over the last six months. But don't worry about getting it perfect because it is easy to add POLCA loops later! In fact, as job characteristics change or new products are added, you will need to modify your POLCA loops from time to time anyway. So, take a first stab at the set of POLCA loops and get started.

To assist you with documenting your POLCA loops, we now return to the convention for drawing the POLCA loops that was introduced in Chapter 2. This convention facilitates unambiguous interpretation of POLCA system diagrams. This will be important during the process of designing your POLCA loops, as well as for documenting the final loops for your employees. The convention is that standard POLCA diagrams must follow three rules:

1. The circulation direction of POLCA loops is always clockwise.
2. The part of the loop that represents jobs going along with cards from the origin cell to the destination cell has arrows on it.
3. The part of the loop that is the route for returning cards does not have arrows on it.

Figure 5.1 shows why this convention is needed. Although in Figure 2.3 in Chapter 2 we displayed the A/B POLCA cards, which helped to indicate that this was an A/B loop, in a more complex diagram of the shop floor with many loops you may not want to crowd the diagram with cards.

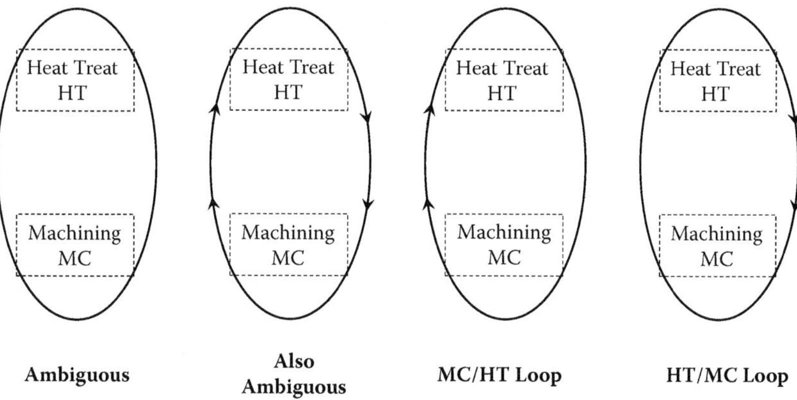

Figure 5.1 Illustration of the convention for clarifying the direction of POLCA loops.

So then, if you see a loop connecting two cells, such as the first loop on the left side of Figure 5.1, it is not clear if this is an HT/MC loop or an MC/HT loop. Also, if you put arrows on both sides of the loop to indicate circulating POLCA cards as in the second loop in Figure 5.1, this representation would be identical for an HT/MC loop and an MC/HT loop, so once again it is ambiguous. On the other hand, the two loops on the right side of Figure 5.1 show how this convention would clearly distinguish between the two situations. Also note that if you are reviewing a complex shop floor diagram with many POLCA loops, and you see a part of a loop without arrows on it, the convention tells you clearly that this is a path for returning cards, and further, even without the arrows on this part of the loop the direction of flow of the cards is clear because it is always clockwise.

The depiction of the actual POLCA loops on a shop floor diagram is an important way of conveying the flow of POLCA cards to the personnel involved. However, for situations involving a large number of POLCA loops on a complex shop floor layout, there is an alternative that keeps the diagram less crowded, and that is to use arrows linking each originating cell with its downstream cells. This approach is used in the descriptions of some of the case studies in Part III of this book; for examples, see Figures 10.6 and 11.1.

We now remind you of the term *POLCA Chain* introduced at the end of Chapter 2. This refers to the sequence of POLCA loops that a job will encounter as it goes through its complete routing, using the metaphor that the POLCA loops are like interlocking links on a chain as shown in Figure 2.9.

As part of determining the POLCA loops, you need to identify cells that are possible ending points for a POLCA Chain. Note that some cells may be in the situation that they are the final steps in some jobs' routings, but they are intermediate steps for other jobs. For example, a machining cell may be the last step for parts that are being shipped out as spares to customers, but it may also deliver parts to an assembly area in the factory. There are some important considerations for cells that serve as the end point of any POLCA Chain: these are discussed in a following section.

It isn't necessary that the POLCA loops only go between cells. You can also include standalone resources. For example, if you have an oven, or a specialized machine, and jobs need to go from various cells to that resource and back, you can include that resource in the POLCA loops as well, just as you would a cell. This situation was shown in the MMC company example in Chapter 2, and in the POLCA Chain of a job at MMC in Figure 2.9.

There is also no problem if jobs go back and forth between two cells (or standalone resources). Using the situation illustrated in Figure 5.1, say a job starts at the Machining Cell (MC) then needs to go to Heat Treatment (HT) for some stress relief, and then returns to MC for additional machining operations. Then you will have an MC/HT loop with MC/HT cards as well as an HT/MC loop with HT/MC cards, and the normal POLCA rules will work just fine for jobs going in either direction. Figure 5.2 shows how you can clearly indicate both loops in your POLCA diagrams. (Note: if your routings can result in a cycle, you need to be careful; this is discussed in detail in Chapter 6.)

If jobs leave your shop floor to go to subcontractors, you can also incorporate this in your POLCA loops. For example, let's say some jobs need to go off-site for a plating operation and then they return to your factory for finishing operations. This situation can be easily managed within the POLCA system, and is discussed later in this chapter.

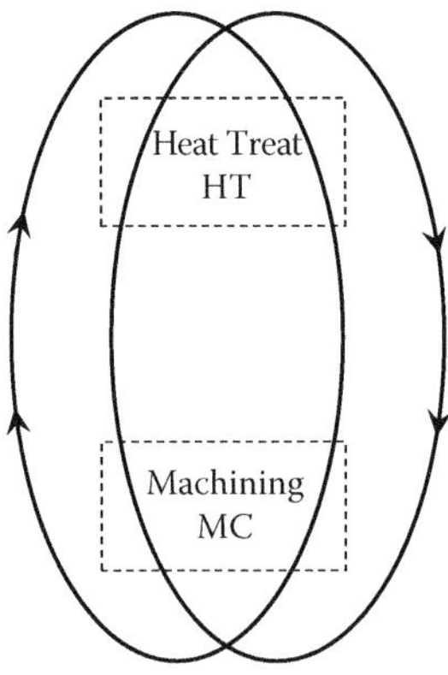

Figure 5.2 It is possible to have loops going between cells in both directions. The figure illustrates how both the loops can be clearly shown in a diagram.

Do Not Include Stocking Points in the POLCA Loops

In the process of identifying the POLCA loops, you must bear in mind a critical rule: *Do not include any stocking points in the POLCA loops*! If jobs sit in staging areas as they are waiting to be moved or to be processed at the next cell, this would be considered work-in-process (WIP) and is okay. However, areas where parts sit in buffers as some kind of intermediate stock or finished goods stock *must not* be included in POLCA loops. Consider a company that manufactures shafts and has a cell that prepares stainless steel bar stock for later operations. Let's call it the Turn and Cut (TC) Cell. TC consists of machines that straighten long bars, then turn their outside surfaces to four different diameters, and next cut the bars into seven different lengths, and finally mill and chamfer the faces of the cut parts. This results in 28 different cut parts. These 28 parts are made ahead of time and stocked in a buffer. When an order is received that uses some quantity of one of these parts, that quantity is picked from the stocking point and proceeds for additional machining operations. This is an example of a buffer of semi-finished goods that has been put in place to reduce the lead time for customer orders. You must *not* instrument a POLCA loop to go into or out of this type buffer, nor into a finished goods buffer.

Working through an example will help to see the reason for this rule. For the shafts company above, let's call the buffer SFS, for semi-finished shafts. If you (incorrectly) put in place a POLCA loop from TC to SFS, consider what will happen to jobs that are placed in SFS if those particular parts wait for several weeks before they are needed for an order. All those jobs will carry TC/SFS POLCA cards that will not be returned to TC. Meanwhile, TC has capacity and could work on jobs to make other parts that might be needed in SFS. But the lack of TC/SFS POLCA cards will shut TC down and it won't work on those jobs. This is obviously not the right course of action, since TC has the capacity to do those jobs that are needed. Hence the correct approach is to *not* have a POLCA loop from TC to SFS.

This example once again serves to highlight a key difference between POLCA and Kanban, discussed in earlier chapters. Kanban is an *inventory* signal, and it could be one of the methods used to manage stock in a buffer such as the one above. However, POLCA is a *capacity* signal, and it would be a mistake to attempt to use POLCA to manage inventory. As discussed in earlier chapters, this is one of the reasons that Kanban is not the right approach for HMLVC environments, while POLCA is effective in these situations.

Note that you can still use POLCA when you have semi-finished or finished goods stocking points in your operations, but you need to be aware of how to use POLCA along with your MRP system to manage these stocks. Essentially, the MRP system—possibly guided by your planners—will create work orders to either use or replenish these stocked items. More specifically, to use a stocked item, the work order will start with the first cell that this stocked item must go to, with the stocked item being part of the material to be picked for the work order, while to replenish a stocked item, the work order will end at the stocking buffer. Then these work orders will simply flow through the POLCA system like any other job. (See the discussion surrounding the definition of a "job" in Chapter 4.) The key point to note here is that the POLCA Chain for the material going into the stocking point is terminated, and a new POLCA Chain is started when the material is picked, so we do not have a POLCA Chain that goes *through* this buffer. Later sections discuss in more detail about the starting and ending points for POLCA Chains, so those sections will also clarify how these work orders would operate.

Implementing POLCA in a Subset of Your Shop Floor

In the case where you have decided to start with a trial of POLCA in a portion of your operation, you need to decide which cells should be included in this initial implementation. To arrive at this decision, there are two main factors to consider:

1. The chosen subset of your operations should include products and processes that are experiencing significant issues described in Chapter 4—such as late deliveries, frequent rescheduling, numerous hot jobs, and other issues listed in that chapter—for which POLCA is a potential solution. The aim is that after implementing POLCA it should be clear if it helped alleviate these problems and make significant improvements in the key metrics for these products and processes.
2. The subset must be relatively self-contained in the sense that interactions with other job flows need to be limited in the following way. If a particular cell will be the downstream cell in any POLCA loop, then ideally, *all* jobs arriving at that cell need to be included in the POLCA system; in other words, all *arriving* jobs need to be carrying POLCA cards. The reason for this is as follows. A POLCA card being sent back to an upstream cell signals that there is available capacity in the

downstream cell. If some jobs are on POLCA and others are not, then the Decision Time rules are no longer clear at the downstream cell. When a job that is not on POLCA arrives at this cell, when should it be started? If you start such a job based on some other rule like "first-come first served," or based on its Authorization Date, then it would use capacity without using a POLCA card and hence invalidate the capacity signals for the POLCA system and could cause bottlenecks. Thus, the test of the POLCA system for your subset would not be fair in these circumstances.

So, during the process of choosing the subset, you should go through the logic of each cell visited by jobs that might be in the subset, and ask the same question as for the previous case: When jobs are completed by this cell, what other cells might those jobs go to next? For each of those destination cells, you have two choices: either the cell will be included in the subset using POLCA, or you can terminate the POLCA operation prior to this cell. For any cell that you decide to include in POLCA, you will then go through the same procedure, until you have identified all the cells that will be included and the boundaries of the POLCA loops.

An example will help to see how to use the process and that it is quite intuitive. Let's revisit the MMC factory described in Chapter 2 and shown again here in Figure 5.3. Let's say the cells at MMC are working well within

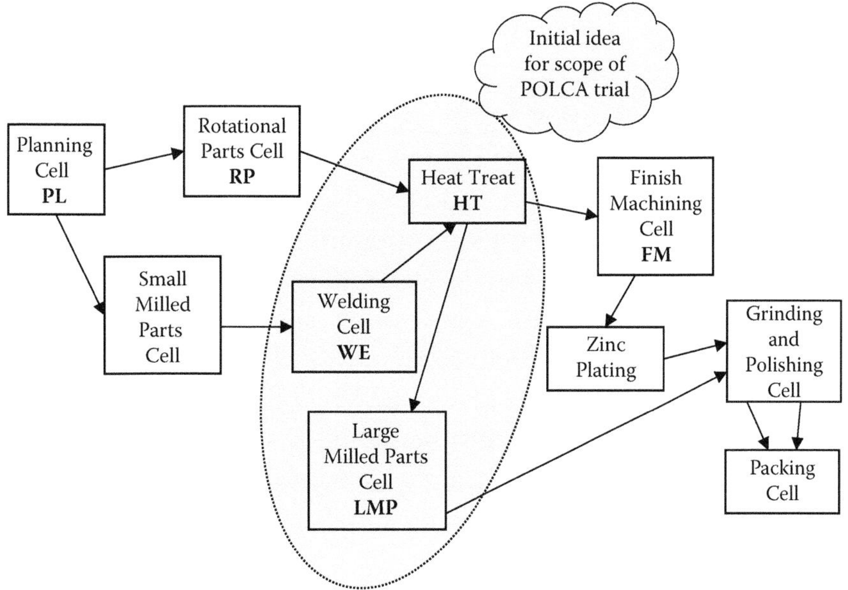

Figure 5.3 Initial idea about the portion of the factory for the POLCA trial at MMC.

themselves, but the teams are complaining that there is a mismatch in the schedules and workload through the Heat Treat (HT) operation, with high variability in the lead times through HT, and as a result, a lot of expediting and rescheduling of jobs through this area of the shop floor. The company's preliminary analysis of various factors (as described in Chapter 4) indicates that POLCA might be a good solution for the company as a whole, but first they would like to try it out in a limited way for this portion of the operation.

Upon further investigation, they find that the Welding Cell before HT and the Large Milled Parts Cell following HT are the most impacted by the expediting and rescheduling. Let's denote these two cells by WE and LMP respectively. So MMC begins by considering a limited POLCA implementation involving only these cells; the POLCA loops they are considering are WE/HT and HT/LMP, see Figure 5.3. (If this seems too small of an implementation, note that Alexandria Industries started with just such a small implementation and immediately experienced substantial results which convinced them to go with a broader implementation, as described in Chapter 10.)

The next step is for the company to go through the "self-contained" tests explained above. Since HT will be a downstream cell, all jobs arriving at HT need to carry POLCA cards. For this example, let's assume that all the possible job flows at MMC are represented by the arrows connecting the boxes in Figure 5.3. From the figure, you see that the Rotational Parts Cell (RP) also sends jobs to HT. This implies that for a successful trial of POLCA, the company must include RP in the subset for the trial. Next, note that LMP will also be a downstream cell. However, since it only receives jobs from HT, and these will be on POLCA, there is no problem. This concludes the logic for verifying that the subset will be self-contained, and the result of this logic is that MMC should, at the minimum, include cells RP, WE, HT and LMP in the subset for their trial POLCA implementation. At this point the implementation team at MMC notices that since RP receives jobs from the Planning Cell (PL) it might help to smooth out RP's operation if they also include a POLCA loop from PL to RP. While this is not required by the logical tests above, it makes sense to boost the potential for success in the trial, and having smoother flow through RP will only help. This points out that some common-sense brainstorming should also be used to accompany the decision on the subset to be chosen. As a result of all this reasoning, the final subset for the trial is shown in Figure 5.4. Also, to be clear, when jobs

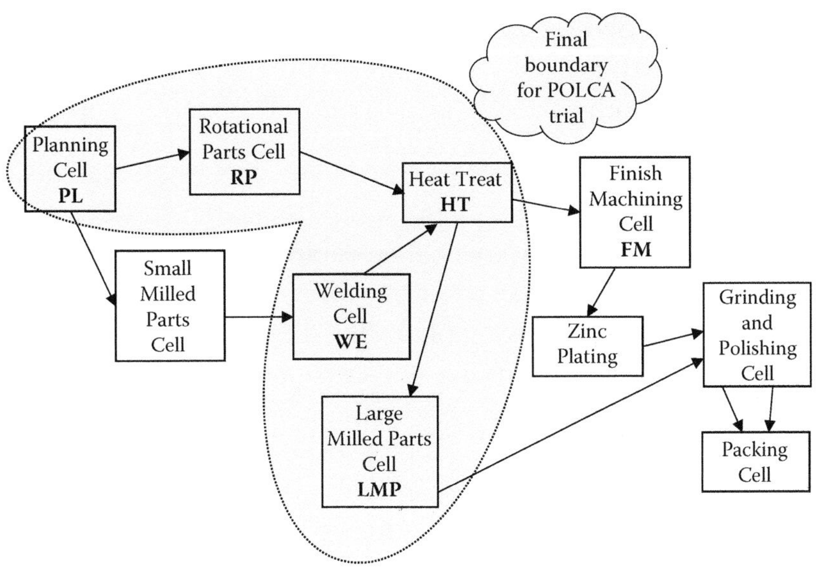

Figure 5.4 Final decision about the boundary for the POLCA trial at MMC.

leave the boundary of the trial area, they will go back to flowing according to the existing system that is in use at the company.

In looking at Figure 5.4, you may be concerned about one aspect, which is that jobs from HT also go to the Finish Machining Cell (FM), which will not be included in POLCA. Perhaps you are already asking the question, "HT will have jobs going to LMP which is on POLCA, and to FM which is not on POLCA, so won't this be a problem?" The key to the answer here is that for a downstream cell, the logic given above is to check that all *arriving* jobs need to be on POLCA. But that test is not needed for *departing* jobs! A later section on "Determining the Ending Points for Your POLCA Chains" will explain how this will work fine for HT. Similarly, PC will be dealing with jobs that are on POLCA and going to RP, as well as jobs not on POLCA and going to the Small Milled Parts cell. Again, note that these are departing jobs, and so for the same reason as for HT, this is not a problem.

On a related note, from Figure 5.4 you see that the POLCA Chains that come from RP to HT will actually start at the Planning Cell, and this process has already been discussed earlier in this chapter. However, the POLCA Chain for jobs going from WE to HT will actually start at WE. So, this is a new situation: *a POLCA Chain starting part-way through the shop floor.* A later section on "Decide on the Starting Points for Your POLCA Chains" will also explain how to do this properly.

Should You Put in Single or Multiple Loops for a Given Area?

A company that manufactured large electrical control systems puzzled over a situation that serves as a good example for this topic. The company fabricated large sheet metal cabinets in one area, and then these cabinets went to several assembly cells in other areas where the cabinets were stuffed with electrical and mechanical equipment. The cabinet area had four fabrication/assembly lines that could put together the cabinets, and then these cabinets were delivered to seven assembly cells. The question was whether to dedicate each cabinet line to serving one or more assembly cells, or to have the cabinet area available to serve any of the assembly cells according to need. The first case would require POLCA loops from each of the lines to the associated assembly cells, as shown in Option 1 in Figure 5.5, while the second case would include the whole cabinets area as one originating cell in the loops (Option 2).

The decision on which way to go is similar to the one encountered in creating cells. Dedicating the lines is beneficial when you can take advantage of similarities in the downstream operations' needs in order to make the upstream operations more effective, but the demand across each of these upstream dedicated lines needs to be fairly even and steady. For the company that we are discussing, it turned out that there was substantial variability in the mix of products being ordered and thus it was better to have the whole cabinet area available to make cabinets for any of the assembly cells. This provided more flexibility and so Option 2 was chosen by the company.

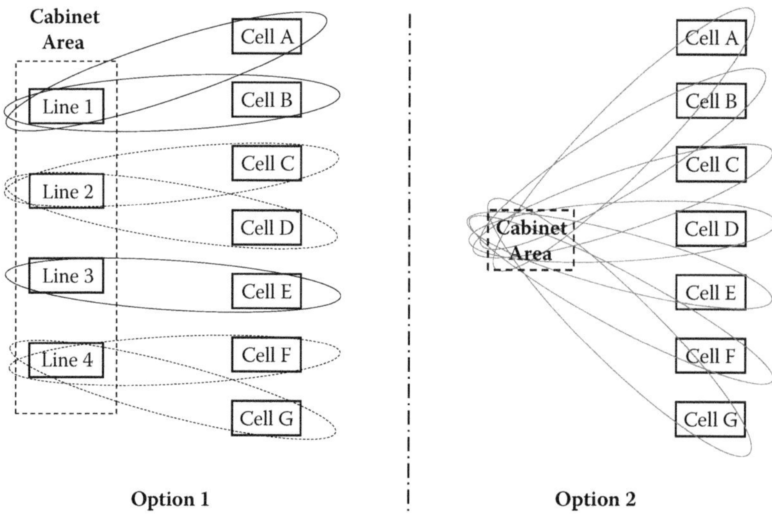

Figure 5.5 Two options for loops from the Cabinet Area to the Assembly Cells.

We have now completed the discussion of items related to POLCA loops. After you have decided on all the potential POLCA loops in your system, you need to determine how these loops will interconnect to form POLCA Chains for the routing of various jobs. This is covered in the next section.

Lay Out the POLCA Chains

The main items that you need to consider in designing your POLCA Chains are the following:

- The starting and ending points for all the POLCA Chains, including some considerations when implementing POLCA only in a subset of your shop floor.
- How to properly include the lead time for the Planning Cell.
- The procedure for incorporating subcontracting into the POLCA system if your routings include such outside operations.
- Deciding whether you will go with the standard POLCA described in Chapter 2, or an even simpler version, *Release-and-Flow* POLCA (explained below). To help with this decision, we identify situations where standard POLCA is recommended.

We begin by discussing the starting point for the POLCA chains.

Starting Point for the POLCA Chains

As mentioned earlier, if you intend to implement POLCA throughout the shop floor, the first loop in a given POLCA Chain should be from the Planning Cell to the first cell on the shop floor for that chain. Using the MMC factory as an example, from Figure 5.4 we see that the first shop floor cell for some jobs is the Rotational Parts Cell (RP). So, the first loop for those jobs will be from the Planning Cell (PL) to RP, as also shown in Figure 5.4. It is important to understand the reason for starting the POLCA Chain at the Planning Cell, and not at the first shop floor cell. Let's use the RP Cell as an example. This cell has a certain capacity, and a major aim of POLCA is to hold off on sending jobs to cells that are overloaded. Suppose it has been determined that there should be four POLCA cards in the PL/RP loop. (The process for calculating the number of cards is explained later in this chapter.) The Planning Cell, being part of the POLCA system, will also need to follow the Decision Time logic. So once PL has released four jobs to RP, and if no POLCA cards have been returned by RP, this indicates to PL that RP is backlogged. Releasing more jobs to RP would not be a good idea, for all the reasons discussed earlier in this book. Hence the POLCA rules kick in at this time, and PL will hold off on releasing another job until a PL/RP card comes back.

In this example, we are assuming that as part of its work in releasing jobs a Planning Cell also sends signals to material handling people to deliver the appropriate raw material to the first cell. Using the previous example, PL will deliver the shop packet with a PL/RP POLCA card to the RP cell, while also arranging for the material to be delivered to this cell. Alternatives to this arrangement are: (i) as part of its job, a Planning Cell team also delivers the material along with the POLCA card to the first cell; and (ii) instead, part of the first cell's job is to retrieve the necessary material for the job, in which case when the job is Authorized, the team should get the material, instead of waiting for it to arrive. You can use one of these methods, or a similar method appropriate for your company, for the operation at the first downstream cell from a Planning Cell.

Having the POLCA Chains start from a Planning Cell to various possible initial cells has another beneficial side-effect. It provides an effective feedback mechanism to the very people who are responsible for planning the

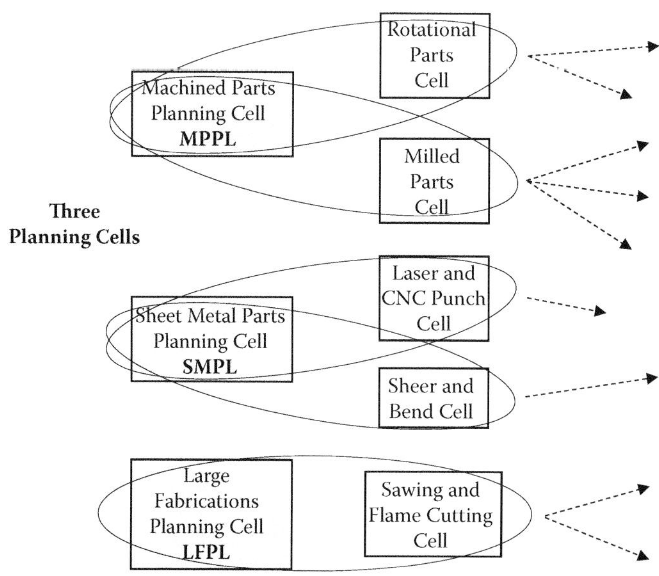

Figure 5.6 Illustration of starting points for POLCA Chains in a company with multiple planning cells.

work in the shop. If the personnel in a Planning Cell find that they are often waiting on a particular type of card, this indicates the need to add capacity at that downstream cell or to consider if jobs can be diverted to a different cell. Conversely, if there always seem to be enough cards of a particular type, then either a card can be removed from that loop (which helps to tighten the WIP control in that direction) or they can rethink how to use the capacity in that downstream cell.

As mentioned, larger companies may have multiple planning cells that are responsible for different areas of the shop floor, or for different segments of the business. In that case, each of those planning cells can be the starting point for the relevant routings. Figure 5.6 shows a company with three planning cells—MPPL, SMPL and LFPL—each of which is the starting point for some POLCA Chains.

Including the Lead Time for the Planning Cell

Because the first step in a job's routing is now a Planning Cell, you also need to include the lead time for this cell in the POLCA system design. Specifically, this will affect the calculations of both the Authorization Dates and (as will be discussed later) the number of cards in the relevant loops. We will show how to determine the lead time for a Planning Cell with a few examples. These examples also illustrate the different types of tasks that might be performed in a Planning Cell.

- Suppose that each day the Planning Cell's tasks include reviewing all the jobs that are Authorized to be released on that day, and for a given job, checking that the needed raw material and necessary components are in stock, making sure that all the paperwork is available, and preparing the work order and shop packet for the job. For instance, the work order could include the number of pieces and the routing with the full sequence of operations, and the shop packet could include detailed drawings and other instructions. Since the Planning Cell is included in a POLCA loop, *it must also follow the Decision Time logic before launching work into its cell.* So, the Planning Cell team will look at the next Authorized job, next check if the material is available, and then make sure that the right POLCA card is available. Only then will it start working on preparing the work order and shop packet for the job. As each job's packet is ready, the Planning Cell will deliver the packet to the first shop floor cell. This task is performed throughout the day for jobs in the list for that day. So, some jobs get released earlier in the day, and some toward the end of the day. But typically, based on past performance, you know that all the jobs for a given day can be handled during that day. Then a Planning Cell lead time of one day would be a good choice. This number should be used to calculate the Authorization Date for when the Planning Cell should review a job for release.

- In the situation where the Planning Cell just does a quick review of each order that is Authorized, with minimal paperwork and other processing involved, and this does not take much time, you could simply assume a lead time of zero. As you will see later in this chapter, there is already a safety margin used in the POLCA system when calculating the number of POLCA cards, so the small amount of time used in the Planning Cell will be contained in this margin, and the value of zero will work just fine.
- In the opposite situation, let's say there is high variability in the arrivals of orders and also there are several tasks to be done for each order. For instance, the Planning Cell may also be responsible for preparing or obtaining the NC programs needed at the machines in each job's routing, as well as issuing orders to get the tooling prepared at each machine. In this case, the Planning Cell could sometimes be backed up with work. So now you should estimate the average time that orders wait for the Planning Cell team to process and release them. If actual data on this is not easily available, a rough estimate based on discussions with the Planning Cell team will be good enough for a starting point. Or if you prefer to have better data, then for two or three weeks during the period that you are designing your POLCA system, you could collect actual data on the time that jobs take to go through the Planning Cell. Even if you have to collect this data manually, you only need to do it for a short period to get the initial lead time estimate. After that, the POLCA system will help you to refine the numbers through performance feedback to the Planning Cell, as explained above.

The decision on lead time for the Planning Cell will then be used in determining the number of cards in each of the POLCA loops originating at the Planning Cell. This calculation is explained later in this chapter.

Starting Points for POLCA Chains When Implementing in a Subset of Your Shop Floor

If you have decided to begin with installing POLCA in only a subset of your factory, there are two possibilities for the starting points of the POLCA Chains:

1. *The first shop floor cell in a job's routing is included in POLCA*. For such a job, the POLCA Chain will start at the Planning Cell and go to the first cell, as explained previously. (An instance of this was the PL/RP loop in the MMC example above.) Note that some jobs will be on the POLCA system and others will not, so the Planning Cell will be dealing with jobs that are both on POLCA and not on POLCA. But this okay, because the rule stated earlier was for *downstream* cells, and the Planning Cell is an *upstream* cell in this loop. Specifically, your lead time calculation for the Planning Cell is based on current operations where they are doing the work to process all the jobs. The only change will be that for jobs that are on POLCA the Planning Cell will need to follow the Decision Time rules to release and move the jobs to the first cell on the shop floor, while for jobs that are not on POLCA, they can just release them in the same way as they are already doing today.
2. *The POLCA Chain starts part-way into a job's routing*. This is the case when the boundary of your POLCA trial is such that some jobs will start at cells that are not on POLCA, and then encounter a POLCA loop at an intermediate routing step. This is the case in Figure 5.4 where you see that the POLCA Chain for some jobs can start at the Welding Cell, in other words, part-way into the routing of a job. In this situation, it is important to train the teams in cells that will be the first step in a POLCA Chain. Based on the logic described previously, a downstream cell in a POLCA loop must have all jobs on POLCA, so the normal Decision Time rules will apply for the team at this cell. However, the first cell in a POLCA Chain is an upstream cell. It could be that this first cell includes jobs that are not on POLCA. To illustrate this, let's extend the MMC example so that the Welding Cell also deals with jobs that can go directly to Packing, which is not included in the boundary of the POLCA trial (see the "Additional routing" in Figure 5.7). This means that there are jobs being processed in the Welding Cell which

do not require POLCA cards. The team in the Welding Cell therefore needs to be aware of the fact that if jobs are going to Heat Treat, then it must use the Decision Time rules for those jobs. We can state this more generally as follows. If an upstream cell in a POLCA loop processes jobs that are going to downstream cells that are not on POLCA, the team should be trained to know that jobs going to cells on POLCA can only be launched using the Decision Time rules, while other jobs can be launched based on Authorization Dates (without requiring a POLCA card). Also, such a cell will need to have a bulletin board for POLCA cards, even though the cell does process some jobs that are not on POLCA. So, for the example in Figure 5.7, with the loop from the Welding Cell (WE) to Heat Treat (HT), the Welding Cell will need to have a board where it keeps the WE/HT cards, even though some jobs in WE are not included in POLCA Chains.

If only part of your factory is on POLCA, then you also have to consider the interface between POLCA and non-POLCA cells at the *end* of a POLCA Chain. In other words, a job's routing may go through a POLCA Chain but there may be additional routing steps at the end that involve non-POLCA cells. Next, we discuss the general issue of ending points for POLCA Chains as well as this situation.

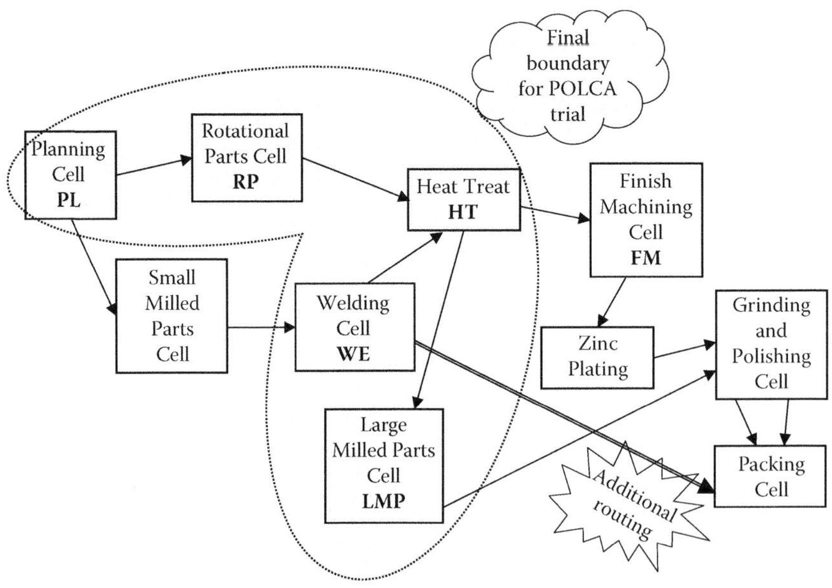

Figure 5.7 Additional routing from Welding to Packing.

Ending Points for the POLCA Chains

In designing your POLCA system, management of the last step in a job's routing also needs some thought. In manufacturing, the three most common ways for a job to finish its routing are:

1. The job ends up at a shipping dock, waiting to be shipped out;
2. It is placed in stock in a warehouse or other stocking point in the factory;
3. It is immediately incorporated into an assembly or into a fabrication such as a weldment, so it loses its identity and becomes a part of this bigger object.

You may think there are other possibilities, but they will typically fit into one of these three categories. For example, if parts needed for assembly arrive at a stocking point near the assembly area where they are stored, and are later picked when it is time for them to go into an assembly, this is an instance of category (2) above. Note that, for this reason, in category (3) we used the qualifier "immediately" to differentiate it from this situation. As another example, consider a long aluminum extrusion that gets cut into 10 shorter pieces. These pieces will now have new part numbers and will be put into a stocking area as these new parts. Once again, this can be seen as an instance of category (2).

We will use these three categories to illustrate the thinking needed for the last steps in a job's routing in POLCA, because this impacts the decision on the last link in a POLCA Chain. If you have a situation that is truly different, you can use the logic below as a model to help you decide on the right approach for your situation. For the first set of descriptions below we will assume that you have decided to implement POLCA throughout your shop floor. Later, we will consider the situation where POLCA is implemented in a portion of your shop floor. Here are the proper ways to end a POLCA Chain for each of the three categories above.

1. *The job will be shipped out when it is completed.* Suppose the last three steps in a job's routing are: Assembly and Test Cell (AT); Packing Cell (PK); and Shipping Dock (SD). AT and PK are capacity constrained. However, when a job gets to SD, it is simply placed in a staging area until it is loaded onto the appropriate transport. Depending on whether the job got completed early and/or whether the relevant transport

arrives as planned, the job will sit here for a variable amount of time; it takes up space, but no machine or labor capacity is involved. Thus, it makes sense to have a POLCA loop from AT to PK, but not from PK to SD. Hence the last link in this job's POLCA Chain will be the AT/PK loop. If this is not clear, an example will help. If you (incorrectly) put in a POLCA loop from PK to SD, consider what will happen if some trucks are delayed, but the jobs destined for those trucks have arrived in SD. Those jobs will all have PK/SD POLCA cards that will not be returned to PK. Meanwhile, PK has capacity and could work on jobs that need to go on other trucks that are expected to arrive on time. But the lack of PK/SD POLCA cards will shut PK down and it won't work on those jobs. This is obviously not the right course of action, since PK has the capacity to do those jobs and there are clear benefits to working on them! Hence the correct approach is to *not* have a POLCA loop from PK to SD, and to end the POLCA Chain at PK.

2. *The job is put into a warehouse or other stocking point.* Usually, jobs are dropped off at a staging area near the stocking point, and then the warehouse operators put the material into storage. Going with a similar example as before, suppose the last three steps of a job's routing are Assembly and Test Cell (AT); Packing Cell (PK); and Warehouse (WH). In this example, we are assuming the job still needs some sort of packaging before being placed in storage, hence the need for PK. If you think it through, the situation for this job's routing is actually no different than for the example that ended with shipping. For the same reasons as before, you don't want POLCA cards sitting with jobs in the staging area for the warehousing, so the last link in this job's POLCA Chain should be the AT/PK loop.
3. *The job is immediately used in an assembly.* Suppose a company makes custom gearboxes, and it has a Small Gearbox Assembly Cell (SGA). SGA receives gears from a particular Gear Cell in the company (say G4) and shafts from a Shafts Cell (say S7). The gears and shafts are customized for each gearbox order so they are not stocked; they are made to order and delivered to the assembly cell as needed. When SGA completes a job, it sends it to a Load Test Cell (LT). In this situation, we will simply use the normal POLCA rules for these cells. (Thus far, we have only discussed jobs which use material coming from just a single upstream cell. It turns out that the POLCA rules work fine even for the case where there is an assembly that uses jobs from multiple upstream cells. We will demonstrate this briefly here, and in more detail in Chapter 7.)

Let's review how the POLCA rules will work in this situation with two arriving parts going into the assembly job at SGA. G4 and S7 will follow the usual Decision Time rules to launch jobs and send them to SGA. In turn, SGA will also follow the normal Decision Time rules. However, when it gets to Step 2, since it is an assembly cell, it needs to check that all needed materials have arrived, which includes the gears and the shafts. And then of course it will need to make sure that it has an SGA/LT card, as in Step 3 of the rules. After the job is launched and then completed in SGA, the gearbox with the SGA/LT card will be sent to LT, while the G4/SGA card will be returned to G4, and the S7/SGA card will be returned to S7. So, these cells are all following the standard POLCA operating procedures, and nothing unusual needs to be done in this case, other than the fact that *two* cards need to be sent back upstream. Also note that the POLCA Chains for both the gears and the shafts *will end* at SGA, but the POLCA Chain for the gearbox will continue to LT. Now let's remind ourselves of the benefits of POLCA in this situation where jobs are delivering parts that are expected to be used in an assembly right away. Suppose a gearbox typically requires six different custom parts being made in upstream operations. Next, if the cell SGA starts getting backed up, then without any POLCA controls you could have a number of such parts being delivered and occupying space in and around SGA. If SGA is behind its planned schedule by eight jobs, there could be 48 different parts placed in various areas near SGA. This would not only take up floor space, and possibly make movement of other materials awkward, but would also require searching and sorting of parts for the next gearbox to be assembled—not to mention the potential for damage or even loss of parts if their location is not properly noted. So POLCA helps in the usual way, by temporarily stopping the upstream operations from making these parts and thus avoiding all these bad consequences.

One more important aspect about the ending point of a POLCA Chain is the Decision Time rule for the last cell in the chain. Let's illustrate this with the example involving the Packing Cell (PK) and Shipping Cell (SD). Since there is no POLCA loop from PK to SD, PK is the last cell in this job's POLCA Chain, and there will be no PK/SD POLCA cards. So, when the team in PK gets to a Decision Time, it will go through Step 1 ("What's the next job on the Authorization List?"), and Step 2 ("Has the material for this job arrived from the upstream cell?"), as usual. Then when it gets to Step 3 ("Do we have the right POLCA card for this job?"), the team will need to

Madison Machinery Corp – Authorization List			
Cell Name: **Cell PK**		Date: **February 14**	
Job ID	**Authorization Date**	**Next Cell**	**Additional Job Data...**
JML	February 12	SD	No card needed
WJO	February 14	LPK	...
NMR	February 14	SD	No card needed
MLL	February 16	LPK	...
RWO	February 17	LPK	...
...	...	...	...
...	...	...	...

Figure 5.8 Authorization List for the last cell in a POLCA Chain.

look at the Authorization List to find the "Next Cell" for this job in order to determine the type of POLCA card needed—at this point, the Authorization List should flag the next cell with a comment such as "No card needed." This can be made more visible by color coding, such as the use of green highlighting to signify: "It's okay to go without a card." In summary, if the answer is "Yes" to Steps 1 and 2, then when the team gets to Step 3 it will immediately be allowed to launch the job into its cell. We illustrate these points with the example of the Packing Cell just described.

In this example, suppose the Packing Cell can either send jobs directly to the Shipping Dock, or else they can be sent to a Large Packing Cell (LPK) where they are combined with other large items for the same customer and put together in a bigger container. Figure 5.8 shows an Authorization List for PK. First, note that based on the date of the list (February 14), only the first three jobs are Authorized and these are highlighted in yellow to draw attention to this part of the list. Second, you see that jobs going to SD have an additional comment on the right and also, the "Next Cell" along with this comment are both highlighted in green as recommended above.

Define the Procedure for Incorporating Outside Operations into the POLCA System

If your job routings involve outside operations such as plating, or if you subcontract some specialized operations such as milling for very large parts, you can easily accommodate such routings within the POLCA system. Let's suppose that the logistics for the outside operations are managed by a team of people at your company; we'll call this team the Subcontracting Cell (SUBC). Typical activities conducted by this cell could include: choosing which supplier to use in the case where multiple options are available; negotiating the price and due date with the supplier; arranging transportation to and from the supplier; packaging the parts for transportation; and unpacking and inspecting the parts when they return. It may be that in your company this is the same team as the Planning Cell, in which case you can replace all references to "Subcontracting Cell" with "Planning Cell" and the procedures below will still work properly. We will further assume that there is a lot of subcontracting involved, and SUBC is capacity constrained, and you would like the POLCA system to both manage and highlight issues with SUBC's operation just as it would with any other shop floor cell. Here's how you should include SUBC in your POLCA system.

We will use the MMC factory again (see Figure 2.1), but now let's say that Zinc Plating (ZP) is an outside operation. Also suppose that we are planning to implement POLCA in the whole factory, not just part of it as we did earlier in this chapter. Let's revisit the POLCA Chain for Job A, which was explained in Chapter 2 and is shown again here in the upper half of Figure 5.9. Note that this chain is for the case when ZP was an in-house operation. Instead, now Job A needs to go from Finish Machining (FM) to SUBC, which will send the job out to ZP. Next, SUBC will receive the job when it returns from ZP, and then send it on to Grinding & Polishing (GP).

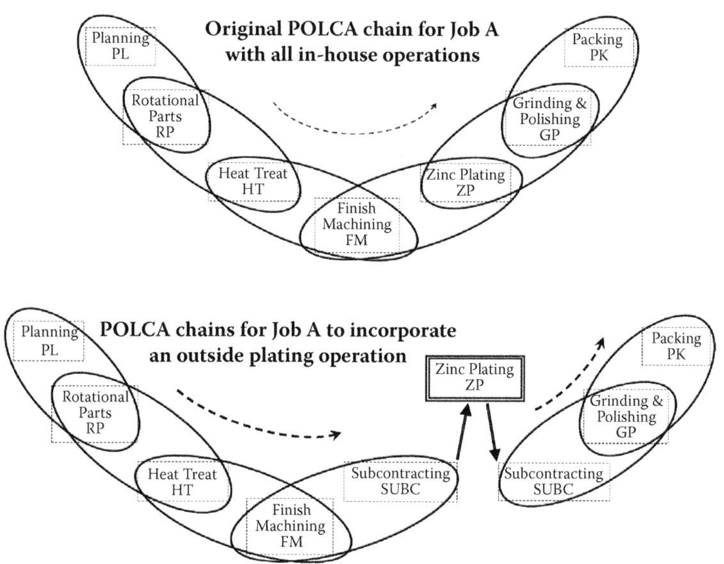

Figure 5.9 POLCA Chains for Job A when operations are in-house (upper diagram) versus when an outside operation needs to be included (lower diagram).

The first change we need to make to the previous POLCA Chain is that there will be a new loop from FM to SUBC. This loop will operate with the normal POLCA rules. However, there will *not* be a loop from SUBC to ZP. It is reasonable to assume that your subcontractors have other customers and your company is not involved in managing the capacity at the subcontractor's facility. So, it does not make sense to install a SUBC/ZP loop. Hence, the second change we will make is that POLCA Chain for Job A will *end* at SUBC. So, for this ending point, SUBC needs to follow the rules for the last cell in a POLCA Chain, as described previously.

Next, when Job A returns from ZP, SUBC needs to process the necessary paperwork or other procedures for receiving this job, and then send it on to GP. Hence, the third change is that we will start a *new* POLCA Chain for Job A from SUBC through to the end of the job's routing. At this point, SUBC is acting in the same way as the Planning Cell did when it released jobs to the shop floor, so SUBC needs to follow the same POLCA procedures that we described for the Planning Cell earlier, and in particular, it will need to follow the Decision Time rules for working on jobs.

Also, you need to accommodate the lead times for these new operations in your Authorization Date calculations. The lead time for ZP will simply be the planned lead time for this subcontracting operation (including transportation time). You also need to include lead times for SUBC in both POLCA Chains. These lead times can be calculated in the same way as was suggested for the Planning Cell earlier in this chapter.

While the preceding set of processes might sound a bit involved, in fact we are following all the standard POLCA rules. All that we did was to split the POLCA Chain into two chains; but within each chain the normal POLCA rules are being used, as well as the usual ways of calculating Authorization Dates. So, there should not be any confusion with the cell teams about how to manage the flow of Job A. The fact that this job goes outside the factory is essentially transparent to the other shop floor teams; this aspect is managed by the POLCA system and the work being done by the team at SUBC.

Consider Whether You Will Use the Simpler Release-and-Flow POLCA (RF-POLCA) Version

Even though the POLCA system is simple—as seen from the overview in Chapter 2—there is an even simpler version of POLCA, which has been used by several companies. This simpler version has still produced significant results at these companies. We call it *Release-and-Flow* POLCA or

RF-POLCA. In this version, the Authorization Date is used only at the first cell in the POLCA Chain for a job. All subsequent cells still use the rules for POLCA cards, but do not need to check Authorization Dates; jobs flow from cell to cell based on POLCA card availability only. Hence the name for this version: you release a job, and then it just flows through the system. Now we describe a few details that are important when implementing RF-POLCA.

First, note that the initial release of a job is no different than for standard POLCA. This release could be at a Planning Cell or, if you are implementing POLCA for a part of your shop floor, it might be at an intermediate cell as previously described. In either case, the standard Decision Time rules will be used for this step, so no new procedure is needed here. The main point to note is that you only need to generate an Authorization Date for this first cell. Let's illustrate this by revisiting the Authorization Date calculation example in Chapter 4. The upper half of Figure 5.10 shows the standard calculation that was explained in Chapter 4, and the lower half shows the process for the same POLCA Chain if it will use RF-POLCA instead. The total lead time for all three cells is 14 days, so we just need to subtract this number from 18 to get the Authorization Date for Cell A, which is December 4. Of course, since the cell lead times are identical in both cases, the Authorization Date for Cell A is the same as before. However, there are no Authorization Dates for subsequent cells.

Authorizations at each cell in standard POLCA

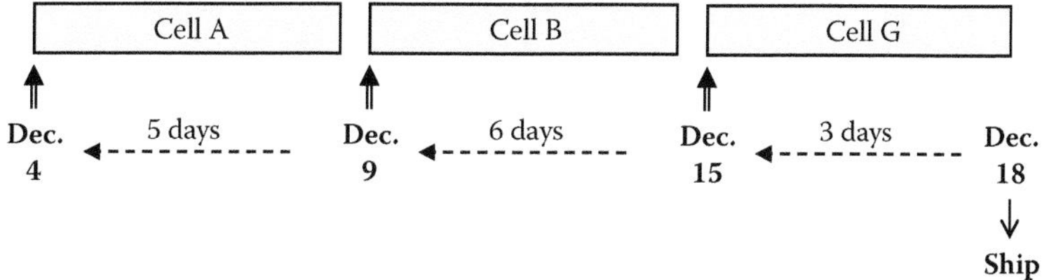

Single Authorization (Release) in R-F POLCA

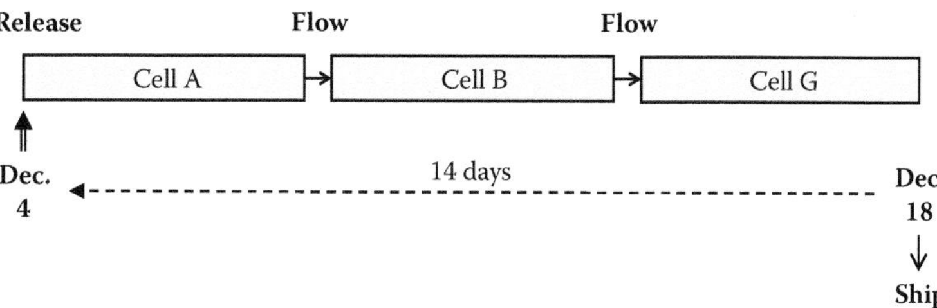

Figure 5.10 Comparison of Authorization Dates in standard POLCA and RF-POLCA.

Next, we discuss the operation at all subsequent cells in the POLCA Chain for a job. An important rule here is that jobs at these cells must be processed using a strict First-Come First-Serve (FCFS) discipline. This implies that as jobs arrive at a cell, the team needs to have a way of queueing them up so it knows which job is next. This could be done physically, by lining them up at a place near the cell, or virtually, by queueing them up on a bulletin board (perhaps a portion of the POLCA Board) using shop packets or other identifiers. Whichever method is used, it should be clear which job is next in line. Let's underscore why using a strict FCFS rule is important. Since there are no Authorization Dates being used, in the absence of such a rule, there could be a return to expediting and changing of job sequences, as supervisors or downstream cells exert their influence on this cell. There could also be the temptation for workers to cherry-pick jobs based on ease of setup or ease of processing. Since RF-POLCA has no further checks on dates once a job is released, it relies on reasonably consistent system behavior in order to achieve its due-date performance. Maintaining the strict FCFS rule helps to keep this consistency.

The next point to note is that the Decision Time rule at subsequent cells is now reduced to just two steps. The first step in the standard rule (Figure 2.6) is to look up the next job on the Authorization List; this is no longer necessary. And the second step is to check if the material for this job has arrived. In RF-POLCA the queue of jobs is based on jobs that have arrived, so this is already the case. In fact, these two steps get replaced by the single question: "What's the next job in line?" The next question in the standard rule now becomes the second question, specifically: "Do we have the right POLCA card to launch this job?" This requires the team to check if there is a POLCA card from their cell to the next cell for this job. Since the cell is not using an Authorization List, the team needs to have a way to look up the "Next Cell" for each job: for example, this could be available on the paperwork accompanying the job, or on a computer screen if team members are logging in their time and job launches. To summarize, in RF-POLCA the Decision Time rules for the first cell in a POLCA Chain are the standard rules (Figure 2.6), while all subsequent cells follow the simplified logic shown in Figure 5.11.

Note that since we recommend starting each POLCA Chain at the Planning Cell, only this cell will need to follow the standard rules starting with checking the Authorization List (Figure 2.6), while all shop floor cells will follow the simplified rules in Figure 5.11. So, this makes it easy to train the shop floor personnel as all the shop floor cells will be following this simplified rule.

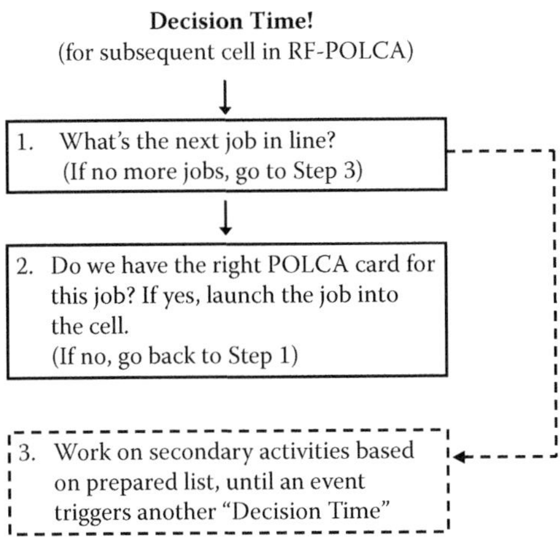

Figure 5.11 Simplified Decision Time flowchart for subsequent cells in RF-POLCA.

Situations Where Standard POLCA Is Recommended

Since RF-POLCA is indeed simpler and doesn't need the generation and lookup of all the intermediate Authorization Dates and Authorization Lists, why don't companies just go this route? The answer is provided by reviewing the benefits of POLCA described in Chapter 3. Many of those benefits derive from the combination of the Authorization process along with the POLCA card signals. Reviewing these benefits also helps us to list the factors that should help you decide on which version of POLCA to use. Specifically, the use of *standard* POLCA is recommended for the following situations:

- *When jobs can have long POLCA Chains.* If a POLCA Chain has a large number of links, there may be a lot of variability in the flow time of jobs and the initial release date may not be sufficiently reliable, which is why the re-checking of Authorization Dates at each cell is important to keep jobs on track for their due dates. For RF-POLCA to work in this case and reliably meet due dates, it would require a lot of padding in the lead time, which will add to the WIP and negate some of the benefits of POLCA. Typically, we recommend that if any of your POLCA Chains could have four or more links, RF-POLCA may not be the way to go.
- *If there is a lot of variability in the workload within each cell.* Even with shorter routings, if there is a lot of variability from job to job, the lead time at a given cell will not be sufficiently reliable for RF-POLCA to

work, or this lead time will require a lot of padding with similar consequences as above. Standard POLCA will help to deal with this situation by reordering jobs that are early or late.
- *When there are many intersecting POLCA Chains.* If jobs arrive at a cell from multiple upstream cells and are following different POLCA Chains through this cell, then the FCFS rule for jobs might not give good enough results. If there is a lot of variability in the upstream cells, as is likely the case if you are considering POLCA, some jobs could be delayed and others could arrive earlier than expected, so it helps to keep resequencing jobs using the Authorization Date rules. This, and the preceding two points, are part of the benefit explained in Chapter 3 in the section "System Is Adaptive."
- *If there are often schedule changes after jobs are released.* If the start-to-finish lead time for your POLCA Chains is long, and there is a possibility that customers and/or salespeople want to make changes in the scheduled dates after jobs are released, there is no way to accommodate this in RF-POLCA, while this can be done seamlessly in standard POLCA as the new dates will just be reflected in the Authorization Dates for remaining cells in a job's POLCA Chain.

Reviewing the case studies in Part III might also help you decide on this option. For example, Alexandria Industries (making aluminum extrusions, see Chapter 10) decided on RF-POLCA because their routings were short. Patheon (Chapter 11) is a larger company in the pharmaceutical industry, but it also went with RF-POLCA because every job has exactly five steps in its routing. On the other hand, BOSCH Hinges (custom hinges, Chapter 12) and Provan (metalworking subcontractor, Chapter 13), both small companies, chose to keep the Authorizations throughout the POLCA Chains since they have jobs with widely different routings and numbers of steps, so they implemented the standard POLCA.

You don't need to decide right at this point if you will opt for RF-POLCA; it might be better to return to this decision after you read the remaining sections of this chapter. However, if you do eventually feel that RF-POLCA is right for your situation, you should review this chapter as well as Chapter 4 to see how that will impact some of the details described in these two chapters (such as calculating and using the Authorization Dates).

Decide on Details Related to the POLCA Cards

This section covers several design decisions that are related to the signals conveyed by the POLCA cards and how they will be communicated through the shop floor. These decisions involve the following topics:

- Design of the POLCA cards.
- The *quantum*, or amount of work to be done per card.
- The process for returning cards to upstream cells.
- The number of cards in each loop.

We now discuss each of these topics in detail.

Design Your POLCA Cards

As part of deciding on the set of POLCA loops, you should assign a color to each cell or standalone resource and design the corresponding POLCA card for each loop. Figure 5.12 shows an example of a POLCA card. The card should be large enough to be visible from a distance and the most important information is the names of the two cells in the loop, such as the "F4/W2" label on the example card, which should also be in a large font. Along with the label, the card background has two colors corresponding to the two cells. The left half of the card has the color of the first cell in the loop (the "Originating Cell") and the right half has the color of the second cell (the "Destination Cell"). The colors help with visual management, allowing cell team members, planners, material handlers, and managers to identify a card from a distance. If you run out of colors, you can use cross-hatching,

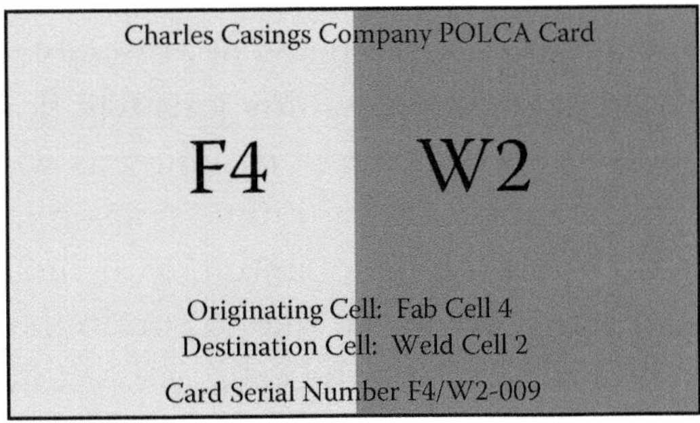

Figure 5.12 Detailed design of a POLCA card.

checkered patterns, or other creative patterns to denote a particular cell. Get your cell teams to help brainstorm these ideas so that they "own" the color or pattern allotted to their cell.

The card also contains detailed explanations of the abbreviations for the cells in case an employee is not familiar with a particular abbreviation. As shown in Figure 5.12, it is also important to put a unique serial number on each card that helps planners keep track of the POLCA cards and—through occasional audits described in Chapter 9—to determine if a card has been misplaced or lost.

Finally, for readers who are familiar with or already using Kanban, you should note that a POLCA card does not have any part numbers, which a Kanban card always has—this results in some of the advantages of POLCA over Kanban for HMLVC products as explained in Chapter 3 and Appendix C.

Decide on the Quantum—The Amount of Work to Be Done per POLCA Card

The next step in your POLCA card design decisions is to determine how much work a POLCA card represents; in a POLCA system this quantity of work is called the *quantum* and the motivation for it is as follows. A POLCA card returning to an upstream cell signals available capacity at the downstream cell. The question then is, how much work should the upstream cell send to the next cell along with this POLCA card?

The tradeoff involved is that, on the one hand, if the quantum is too large it will result in infrequent and lumpy flow; no work will be sent for a while but then a large amount of work will be sent to the next cell. Both these factors—variability in arrivals of jobs, and large batches—exacerbate queuing effects and waiting times, so this is not desirable. On the other hand, if the quantum is too small, many tiny jobs will flow between the cells and this will require a lot of material handling as well as a large number of POLCA cards in the loop, making it time-consuming to manage and keep track of them. In setting the quantum, you should also take into account the ideal batch sizes you would like to run in the cells. Finally, consider how material is moved between the cells: we will illustrate this with some examples below.

The quantum should be specified in an easy-to-use unit. The simplest quantum is a job (or work order). If jobs usually don't get too big in terms of their workload, then assigning a POLCA card to each job is the ideal situation and easy to implement because the rule is so clear: "one POLCA card = one job." Practical examples illustrating this choice are the implementations at the pharmaceutical company Patheon (see Chapter 11) and the custom hinge manufacturer BOSCH Hinges (Chapter 12). Another option is to specify the quantum in terms of work content, such as 20 hours of machining. (An approach recently proposed is to specify the quantum in terms of an innovative concept called *Capacity Clusters*: this new approach is described in Appendix D.) You can also specify a quantum in pieces. Combining this option with the consideration of your material-handling method can result in a good practical solution. For example, if housings are typically placed on a pallet that is moved by a forklift truck, then the quantum could be the number of housings that will fit on a pallet. This would work well with the material handling process as each pallet could carry a POLCA card attached to it while it is being moved. For a large housing, if only four can be placed on a pallet the quantum becomes four housings, while for a small housing, if nine fit on the pallet then the quantum is nine. The rule for this example is "quantum = one pallet." Chapter 8 explains why this type of reasoning proved helpful in deciding that "quantum = one cart" at a company making control system cabinets of varying sizes. Companies have used similar rules for other means of material handling to conform to what is typical in their operation. For instance, other rules that have been used are "quantum = one basket" and "quantum = one crate." Figure 5.13 shows an example of this last option; notice how the POLCA cards are

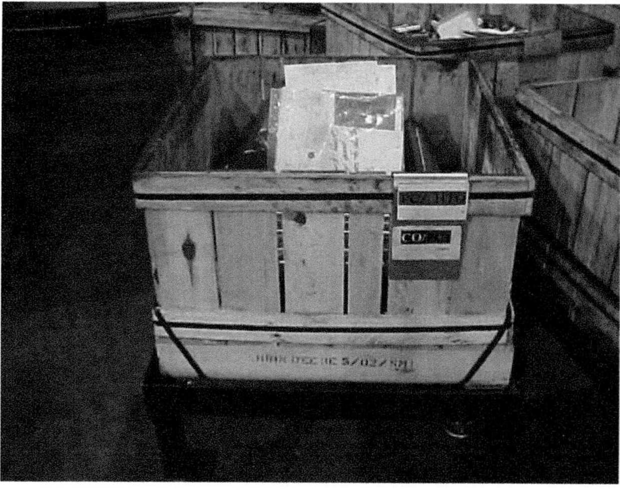

Figure 5.13 Material-handling method of using crates also works as a quantum here. In addition, the crate is being used as a holder of (up to) two POLCA cards.

prominently displayed on the material handling crate, and there is space for two POLCA cards since both are needed when the job is in process in a cell.

When the quantum is the same as a job or work order then POLCA works as described in Chapter 2. However, if the quantum is pieces or hours, here's how the quantum rule is used by the cell team. When a job is Authorized and the right POLCA card is available, if the workload in the job is less than or equal to the quantum, then the job is launched into the cell. If the workload exceeds the quantum, then one of two methods can be used: in the case that the job can be split, then the amount allowed by the quantum is launched into the cell along with one POLCA card. If the job cannot be split, then the team must wait for a sufficient number of POLCA cards of this type to arrive before the job can be launched at all. Let's go over these two options with an example.

If there are seven housings in a job and the quantum is four, then in the first option, four housings can be started in the cell. The remaining three housings will need to wait for the next POLCA card of the same type to arrive. Now let's consider the second option. Suppose we are in Cell A, the housings are going on to Cell D next, and the planners don't want jobs to be split (this could be for quality control or lot integrity reasons). Then the Cell A team needs to have two A/D POLCA cards before it can start work on the housings. Until the two cards are available, the team will need to skip this job just like it did when it was waiting for the one card to show up, and it can work on a job destined for a different cell (any cell other than D) if that job is authorized and has a POLCA card. In particular, it is important to note the qualification in the previous sentence: if there is another authorized job for Cell D that only needs one A/D POLCA card, the team *cannot* launch this job! Doing this could lead to an ongoing cycle where jobs "steal" A/D cards and the job needing two cards might not get launched for a long time! As part of designing the POLCA system for your operation, you need to decide which of the two options will be used so that the rules are clear for the cell teams.

One other point to note here is with regard to the number of cards in a loop. In a following section, we will go over the calculation of this number. However, keep in mind that if you go with the second option above (that is, that the full set of POLCA cards must be present before the job can be launched), then you should ensure that each POLCA loop has enough cards to accommodate the largest non-split jobs that you expect to release to the shop floor. For example, using the same housings example as above with a quantum of four, if the planner will not release any jobs with more than

20 housings, then you need to have at least five POLCA cards in any loops associated with the routings for the housings. If the calculation below results in a number that is five or more POLCA cards, then this is already accounted for. On the other hand, if the calculation results in less than five cards, then you need to go with five cards. However, if the calculation recommended a much smaller number, such as two cards, you might worry that going with five would often lead to too many jobs in the loop. If the larger jobs are relatively rare, you can go with the recommendation of two cards, and then add a "Safety Card" for the occasional very large jobs. The use of such exceptions is described in the next chapter. Just such a procedure was used by the pharmaceutical company Patheon to deal with occasional large batches that needed to be run all at once (see Chapter 11). A final point to be made here is that if large jobs are relatively rare, you should consider imposing a strict limit on your batch sizes and have your Planning Cell split very large jobs into two separate jobs (work orders). This will also be beneficial to your overall flow as it will reduce the potential for queueing at all cells that are part of this job's routing.

The quantum doesn't need to be the same for all POLCA loops. Based on the characteristics of each of the flows, you could have different quantum rules for different parts of your shop floor, as long as this doesn't cause confusion or too much job-splitting.

Determine the Process for Returning the POLCA Cards

Having decided on which cells will be connected by POLCA loops, as well as the quantum (or quanta, if there are several different ones), you need to design the process that will be used to return POLCA cards from downstream cells to the relevant upstream cells. It may be advantageous to design this procedure in parallel with the decisions on the various quanta, since those decisions will impact the number of POLCA cards that will flow between cells. Also, as mentioned in the previous section, sometimes a particular quantum decision can be based on the material handling method in use for the associated jobs.

Thus, it makes sense to also consider how jobs will be moved from the upstream cell to the downstream cell, because the factors involved actually impact both decisions: moving the job downstream and returning the card upstream.

As a reminder, let's review the POLCA-related tasks involved in the moving of jobs and cards between cells. When an upstream cell completes a job, that job and the associated POLCA card need to be moved to the downstream cell. When a downstream cell completes a job, the POLCA card which had arrived with that job from the upstream cell needs to be returned to the upstream cell. Note that since many cells will be contained in overlapping loops, for such cells when a job is completed we need to take care of both these tasks. Let's illustrate this with an example to make it clear. Suppose Cell B is part of an A/B POLCA loop as well as a B/C POLCA loop, and it is currently processing a job that was routed to Cell A, then to Cell B, and next to Cell C. Since the job came from Cell A, it has an A/B card attached to it, and since Cell B is processing this job, through following the Decision Time rules it must have a B/C card also attached to it that allowed Cell B to launch the job into the cell. So, when Cell B completes the job, two tasks must be performed: moving the job along with the B/C card to Cell C; and returning the A/B card to Cell A.

There are many options for how these two tasks can be carried out. The best way to illustrate the choices is to describe various procedures that have been implemented by companies that are using POLCA:

- If the two cells in a POLCA loop are closely located and the parts being moved are not too heavy and can be manually moved—for example, in baskets or using light carts—then companies have made it the responsibility of the cell teams to move both the jobs and the cards. So, when a cell completes a job, the cell team is responsible for moving the job and the associated POLCA card to the downstream cell, and the team is also responsible for returning the POLCA card that had arrived from the upstream cell back to the upstream cell. Of course, for the last cell in a POLCA Chain there will be no card to be delivered downstream, and likewise for the first cell in a POLCA Chain there is no upstream cell, so the relevant actions are not needed in those cases.
- In the case where your production deals with heavy parts and materials, and jobs are typically moved by specially trained material-handling (MH) operators, both these tasks could be made the responsibility of the MH staff. This could also be the decision in the case where parts are not necessarily heavy, but distances between cells are large and the task of moving jobs is performed by separate MH people, not the cell operators. In this case, completed jobs should be placed in a designated location for each cell; one that is clearly visible to MH operators. Similarly, cards that need to be returned should be put on a board in or near the cell, that is also clearly visible to MH operators. The MH staff

should be made aware that the job of returning cards is just as important as moving material; without cards, upstream cells will not be able to start jobs even if they have capacity and the jobs are Authorized, so this will result in unnecessary delays and longer lead times.

- At one company that makes large electromechanical products, it was necessary for the jobs to be moved by forklifts operated by specialized MH staff. However, the POLCA implementation team decided that returning the cards would be "everyone's job." What this meant was that anyone who walked by a bulletin board and saw POLCA cards that were waiting to be returned could decide, if they were headed in the direction of the upstream cell, to carry the POLCA cards back to that cell. This included managers, supervisors, MH staff, or an operator who happened to be walking through the shop floor for any reason. The company found that there were no problems with this arrangement; on the contrary, it had the benefit that it made everyone feel they were part of the overall POLCA process and cards were returned with minimal delays. The process of returning cards was also made everyone's job at Patheon, which experienced similar benefits of giving employees a sense of involvement and even ownership of the POLCA implementation (see Chapter 11).

A different way to tackle the issue of returning POLCA cards is to do away with physical cards altogether, and go for an electronic implementation of POLCA. The case study of BOSCH Hinges in Chapter 12 describes why the company decided to go this route and it helped in the development of a "Digital POLCA" system. This system is now in use at several other companies, including Provan (see Chapter 13). Another example of an electronic POLCA system is in Chapter 15. Indeed, these electronic implementations do away with having to manage the physical POLCA cards, and they have also been well received. Electronic POLCA systems also allow the use of a smaller quantum since the number of cards to be managed and returned is not an issue, and a smaller quantum can help smooth out the flow. However, in keeping with the aims of Visual Management, many companies still prefer the visual impact of the physical POLCA cards and POLCA boards, and even though managing the physical cards takes effort, their management believes the benefits are worth this effort. So, at the end of the day, this comes down to a choice by your POLCA Implementation Team as to whether you want to invest in an electronic POLCA system, or if you prefer some of the benefits of using the physical cards. Part III of this book has case studies of companies that chose one or the other of these options; reviewing these case studies will help you with your decision. In fact, some electronic POLCA implementations described in Part III aim to get the best of both worlds by retaining some of the visual benefits through the use of electronic POLCA boards.

As you can expect with physical POLCA cards being handled and moved in a complex shop floor environment, cards will occasionally be misplaced or lost altogether. You need to keep on top of this situation, because missing cards mean that a particular loop has fewer cards than needed, and this will impact your production performance. Chapter 9 will explain the recommended POLCA audit process which assists you in monitoring the availability of POLCA cards.

After you have decided on the POLCA loops, the quanta, the rules for starting jobs, and the procedures for moving jobs and returning the POLCA cards, you should document all these in a flowchart. In addition to being a handy reference for cell teams, this flowchart serves as a valuable tool in the initial training of the cell teams, planners, schedulers, and material handlers as well as any new employees in the future. This will be discussed further in Chapter 7, along with an example of such a flowchart.

Calculate the Initial Number of POLCA Cards in Each Loop

Now that you have decided on the POLCA loops and the quanta, there is a straightforward formula to help you calculate the initial number of POLCA cards you should have in each loop. We deliberately use the term "initial number" because this allocation of cards is expected to change as the system starts working, but as mentioned several times already, it is easy to determine when to add or remove POLCA cards, and equally easy to physically add or remove the cards from any loop. We will discuss this below and also in Chapter 7 in the section on POLCA audits.

We will use an example to explain each of the terms in the formula for the number of cards. Let's say we are planning a POLCA loop that will go from Cell A to Cell B, in other words an A/B POLCA loop. First, we will need the planned lead times for these two cells. These should already be available since you will be using them to calculate the Authorization Dates (as explained in Chapter 4). Let's say these lead times are LA for Cell A and LB for Cell B. For this example, we will assume these lead times are in working days, but any units can be used as long as they are consistent with the units used for other items below.

We will also assume that these lead times are internal lead times for the cells and do not include move times. So, next you need to estimate the average time for jobs to be physically moved from Cell A to Cell B. Let's call this time MovAB. An example for calculating the move time will help you understand the point about using consistent time units. Let's say that the two cells are not close to each other, and your products are heavy and need to be transported via forklift by special material-handling operators. Based on current operations, you estimate this move will take half a shift on average. Next, say your factory has a two-shift operation. This means that a working day equals two shifts. Suppose the LA and LB values are in working days, and we want to keep MovAB consistent with these units. Then half a shift equals a quarter of a working day, so you will need to use the value MovAB = 0.25.

In a similar fashion, you need to estimate the average time for a POLCA card to be returned from Cell B to Cell A, after Cell B has completed the job attached to that card. You should review the different options for how the card will be returned as explained in the preceding section, and based on your decision, you should estimate the return time. Let's call this time RetBA; again, you need to be sure it is in consistent units as in the preceding example.

Typically, you calculate your POLCA cards for a planning period such as a month or a quarter. The time horizon for your planning period should be long enough to allow for some aggregate planning through averaging of demand, but at the same time, you should not average widely differing demand periods into one planning period. For example, if there are three months in your planning period, then those months should be expected to be roughly the same on average. On the other hand, if one of those months is expected to have much higher (or lower) demand, then it should be a separate planning period. We will return to this point at the end of this chapter in the section on "Hedging with Lead Time or with Cards?"

To continue the explanation in consistent units as used above, we will stay with working days. Suppose the number of working days in your planning period is D. Now you need to estimate the number of times an A/B POLCA card will flow from A to B during this planning period. Note that this number depends on your choice of quantum: for example, if a particular work order requires two POLCA cards then you need to count this as two cards flowing. The first few rows of Table 5.1 show you how to do this calculation easily, using a situation where the quantum is five housings. Observe from Table 5.1 that this calculation requires that you have rough estimates for the number and types of jobs you can expect during

Table 5.1 Calculation of Number of Cards for A/B POLCA Loop with Quantum = 5

Order Size	*Estimated No. of Orders*	*No. of POLCA Cards Needed Per Order*	*No. of Times a POLCA Card Flows Between the Cells*
1–5 Housings	70	1	70
6–10 Housings	20	2	40
11–15 Housings	6	3	18
FlowAB (Total no. of times a card flows from A to B)			128
Additional data: Planned lead time for Cell A, LA = 3 days Move time from A to B, MovAB = 0.25 days Planned lead time for Cell B, LB = 1 day Return time from B to A, RetBA = 0.25 days Planning period of one quarter, D = 63 days Safety margin, S = 0.1 *POLCA card calculation using formula in text*: No. of A/B POLCA Cards = (3 + 0.25 + 1 + 0.25) × (128/63) × 1.1 = 10.1 Rounding up to the next integer = **11 A/B POLCA Cards.**			

the planning period. Even if you have high variability in demand as well as custom jobs that may also vary, you need these types of estimates for your rough-cut capacity planning anyway, which is a prerequisite for POLCA as explained in Chapter 4. So, based on historical demand and/or input from your salespeople, you should have a process for aggregating jobs into categories even without knowing the details of the actual jobs that will arrive, in order to do your rough-cut capacity planning. That same process will enable you to get the estimates required for Table 5.1. Using the approach in Table 5.1, let's say you estimate that the total number of times a card will flow from A to B is FlowAB.

The last item that you need for the formula is a choice of safety margin. There are two reasons to include a safety margin for the initial calculation of the number of cards. The first is that the data used to calculate the various quantities in the formula, such as move times or demand numbers, are estimates and we need to allow for a margin of error. The second reason is that the formula is derived using theory based on idealistic average behavior, and in practice the variability and dynamics of the shop floor may require slightly more cards than the formula predicts. (Readers interested in the theory behind the formula can read the technical details in Appendix G.)

In practice, from several implementations we have found that a safety margin of 10% along with rounding up the result to the next integer has provided a good starting point for the number of cards in each loop. We will use S to denote the safety margin, expressed as a decimal. Thus, a safety margin of 10% implies that S = 0.1.

You now have all the data you need to calculate the number of A/B POLCA cards—let's call this number NumAB. The formula for this number of cards is simply:

$$\text{NumAB} = (\text{LA} + \text{MovAB} + \text{LB} + \text{RetBA}) \times (\text{FlowAB}/\text{D}) \times (1 + \text{S})$$

You can now follow the numerical example in Table 5.1 to see this formula in use, including the rounding off to the next higher integer, to get the value of 11 for the initial number of cards in the A/B POLCA loop.

Once you get your POLCA system operating, you can fine-tune the number of cards based on the actual experiences of the cells. If a cell team complains that it is always waiting for a particular type of card, and there is no obvious problem in the downstream cell, you can consider adding a card to that loop. Conversely, if you frequently see several cards of a given type posted on a cell team's bulletin board, you can consider taking a card out of that loop. The POLCA audit process described in Chapter 7 will help with these decisions.

Add the Finishing Touches with Some Final Details

We end this chapter with two items that apply across the whole POLCA system. Having thought through all the other elements of the system design, you now have the foundations to finish up with these two final touches:

- Thinking through the instances when a Decision Time should occur at a given cell.
- Creating a safety valve for extreme variability: whether to hedge with lead time or with cards.

Identify the Triggers for Decision Time

Thus far, we have described what a cell team needs to do when a Decision Time occurs. Specifically, the team's process was delineated in the Decision

Time Flowchart (henceforth "DT Flowchart") in Figure 2.6, which was explained in Chapter 2. For ease of reference, we reproduce the DT Flowchart here in Figure 5.14. However, we haven't as yet identified what determines that a particular moment should become a Decision Time. This section recommends the events that should initiate a Decision Time. You can modify or enhance this list to suit your situation.

Let's start by understanding the motivation for having a trigger for a Decision Time at a particular cell. The aim is to check if the status in the cell has changed in any way that is related to the POLCA system and which would affect the cell's activities. Looking at the DT Flowchart in Figure 5.14 immediately suggests some examples of status changes that could occur and affect the cell, for instance: a job moves from being Not Authorized to being Authorized; or a job arrives from an upstream cell; or a POLCA card is returned by a downstream cell. Since the cell team members might all be busy and involved with current fabrication or assembly work (and since factories can be noisy environments!) these status changes might not be immediately noticed. So, the first set of triggers is designed to force the cell team to review their POLCA-related status. Hence the following events should trigger a Decision Time for a cell team:

- The start of a shift.
- Returning from a lunch break or any other mid-shift break or meeting.

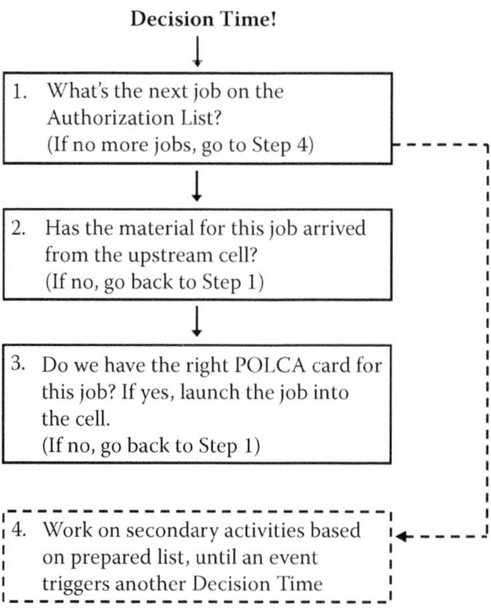

Figure 5.14 Decision Time flowchart.

If a normal eight-hour shift at your company has a morning and afternoon break along with a lunch break, then just these events alone will ensure that a Decision Time occurs at least once every two hours or so. The next two triggers are based on activities by the team members, so they will know when one of these events occurs and can follow it immediately with a Decision Time:

- The team completes a job and delivers it to the next cell.
- A team member completes some work and is available to take on new work.

These are the most obvious triggers, but there are other important ones too, as we now elaborate. According to the DT Flowchart, if a cell team cannot launch a job then it goes to Step 4 of the flowchart: "Work on secondary activities based on prepared list, until an event triggers another Decision Time." The justification for these secondary activities, as well as examples of them—such as cross-training or other improvement tasks—were discussed in Chapter 3. Note that these activities are intended to make use of the time when there are no jobs to be launched, but these activities should not significantly hold up a job if it is available to be launched. So, while the team is engaged in such secondary activities, it needs to be vigilant about whether a job can now be launched. To support this aim, the following additional event should trigger a Decision Time when secondary activities are being conducted:

- The team completes a logical phase of a secondary activity (such as, completion of a short cross-training exercise for a given skill). At this stage, it should stop the secondary activity and revisit the Decision Time process to see if a job can be launched.

If your time horizons are short and you are concerned about the time it takes for a cell team to react to a change in status (in other words, jobs might be ready to be launched but the team has not noticed this), then you can consider putting in a signaling mechanism such as a bell, buzzer, or blinking light, that goes off when any of the following events occur:

- A job arrives from an upstream cell. (This job might already be Authorized on this cell's Authorization List, but it is waiting for the material to arrive from upstream—see Step 2 of the DT Flowchart—so it is good for the team to check on this.) The person delivering the job should be instructed to turn on the signal (e.g., one of the items listed above).

- A POLCA card arrives from a downstream cell. (There might be an Authorized job waiting for this particular card—see Step 3 of the DT Flowchart.) Similarly, the person returning the card should be instructed to turn on the signal.
- A job moves from being Not Authorized to being Authorized. Your IT personnel can program the system that generates the Authorization List to turn on the signal when this happens. (But if you don't want to get into more dependence on software, or this is too complicated to program, it's fine, because as we explained above, other events like returning from a break will trigger a Decision Time fairly often.)
- If jobs arrive frequently at cells, your shop floor environment operates on short time scales, and you are still concerned that jobs might wait, there is another simple solution: you can set the signal to go off (say) every hour, ensuring that the team will visit the Decision Time logic at least once an hour.

Needless to say, when the signal is turned on, any team member that initiates the resulting Decision Time process should be empowered to turn off the signal. We should also add an alternative perspective here. Some teams may find the signals invasive or disturbing to their work, so it is not necessary for you to use the signaling methods just listed. We mention these ideas here as a possible way to go, but at the same time many of the practical POLCA implementations have worked well without a need for such active signals. The ownership and vigilance of the POLCA-enabled cell teams has been enough to keep jobs moving sufficiently well.

We end this list of triggers with one final item. The following is also an acceptable trigger:

- Any other time that a team member elects to go through the Decision Time logic. (This is fine, because it never hurts to go through the Decision Time logic; it can only be helpful.)

As mentioned at the start of this section, above we have listed the recommended triggers for a Decision Time, but you can modify or expand these to suit your particular manufacturing environment.

Hedging with Lead Time or with Cards?

We end this chapter with some mechanisms to mitigate management concerns of POLCA's ability to operate successfully in a highly variable environment. The POLCA system by its very nature has a certain amount of flexibility as well as ability to adapt and adjust for deviations from the production plan. So, a degree of variability is already accommodated in the combination of planned lead times and card calculations. However, if your company, or certain portions of your operations, experience very high variability, then you need to examine these areas more carefully and consider one of two options to enable POLCA to work effectively. These options are: adding safety through more lead time, or adding safety through more cards. Which way should you go? It turns out that the best option for a company is easy to determine; we will explain the approach in the following sections with examples.

One caveat here. The approach below is *not* intended for use with seasonal variations. Large variations due to seasonal effects (for example, at a company making snow blowers) should be accommodated through the choice of planning period and corresponding calculation of POLCA cards as explained above. In other words, you should have your planning periods separated into segments with significantly different demand, and calculate the POLCA cards for each of those segments as they occur. The analysis below is for high unanticipated variability that occurs *within* a planning period. We now discuss the two main options for such situations, and when it is to appropriate to use one or the other.

Adding Safety through Lead Time

If you have a cell that is capacity constrained, and there is no easy way for you to add capacity in the near future, then you have no choice but to go with increasing the planned lead time for that cell. For example, let's say you have a machining cell with highly specialized and expensive machines, and it is already running two shifts and is still capacity-constrained. In Chapter 4, the section on "Engage in Effective Planning Before Control" explained that companies should aim to have their resources running at no more than 85% utilization on average, for POLCA to be effective. Suppose this cell is already running at 90% utilization—and that's just on average: some weeks it is running full-tilt at 100% for several days in a row. Although you recognize this is not ideal for POLCA, at this time you do not have the resources to create more capacity. Adding a third shift or weekend shifts would be too expensive or unpopular with the employees, and your company does not have the finances to invest in more machines at the moment. In this case, the only solution for POLCA to work is to add some safety time in the lead time, in other words, the planned lead time for this cell should be increased.

Why will this work? As long as the *average* utilization of the cell over the planning period is still significantly below 100%, this means that while the cell may be bottlenecked at some points in time, at other points in time it will have capacity to catch up. However, this means that jobs will often wait in queue as the cell goes through the dynamics of being backed up and then working off the backlog. So, this needs to be anticipated in the planning process via a longer lead time for the cell. (This approach also reinforces the logic, now well-recognized in manufacturing system design, that a critical-bottleneck resource should not be starved; this method will indeed put more WIP in front of this bottleneck.)

We should, however, insert a strong caveat here. The mechanism of adding some more lead time, and possibly creating more WIP in front of the bottleneck, is *not* to be seen as a long-term solution. Worse yet, do *not* see this approach as a way for you to increase the utilization of all your resources to more than 90%! To meet the goals set for the POLCA system, and to overcome the manifold symptoms listed at the beginning of this chapter, it is essential that you work towards bringing the utilization of this bottleneck more in line with the 85% goal. This might mean adding more capacity, or outsourcing some work, or even rethinking which orders you want to accept; regardless of the solution you choose, do *not* let the short-term fix make you lose sight of the longer-term goal.

Next, after deciding on the increase in lead time, if you review the formula for calculating the number of POLCA cards, you will see that if you increase the lead time for a cell (in the formula, say LB is increased by 20%), then the number of cards for all loops going to that cell will also increase! Don't be confused by this, because in this case the *driver* is the decision to increase the planned lead time, and the change in the number of cards is just the logical result of that decision. The next example will show a different situation.

Adding Safety through Number of Cards

For a cell that has spare capacity, or the ability to create sufficient spare capacity, the approach to deal with high variability should be to stick to the planned lead time, but increase the number of cards by using a higher value for the safety margin (S) in the formula. Again, let's go over an example, and you'll also understand what we mean by the *ability* to create more capacity.

Consider an assembly cell with a team of five people assembling large electromechanical products. The company's management has a firm belief in the strategy of maintaining spare capacity, so the team's workload is planned for 80% of capacity. At this company, a normal work week is 40 hours. So, the five-person team has an aggregate of 200 working hours of capacity per week, but with the 80% planning target, their workload averages 160 hours in a week. (As an aside, the management of this company also strongly believes in continuous improvement, so when there are no assembly jobs for the cell, the team works on activities such as cross-training, quality improvement, process improvement, and other tasks, as discussed in Chapters 2 and 3. The company has witnessed substantial productivity gains from these activities over the long term, so there is no concern in upper management about the fact that people are not assembling jobs for 20% of their time.)

This company has instituted several mechanisms to be able to rapidly increase the capacity of any given cell. The first is that just as a baseline, if a cell gets a lot of work, the team can spend the full week doing assembly work and no other activities. This provides an additional 40 hours over the planned capacity of 160 hours—in other words, a 25% increase in capacity without having to do anything out of the ordinary! Next, in a high-demand situation, the team can request the addition of a temporary worker from an agency. True, short-term employees are not trained and skilled up to the level of the cell team. But the team has already analyzed its work and identified a number of simpler jobs that can be handled by a new employee with just a little initial training. Examples of such jobs are cutting insulation foam or picking parts from storage racks. The team has found that it can accommodate and fully utilize one temporary employee without any degradation of quality. This can add up to 40 hours of capacity if needed, or another 25%. As a final option, in an unusually high demand period, management has the option to institute a Saturday shift—an option that has been previously negotiated with the employees. With five people each working an extra eight hours each, this provides an additional 40 hours, or another 25%. If the temporary employee is also included in the Saturday shift, this adds another eight hours, or 5%. Adding up all the possible options, you come to an amazing realization—the management of this company has found a way to quickly increase assembly capacity by a full 80% over its baseline capacity plan!

If maintaining short lead time is strategically important for your company, and if you have cells with spare capacity or the ability for spare capacity, then the best option is to stick with your planned lead times but add more cards. You do this by increasing the value of S in the formula above. For example, instead of the normal choice of 10% of safety margin (S = 1.1) you decide that the team can find ways to handle a surge of work and you choose to go with 40% (S = 1.4).

Again, we can ask, why go this route and why will this work? Let's say this is a downstream cell for a POLCA loop, and if you start with no safety margin (S = 1.0), the formula gives the number 4.4. If you used the commonly recommended value of S = 1.1, the formula would give 4.8, so you would put five cards in the POLCA loop. If you go with S = 1.4, then the new answer is seven cards (after rounding up to the next integer). Now what will happen is as follows. With its normal planned capacity and lead time, the cell can handle the rate at which jobs arrive when there are five cards in the loop. So, in an "average" week, typically there will be two more cards in the loop than are needed. However, when demand suddenly goes up more than five cards may get used, so more jobs will arrive at the cell, and the cell team will notice that it is getting backlogged. At this time, the team can kick into high gear and start adopting some of the capacity increasing strategies. For example, it could begin by just eliminating improvement activities and adding the first 40 hours of weekly capacity. If jobs continue to arrive at a fast pace and the cell is still backlogged, it can go on to some of the other strategies listed above. The main point is that by using its spare capacity, the cell can get jobs completed without lengthening its lead time. Similarly, upstream cells can continue to send jobs without being temporarily shut down due to lack of cards, again avoiding the addition of any lead time.

Note that *POLCA acts as an effective signaling system* to tell the team when it needs to kick in these additional strategies, as well as for management to see and approve, for example, a request for a temporary worker. Suppose there is a surge in demand and jobs are scheduled for the assembly cell at a higher than normal rate. Then all seven POLCA cards will get used as described above, and the cell team will notice that it is backlogged. It will also notice that no sooner that it finishes a job and sends a card back upstream, another job arrives with that card! If this continues for a couple of days, and the team looks into the schedule and sees that there is no let up for at least another week or more, it can decide which of the capacity strategies it wants to implement. Conversely, when the surge subsides and the team has caught up, there will be unused POLCA cards in the upstream cell and fewer than seven jobs at the assembly cell. If this also continues for a couple of days, the team can decide that it no longer needs the spare capacity.

This completes the description of the major aspects of the design of your POLCA system. Realizing that, in the real world, there will occasionally be some exceptional situations that arise, the next chapter will show you how to deal with such exceptions while still staying within the POLCA system rules.

Chapter 6

Accommodating Exceptions within the POLCA Framework

Chapter 2 described how POLCA works, and in particular, it explained the Decision Time Flowchart, which is the kernel of the system's operation. Chapter 5 then provided additional details needed for the design of a system to suit your particular needs. Even with all these details thought out and in place, during the operation of POLCA further exceptions can arise due to issues such as component part shortages, machine breakdowns, expediting of rush jobs, and so on. Although these exceptions are undesirable, they exist in almost all real-world manufacturing environments and, almost certainly, in your factory as well. Thus, in the process of deciding to implement POLCA, you may worry whether having to deal with these exceptions will lead to frequent violation of the POLCA rules. Since the success of POLCA depends on the discipline of both workers and managers sticking with the rules, if you start breaking some rules it can become a slippery slope, eventually resulting in many people ignoring the rules and finally rendering the POLCA system ineffective. But do not despair! In this chapter, we will show you how several of the commonly occurring exceptions in manufacturing can indeed be handled within the POLCA framework. You will see that with a few clever tweaks, you can adjust the POLCA system to accommodate some real-world situations while still keeping the key POLCA rules intact.

The remainder of this chapter will address how your POLCA implementation can include procedures to deal with the following exceptional situations, which are among the most common in factories:

- An unexpected shortage of a component part.
- A quality problem that only surfaces after an operation has been started.
- Unplanned machine downtime.
- Expediting a rush job for an urgent customer need.
- Changes in schedule after a job has been released to production.
- Holdups in assembly due to non-synchronized arrivals of components.
- Preventing gridlock in some rare cases when loops form cycles.

This is certainly not a comprehensive list and it is more than likely that you will have other exceptional situations that are important to deal with in your factory. However, it is hoped that the methods below will give you a pool of ideas from which to extend this thinking to other types of exceptions, and your POLCA Implementation Team can devise similar mechanisms to deal with those exceptions while maintaining the integrity of POLCA.

Dealing with an Unexpected Shortage of a Component Part

Since no supply chain can be 100% reliable, material shortages are a fact of life in any manufacturing facility. We will divide these shortages into two main categories. The first category is for the case when there is missing raw material to *start* a job at the first shop floor cell in its routing—for example, missing bar stock, sheet metal, castings, and so on. In Chapter 5 we explained how the Planning Cell needs to check the availability of the needed materials for the first shop floor cell before releasing a job to this cell. So, for this category of shortages, we can assume that a job will not enter the POLCA system until the material is available. Once the material arrives, the job can simply flow through the shop floor with the normal POLCA rules. If the delay due to the shortage was relatively short, POLCA will automatically attempt to help this job catch up with its schedule, because once it is released, based on its original Authorization Dates it will most likely be on the top of the Authorization List (or close to the top) for all the cells in its POLCA chain. So POLCA will help to alleviate the delay from the shortage. On the other hand, if the shortage ends up causing a much longer starting delay, this will be known to the Planning Cell, and at this point the planners along with salespeople can decide if they need to reschedule a delivery date to the customer; if so, then there will be new Authorization Dates in POLCA and when the material arrives, again the job can follow POLCA in the normal way. The preceding discussion shows that for this first category of material shortages, we do not need to change anything within the POLCA system.

The second category of shortages is when purchased components that are needed part-way through a job's routing are not available on time. For example, a welding operation might require a purchased lug or pin to be welded to the main part; an electrical product may require purchased components like switches; an electromechanical product may need a purchased motor to be installed; and so on. Here we will distinguish these situations into two further types: *anticipated* shortages and *unanticipated* shortages. By anticipated shortages we mean those that are known in advance by the relevant planner *before* the job is released to the shop floor: for example, a supplier of switches informs the factory that they are unable to deliver the components on time. For such shortages, once again the Planning Cell can take this into account as explained above, and no special action is needed within the POLCA system.

Unanticipated shortages are those that occur without advance information and are realized *after* the job is already released to the factory floor. For example, the switches were expected to arrive by the planned Authorized Date for the assembly operation, which occurs much later and towards the end of the POLCA Chain, but after the job was released to make its way through the first few cells, there was a transportation issue and the switches are delayed. Now we have a job that is making its way through its POLCA Chain but it will get stuck at some point because the operations on this job cannot be completed. Of course, this happens in most manufacturing companies, so why is this a special problem as far as POLCA is concerned? The answer lies in similar reasoning as we gave in Chapter 5 for jobs that could be sitting on a shipping dock and holding on to POLCA cards. We will explain the detailed reasons with an example.

Let's use the example of the POLCA loops connecting Cells A, B and G from Chapter 2. The related figure from Chapter 2 is reproduced here, along with a callout balloon indicating that a shortage has occurred (Figure 6.1). In the worst case, suppose the job has already been launched into Cell B before the shortage is discovered. At this point, since the job is in the cell, the figure shows that it will be carrying two POLCA cards: the A/B card

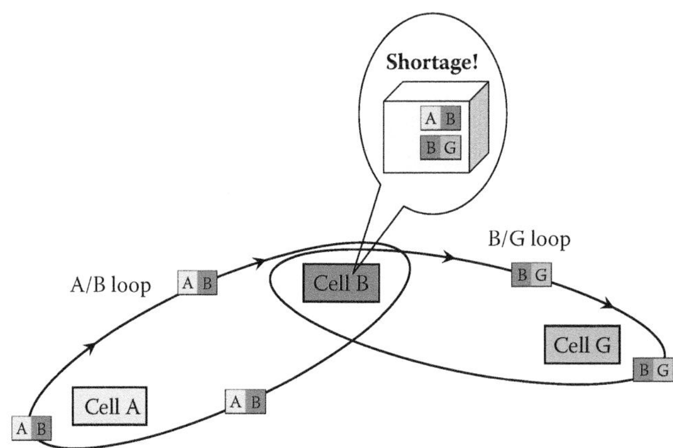

Figure 6.1 Job with routing through Cells A, B, and G, with shortage detected after job is launched into Cell B.

and the B/G card. Both these cards will be stuck with the job for a while. However, since the Cell B team cannot work on this job, it has capacity to work on another job. In fact, for now the situation is the same as it would be if Cell B had finished a job, because it has stopped working on a job so it should signal that it can take on more work. But Cell A will not get a signal that Cell B has capacity. On the other side of the picture, the B/G card is also stuck. Perhaps there are other jobs destined for Cell G that are waiting to be started in Cell B. But if there are no more B/G cards on Cell B's POLCA Board, then those jobs cannot be launched. Again, this is a pity since Cell B does have capacity to work on those jobs, and that would soon benefit Cell G as well. So, this explains the problems that occur in the POLCA operations when there is a shortage detected after a job hits the shop floor. (Even if the shortage is noticed by the Cell B team *before* the job is launched into the cell, and the team holds off on launching it, the job will still be carrying an A/B card that will be stuck, as above.)

Introducing the Safety Card Concept

We will rectify these problems by using a concept called a *Safety Card.* A Safety Card is visually different from the normal POLCA cards and kept with a designated person (the choice of this person will be discussed below). We will explain how this card is used using the Cell B example, and then discuss more details of the Safety Card and associated processes. When the Cell B team identifies that a shortage has occurred and the job cannot be processed further, it will request a Safety Card. This Safety Card will be placed with the job (or its shop packet), and the A/B card will be removed from the job (or packet) and returned to Cell A, enabling Cell A to work on another A/B job if needed. Also, the B/G card will be removed and placed back on Cell B's POLCA Board. This will enable Cell B to launch a different B/G job if so indicated by its Decision Time rules. We will now go into more details on the use of the Safety Card.

The aim of the Safety Card is to ensure that the problem with one job does not snowball into problems with other jobs. The situation with Cell B (before deploying the Safety Card) is an example of this snowballing, because both upstream and downstream cells could be impacted, along with the progress of other jobs. At the same time, we don't want the Safety Card to be an enabler of ongoing problems. So, the first rule is to limit the number of Safety Cards in each loop (we'll discuss this further below). Typically, companies have found that using a rule of 10% of the total number of cards is a good choice. If some loops have very few cards, you can assign one Safety Card to a collection of loops. Also, the Safety Card should be clearly visible, so a bright color should be used. Orange is a common choice since it is often associated with safety signs and safety jackets. Figure 6.2 displays two examples of Safety Cards.

The second rule is that the Safety Cards should be kept with a manager who has responsibility for and/or could impact the shortage issue. We will call this person the Safety Card Owner. The idea is that each request for a Safety Card will highlight the shortage problem to the very person that might be able to rectify it in the future.

Figure 6.2 Examples of Safety Cards. The upper card is dedicated to a particular loop, while the lower card is common to several loops.

Procedure When a Safety Card Is Available

Next are the rules for deployment of the Safety Card. When an unanticipated shortage occurs, as explained previously, a job could have two POLCA cards attached to it (except for the first and last cell in the POLCA chain, when it would have only one). So, in general, when this job is held up, we need to deal with both the POLCA cards. The first step is for the cell team to request a Safety Card from the Safety Card Owner. The team needs to specify for which loop the card will be used. We'll assume for now that a Safety Card is available for that loop. The cell team will attach the Safety Card to the job (or its packet), return the upstream POLCA card to the upstream cell, and place the downstream POLCA card on this cell's POLCA Board. (If the job was not yet released into the current cell, there will be no downstream card with it. Also in the case of the first cell in a POLCA chain there will be no upstream card, and for the last cell, no downstream card, so the related actions can be omitted.)

For now, let's skip forward in time to when the shortage is resolved; in other words, the missing component arrives at the cell and the cell team can continue with working on the job. Remember that this job is going to Cell G, so according to the POLCA rules, it should have a B/G card in order to be processed. However, since the job is already delayed, we don't want it to have to potentially wait for a card to be available. So, the rule here is that the Safety Card acts as the B/G card and the team can work on the job. If in fact no B/G cards were available in Cell B, this means that an extra job will be sent to Cell G (beyond the normal limit), but this will be a one-time temporary overload at Cell G (as explained below), and is to the benefit of keeping the delayed job moving.

When the job gets to Cell G, the standard Decision Time rules will be applied. (In other words, it still needs to be Authorized and it will need a downstream card to be launched.) Since the job has been delayed, the chances are it will be high up on the Authorization List and will be launched soon. When the job is completed, again the standard procedures are used, with one change: the job and downstream card are sent on to the downstream cell as usual, but the upstream card is actually the Safety Card, *which must be returned to the Safety Card Owner*, not to the upstream cell. Hence, the number of cards in this POLCA loop will revert to the normal amount, and the one-time overload of Cell G will be over.

You might be concerned about the Decision Time rules being applied to this job at Cell G. What if the job is going to Cell M next, but no G/M card is available at Cell G? "Since the job has been delayed," you might well ask, "shouldn't we launch it even if the G/M card is not available?" The fact is, if no G/M cards are available it means that Cell M is backed up, so rushing the job to Cell M won't help anyway—it will be a classic "Hurry up and wait!" situation. Also, when the G/M card becomes available and the job is processed and sent on to Cell M, it will be high up on the Authorization List once again, and so on all the way down to the end of its POLCA chain. So, it's best to let the standard POLCA rules be applied for the rest of the job's routing, and this minimizes the extent of exceptions that need to be managed.

This completes the basic procedure for use of the Safety Card in the case where such a card is available. However, companies have found it useful to enhance this procedure in order to improve their operations, as we explain next.

Adding a Shortage Tracker

The Safety Card procedure provides a platform to support continuous improvement efforts. The long-term solution to the component shortage problem is, of course, to prevent the occurrence of such shortages. By enhancing the Safety Card procedure, companies can gather data to help them work toward this goal. When a Safety Card is deployed, it provides an opportunity to track information about the specific shortage that has occurred. To this purpose, companies have designed a *Shortage Tracker* that can be made part of the Safety Card procedure. The Shortage Tracker is a sheet that helps record the details of the shortage incident, such as the part number that was missing, the time and location of the shortage, reason (if known), and also, the time when the shortage was finally resolved. Figure 6.3 has an example of a Shortage Tracker used by one company. An additional idea used by some companies is to print the Shortage Tracker with a sticky back (like a sticky note from a notepad), which allows it to be stuck onto the Safety Card, and then removed once the shortage is resolved. For the example of Cell B above, the enhanced procedure thus includes two more processes for Cell B: first, when the shortage is detected, the team should also fill out the relevant data on the Shortage Tracker sheet; and second, when the shortage is resolved (the job of course is still at Cell B), the team should complete the Shortage Tracker with the final data, detach the Tracker sheet and deliver this sheet to the Safety Card Owner. Over time, as such sheets are collected, the Safety Card Owner can analyze the data and

Charles Casings Company SHORTAGE TRACKER							
Date/Time shortage was discovered	Location where shortage occurred	Part# and quantity short	Reason for shortage (if known)	Date/Time Safety Card was requested	Was Safety Card available (Y/N)	Date/Time Part(s) arrived	Comments

Figure 6.3 Shortage Tracker sheet.

apply the usual statistical and continuous improvement techniques to look for the root causes and come up with solutions.

Procedure When a Safety Card Is Not Available

Now we discuss what happens when a cell team requests a Safety Card for a particular loop, but no more cards are available for that loop. In this case, the job and both its associated POLCA cards will be stuck in the cell. Although this is not a desirable situation, it will serve to raise the visibility of the shortage. If no Safety Cards are available, it means that there have been several recent shortage occurrences and management should be aware of this. The cell team should still fill out a Shortage Tracker sheet and keep it with the job (if it has a sticky back, it could be stuck onto one of the POLCA cards associated with the job). Also, the Safety Card Owner should note down the team's request for a Safety Card and queue it up (in case there are other requests that are also waiting). When a Safety Card is finally returned from a downstream cell, the Safety Card Owner can issue this Safety Card to the next job in the queue, and the team working on that job can then follow the procedure as before. It might happen that before a Safety Card becomes available, the part(s) arrive and the shortage is resolved. In that case, the team can complete the Shortage Tracker sheet, and proceed with the job following the standard POLCA rules. (This job still has two POLCA cards attached, so the usual procedures will apply from here on.)

Limiting the number of Safety Cards helps to emphasize that the Safety Card Process is intended to be used infrequently. As mentioned, if there are frequent requests so that Safety Cards are often unavailable, this will raise the visibility of the problem and require management to devote resources to resolving the shortage issues instead of just living with them, as they might have in the past.

Dealing with an In-Process Quality Problem

The next issue we consider is when a quality problem surfaces after a cell has started working on a job. For example, a casting may have a blowhole, which is only detected after some machining has been done. Or, to take the example of the electrical switches in the previous section, the switches may have been available in stock and delivered to an assembly cell, but when assembled on the unit and tested it is found that they are all from a defective batch, and thus there are no acceptable switches available in stock at this time.

Typically, when such a situation occurs, the cell team needs to stop the job and then wait for an appropriate expert to make an evaluation of the situation. For a minor quality problem, a manufacturing engineer may find a simple fix or even decide that the problem will not impact the product. For a major problem like a blowhole or the switches, a new casting or new batch of switches will need to be ordered. In any case, the job involved will be stuck in the cell for some period of time with the same consequences for operation of the POLCA system as described in the previous section.

We will consider two possible outcomes of the evaluation of the problem by the quality expert involved. In the first case, it is felt that the problem can be resolved but some time is needed. An example is when a new batch of switches is ordered from the supplier. This is very similar to the situation discussed in the previous section! So, the Safety Card concept can be deployed here as well. All the procedures of the previous section can be used as before, including the use of a tracker, now called a *Quality Tracker*, modeled on the Shortage Tracker in Figure 6.3 but with modified headings based on the quality systems in use at the company.

The second case is when the problem cannot be easily resolved and a whole new job needs to be started from the beginning. For instance, when the blowhole is discovered, it could be that the casting has already been through several operations before coming to this cell. So, when a new casting arrives it would have to be started from the beginning of the routing. In this case the Planning Cell needs to get involved and the job and all its associated work orders will have to be cancelled. From the point of view of POLCA procedures, the cell team's responsibility in this situation would be to detach the two POLCA cards, return the upstream POLCA card to the upstream cell and place the downstream POLCA card back on the cell team's POLCA Board.

One decision to be made for this and the previous procedure is whether there will be one pool of Safety Cards for both shortage issues and quality issues, and thus also one Safety Card Owner who is responsible for overseeing both types of issues, or if there will be two separate owners, which would then also require two pools of cards. An advantage of the former choice is that there is one person who is aware of the external issues (i.e., not related to POLCA itself) that are affecting the POLCA system; with this choice you could expand the pool of Safety Cards to (say) 20% of the nominal number in the loop. If you go with the latter choice, the two types of cards should be given different names and colors, and each owner could have a number of Safety Cards that is 10% of the nominal number in a loop.

Note that here, as well, the limited number of Safety Cards will raise the visibility of the occurrences of these quality problems. It could be that many of these problems have been occurring and being resolved without management being aware of the frequency of them. The use of the Quality Tracker will help to gather data on the frequency and root causes, and the limit on the number of Safety Cards will help provide visibility on these problems.

Dealing with Machine Downtimes

Another common occurrence in any factory is machine downtimes, by which we mean unexpected failures. If there are planned downtimes for scheduled maintenance or equipment upgrades, these will be known ahead of time and taken into account during the planning process. So, in such cases we can assume that the jobs and their Authorization Dates would already have been planned around these dates and this does not impact how POLCA works. However, if there are unexpected failures while jobs are being processed, we need to consider how they would impact the POLCA system and what to do in these situations.

Again, we will divide the failures into two categories: short-term failures and long-term failures. A short-term failure at a machine could be as simple as a tool breakage that requires waiting for a new tool to be delivered, or the failure of an electrical component that can be replaced with a spare part relatively soon. In such cases, nothing different needs to be done with the POLCA system. The short-term capacity shortage will create a backlog in the cell. This will limit the jobs being sent from upstream cells. This is as it should be, because those cells will then use their capacity to work on jobs going to other cells that can make use of those jobs. When the repair is completed, the cell team and management can decide if the cell can catch up or if some overtime or temporary workers are needed. The team may also receive offers of help from other teams, as described in Chapter 3. In addition, jobs that were held up in this cell will be high up in the Authorization Lists in downstream cells, and may be able to catch up with their delivery date. In summary, the short-term feedback controls in POLCA will work as expected.

You might have noticed that in the preceding discussion we did *not* advocate the use of Safety Cards. The reason is that in the previous two sections (shortages and quality problems), the job was stuck but the cell still had capacity. So, the Safety Cards allowed for other jobs to take the place of the stuck job. But in this section, the machine failure actually impacts the cell's capacity. So, we do want to limit the rate at which jobs come to the cell. This will be accomplished by the POLCA cards and loops already in place, and there is no need for deploying Safety Cards.

On the other hand, if it is determined that it is going to take a while to fix the failure—for example, if a large bearing has to be replaced and needs to be ordered first—then the Planning Cell along with management will have to decide what to do with jobs in process. They could be outsourced, moved to other cells, or have their delivery dates changed if possible. All these actions would need to be considered whether or not a company is using POLCA, so once again, there is nothing different needed here in terms of the POLCA system. The only special action required would be to detach any POLCA cards associated with jobs being moved and deliver them back to the appropriate POLCA Boards. Once the machine is running again, there will be new Authorization Dates and possibly whole new jobs being released by the Planning Cell, and then POLCA will just be used as normal.

Expediting a Rush Job for an Urgent Customer Need

Despite the best plans, the real world intervenes and all factories experience rush jobs to some extent. Even with excellent planning, a customer situation may call for an expedited order—for instance, if a customer's machine breaks down, their production line is stopped, and your company has the ability to supply the spare part needed; or if the customer's forecast was inaccurate and they need more parts from you than anticipated in order to serve their end demand. Your knack in servicing customers in unusual situations like these helps you to preserve customer loyalty.

Thus, we arrive at the classic dilemma faced by most companies: on the one hand your ability to successfully expedite jobs can give you competitive advantage; on the other, as mentioned in several places in this book, expediting can create a chaotic situation on your factory floor plus an increasing cycle of hot jobs that make people ignore the schedules in place, and eventually hurt all your deliveries—thus undermining the very competitive advantage you were trying to establish! In addition, it has been underscored in the preceding chapters that for POLCA to work, it is important that discipline is maintained and the system rules are obeyed. So how do we deal with expedites within the POLCA system?

Before we answer this question, it should be noted that with POLCA in place you might be pleasantly surprised to find that the number of expedites is greatly reduced in the first place! Patheon, a multinational manufacturer of pharmaceuticals, experienced a significant reduction in expedites at its Canadian factory after it implemented POLCA (see Chapter 11).

Moving on to answer the previous question, once again the approach is to have a safety valve, like the Safety Card described already, but to also limit the use of this process. The safety valve in this case is called a *Bullet Card*. (The name was suggested by an early implementer of POLCA based on a popular superhero who was "Faster than a speeding bullet"; other users liked this name, so it has stuck!) The Bullet Cards are also designed to be visually different from POLCA cards (see Figure 6.4). A Bullet Card can be assigned to a job, and then the procedure associated with this card is simple: the job does not need to wait at any cell for Authorization or for a downstream POLCA card. The cell can launch the job as soon as it has capacity, and in fact it should launch any job with a Bullet Card before any other jobs that are Authorized and have POLCA cards. Thus, the Bullet Card goes

Figure 6.4 Bullet Card used by Alexandria Industries (QROC refers to the Quick Response Office Cell responsible for a set of products, and which is the owner of this card. This is described further in Chapter 10).

with the job all the way through the end of its POLCA chain, and then it is returned to the Bullet Card Owner (discussed next).

In order to properly use (and not misuse) the Bullet Cards, the following processes should be in place. First, the Bullet Card should only be deployed by a senior manager or an approved group that is the Bullet Card Owner. Second, the number of such cards should be severely limited, as their deployment is highly disruptive to existing operations. In fact, companies have found that since POLCA helps to eliminate many of the expedites that were there in the past, even just *two* Bullet Cards can suffice! The limited number also forces the Bullet Card Owner to think carefully before "playing" their card. It is possible to have multiple groups authorized to use Bullet Cards. As an example, from the case study in Chapter 10, Alexandria Industries gave one Bullet Card to each of its four customer service teams (serving four customer segments). Thus, each team had only one card, which made it think carefully before using the card—the teams soon learned that once a team had deployed its card, it couldn't expedite any other job until that card was returned! This example has an interesting ending, once again demonstrating the impact of POLCA. As described in Chapter 10, the company found that after POLCA had been in place for a few months, instead of the total of four Bullet Cards, even two cards shared among the four teams would have been sufficient, because the number of expedites had become so rare!

Accommodating Schedule Changes Seamlessly

Schedule changes are another fact of life in manufacturing companies. The most common reasons are customer requests for delivery date changes, management decisions to move existing jobs in order to accommodate an urgent order, and supply chain delays (if they are known in advance, as discussed in an earlier section); of course, many other reasons exist too. A strength of POLCA is that the logic of the system will simply take care of these changes and they can be accommodated seamlessly. We will divide the circumstances under which a schedule change occurs into three categories:

1. *The change is known before a job is released to the shop floor and is still feasible within the normal lead times.* In this case the new due date will result in recalculation of all the Authorization Dates for the POLCA

chain (using the standard process described in Chapter 4). If the due date has been pushed back, the job will not be released until later. If the due date has been moved up, the Planning Cell will need to check if it is still achievable with the normal planned lead times for the job's POLCA chain. For example, if the new calculation shows that the job should have been released two days ago, but the total planned lead time is 15 days, this might still be okay, because the job will have early Authorization Dates throughout its POLCA chain and hence might make it through all the cells a bit faster than 15 days. If, however, it seems highly unlikely that the new due date can be achieved with normal operations, the next approach should be considered.

2. *The change is known before a job is released to the shop floor but it requires expediting to meet the due date.* For the preceding example with a planned lead time of 15 days, if the recalculation shows that the job should have been started a week ago and the Planning Cell doesn't think this amount of time can be caught up during the job's routing, then this would require either renegotiating the delivery date or considering the use of a Bullet Card described earlier. As already explained, the use of Bullet Cards should be severely limited, so other alternatives should be explored first and this option should be used as a last resort.
3. *The change only becomes known after a job is released to the shop floor.* Here, the job will be part-way through its routing when the change becomes known. At this point the Authorization Dates for all the remaining cells in its POLCA chain will be recalculated. The next time that the job is part of a Decision Time check, these changes will make the right impact on how the job is treated: either it will be started earlier, or it will be held back till later. (If the job is in process at a cell, this will happen when it gets to the next cell. If it is in the queue at a cell, it will occur when the cell team gets to its next Decision Time.) However, the Planning Cell also needs to consider two additional options based on the direction of the due date change:
   - 3.1. If the due date has been moved to much earlier, and—similar to item (2) above—it is determined that the new date can only be achieved by expediting, and if other options are not possible, then a Bullet Card must be used. In this case, a Planning Cell member should take the Bullet Card to the job (wherever it is), remove the upstream card and return it to the upstream cell's POLCA Board (a job will always have an upstream card during its journey through the POLCA chain), and if there is a downstream card too (in other

words, the job is in process at a cell) this should be removed and placed on the POLCA Board at that cell. While doing this, the Planning Cell member should alert the cell team to the fact that the job now has a Bullet Card attached.

3.2. If the due date has been moved to much later, the Planning Cell should consider whether the job should be completely removed from the POLCA loops and started at some future time. This is important because once a job is released it will always be carrying an upstream POLCA card. If the date change is substantial, for example, the customer has requested a delay of several weeks or months, this card will be stuck with the job for this entire time. This will stop the upstream cell from receiving a signal to work on other jobs (which it should do, since this job will not be using capacity at any cell for a while). So, it may be best to move the job aside (or to another storage location) and return any POLCA cards to their respective POLCA Boards, as above. The Planning Cell can then re-launch the job at a future date based on the new Authorization Date for the next cell, using its normal launching procedures explained in Chapter 5. In other words, the fist loop for re-launching the job would be from the Planning Cell to the relevant cell where this job needs to restart.

Although the descriptions above may appear to be lengthy, in fact all the processes follow common sense and still use the POLCA rules that have already been established. These processes should be part of the team training described in the next chapter, so it should not be difficult for the Planning Cell team and the other cell teams to seamlessly accommodate these schedule changes as part of the POLCA system operation.

Adjusting for Non-Synchronized Delivery of In-House Manufactured Components to Assembly

This section considers the situation where an assembly cell requires multiple components arriving from other in-house operations. The difference between this and the first section of this chapter is that the earlier section was concerned with delays in arrival of components from *external* suppliers. Here, since the components are coming from *in-house* operations, there will be POLCA loops from all the supplying cells to the assembly cell (see Figure 6.5).

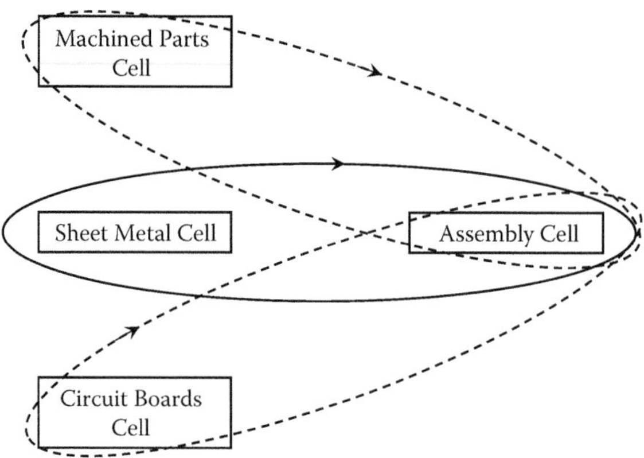

Figure 6.5 POLCA loops from multiple in-house supplying cells to an assembly cell.

We now discuss how POLCA deals with the situation where one or more of the supplying cells is behind schedule and those components are delayed in arriving at the assembly cell. Suppose a job that needs those components is now Authorized at the assembly cell. The first point to note is that when the team at the assembly cell goes through its Decision Time logic, if this job is next on the Authorization List, the team needs to check if the material for this job has arrived from upstream cells. In the case of multiple components being needed (which would usually be shown in the work order or on the computer screen), the team must make sure *all* component jobs have arrived before this job is launched. Thus, if some components are missing, this job will be skipped and the team will move on to the next job. This is what we want, because if the team starts this job then it will get stuck in the cell and also hold up a POLCA card while it waits.

If the delayed components are substantially late, then as explained in Chapter 5 (in the section on "Determine the Ending Points for the POLCA Chains") this will result in the other feeder cells (making components that are arriving on time) being stopped from sending more parts to the assembly cell. This is good, because otherwise the assembly cell will be cluttered with lots of parts that it cannot yet use. Not only do these excess parts take up floor space, but they could cause safety problems or even be damaged during the wait. At the same time, the team in the assembly cell will realize that it cannot start several jobs because of missing components, and so if it has unused capacity it could offer to send people to help the upstream cell (or cells). Again, this is a desirable outcome. In other words, the POLCA system will result in actions that are reasonable in this situation of delays.

Preventing Gridlock When Loops Form Cycles

In rare cases, if a set of loops can form a full cycle, it is possible for jobs using POLCA to get into a gridlock situation. This is similar to a traffic jam that goes all around a city block: everyone needs someone else to move first, but since no one can move anywhere, the traffic is completely halted. Although we have not heard of such gridlock happening in practical POLCA implementations, it is theoretically possible and has been observed by researchers during simulation runs involving complex routings (see Appendix H). So, it is important that in this book we at least address this possibility. Further, you will see that the situation can be *completely prevented* with yet another simple tweak similar to the Safety Card described earlier.

Let us define the setting in which a gridlock can occur. If, after reviewing this definition, it is clear that your factory does not have this setting, *you can skip the rest of this section.* If your factory could have a gridlock, then in the rest of this section we first explain how this gridlock could occur with POLCA, and then we provide the simple solution to prevent it from ever happening.

For a gridlock to occur, there has to be a set of POLCA loops that form a complete cycle. In other words, it is possible to start at a given cell, then follow a POLCA loop to a downstream cell, and follow another loop to a downstream cell, and so on, and eventually end up at the cell where you started. So, if there exists a set of loops that enable you to return to a cell (always going in the downstream direction of each loop) then you have a setting where gridlock could occur. If there is no such cycle possible with your loops, you don't need to worry about the gridlock phenomenon. We will illustrate how you can check this with an example.

There is a simple visual way to check for a cycle: in a diagram of your shop floor, replace each loop with a directed arrow, as shown in Figure 6.6. The upper portion of the figure has all the POLCA loops, with each loop containing one arrow to show the direction of job flow. However, with all these loops the diagram is crowded. In the lower part of the figure we have replaced the loops with directed arrows. Now it is much easier to see if you can follow any arrows around to complete a cycle. If you look carefully, indeed there is a cycle that goes A → B → G → A. So, starting at Cell A, you can follow a set of arrows all the way back to Cell A.

Note that in conducting this check, you shouldn't think in terms of the POLCA Chains for individual jobs. You need to look at all possible POLCA

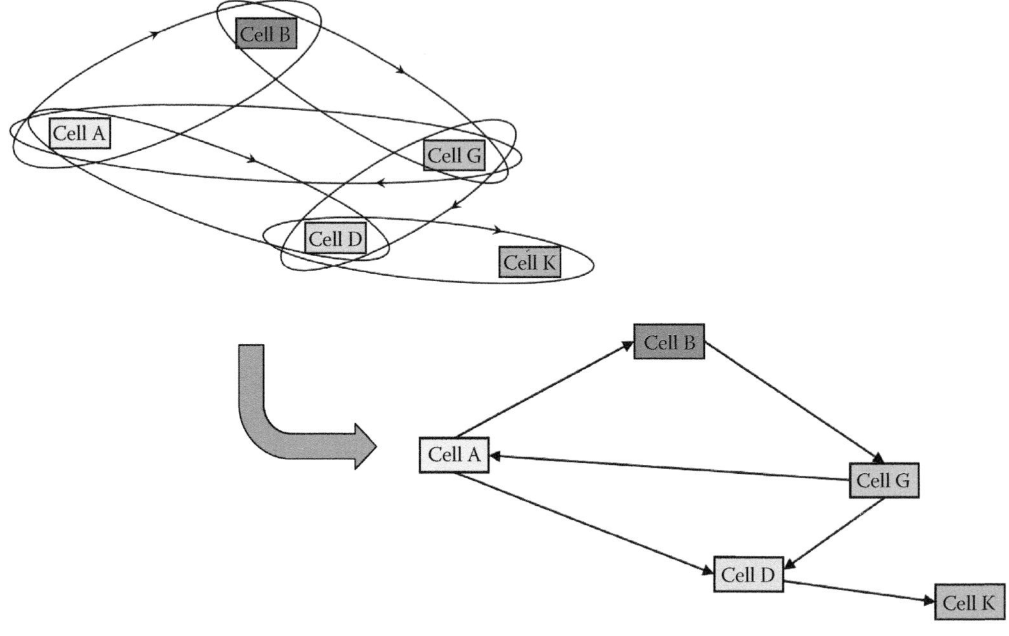

Figure 6.6 Method for visually checking for potential cycles in POLCA loops.

loops that might be in place. In the cycle in Figure 6.6, it is possible that one job is going from Cell A to B, a different job from B to G, and yet another job from G to A. So, the loops in the cycle are not necessarily related to the same job. As long as a cycle can be found using the procedure just explained, there is a possibility of gridlock.

So, at this point if you can verify that you do *not* have any cycles in your POLCA system design, *you may skip the rest of this section*—unless of course you have an intellectual interest in reading about the details of how the gridlock might occur and how to prevent it.

To understand how a gridlock might occur we will use the cycle in Figure 6.6. We will start with the extreme case where there is only *one* card in each of the loops A/B, B/G, and G/A. Suppose Job X came from Cell A to the queue at Cell B, and is now Authorized and needs to go to Cell G next. Since it came from Cell A, it is carrying an A/B card (see Figure 6.7). To be launched into Cell B, it needs a B/G card. However, let's say Job Y went from Cell B to G earlier, and so it took the B/G card with it and is waiting in the queue at Cell G, as also shown in Figure 6.7. Suppose Job Y is now also Authorized and needs to go to Cell A next. So, it needs a G/A card to be launched. But in a similar way, Job Z has taken the G/A card and is waiting in the queue at Cell A. To complete the cycle, let's say Job Z is Authorized

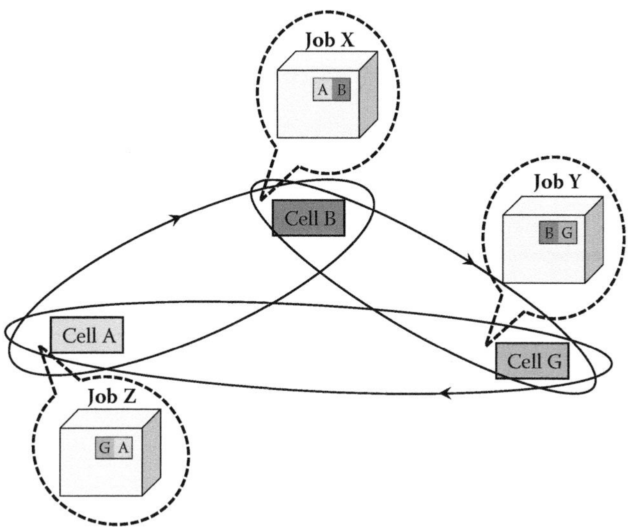

Figure 6.7 Details of situation causing gridlock.

and needs to go to Cell B next. In order to be launched, it needs an A/B card. But this card is with Job X. So, we have our cycle causing the gridlock: Job X is waiting for a card that is with Job Y, which is waiting for a card that is with Job Z, which is waiting for a card that is with Job X! There is a stalemate and none of the jobs can move.

Your first instinct might be to increase the number of cards in each loop. After all, in this example there was only one card in each of the loops. In fact, increasing the number of cards reduces the chances of a gridlock, but does not eliminate it. If there are three cards in each loop, there is a chance that there will be three jobs in each of the queues in Figure 6.7, each job in exactly the same status as the jobs in the figure, so none of the nine jobs can move. The chance of this happening is much less, because now you need nine jobs to simultaneously end up in exactly this situation, but nevertheless it is theoretically possible.

So, what is the solution to this potential stoppage? The solution is actually quite simple, and completely eliminates the possibility of gridlock. All we need to do is to introduce a single card, which we will call a *Cycle Card.* A Cycle Card is particular to the cycle that could cause a deadlock. (In case there are multiple cycles possible in a factory, each would have its own Cycle Card.) For the above example, we will create a Cycle Card for the A → B → G → A cycle (see Figure 6.8).

Here's how the Cycle Card works. First, analogous to a Wildcard (also known as a Joker) in a deck of playing cards, it can act as different cards as

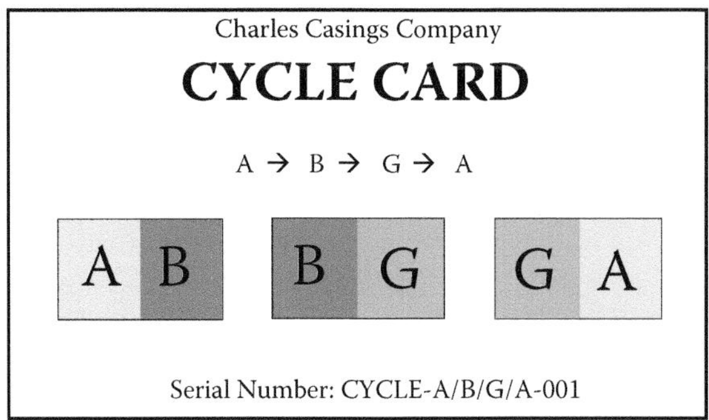

Figure 6.8 Cycle Card.

needed, and second, *it is never returned*; it always moves in a forward direction in its particular cycle. Let's work through an example. We will start by placing the Cycle Card on the POLCA Board in Cell B, which was the starting point of our earlier description of the gridlock. The rule for a Cycle Card is that when it is on the POLCA Board for any cell in the cycle, it becomes the card needed for the *next* cell in the cycle. So, when it is in Cell B, it becomes a B/G card. So it should be placed on Cell B's POLCA Board in the location for B/G cards. Now, since we have an (acting) B/G card, Job X can be launched into Cell B along with the Cycle Card (which is currently a B/G card). So the stalemate is ended. Next, after Job X is completed, the A/B card is returned to Cell A, which will also allow Job Z to be launched.

Meanwhile, let's follow the Cycle Card. This goes with Job X to Cell G. It is still acting as a B/G card. Suppose Job X is going to Cell D next. At some point, Job X will be Authorized and a G/D card will be available, and it will be launched into Cell G along with both the Cycle Card and the G/D card. When Job X is completed in Cell G, the job and the G/D card will move on to Cell D as usual. But note what happens to the Cycle Card. It is *not* returned to the previous cell. Now the Cycle Card is kept in Cell G, and placed on Cell G's POLCA Board. According to the rule stated, if the card is in Cell G, it now assumes the role of a G/A card. So it should be placed in the location corresponding to the G/A cards. This means that at some future point in time, if three jobs are in the potential gridlock situation previously described, this card will act as a G/A card and allow the job in Cell G to be launched, and the deadlock will be averted before it even happens.

In case you are wondering, there is a reason why a single Cycle Card works, when adding a second B/G and/or G/A card does not avert the gridlock possibility. The key is that the Cycle Card *does not get returned* to the upstream cell. It keeps moving forward, and, like a chameleon, it changes character to adapt to its current surroundings. This means that, wherever it is, the Cycle Card becomes the appropriate release valve to keep jobs moving. To emphasize its chameleon character, and also to make the role of the Cycle Card visually clear, we recommend printing the symbols of the possible cards right on the Cycle Card, as also shown in Figure 6.8.

The Cycle Card should be used freely, just like it was an available POLCA card of the particular type. The purpose of the Cycle Card is to act like a lubricant in the cycle. Just like you would grease a wheel that is jamming on a cart, in order to keep it moving, the Cycle Card circulates around the cycle, keeping jobs moving and preventing a gridlock from ever occurring. So, there is no need to hold back on using the card; its use is beneficial to all potential jobs in the cycle. Hence, when such a card arrives at a given cell with a job, after the job is completed the Cycle Card is placed on the current cell's POLCA Board, in the appropriate slot for the card that it is next mimicking. In the previous example, when Job X arrived with the Cycle Card at Cell G, after Job X was completed and moved on the Cycle Card would be placed in the G/A slot on Cell G's POLCA Board.

Having a freely used Cycle Card means that there is one more card in the set of loops that comprise the cycle. This should not be a significant concern. If the cycle has three loops, as in the above example, this means that on average each loop has 0.33 more cards in it. If you consider that the POLCA card calculation already has a safety margin and the number is rounded up to the next integer, an additional amount of 0.33 is not worrisome. For example, if the card calculation recommends 6.7 cards and this has been rounded up to 7, then an average of 7.33 cards will not make much of a difference. In any case, the number of POLCA cards is typically fine-tuned as an implementation progresses, as explained in Chapter 5.

We have one more simple recommendation to keep jobs flowing where there is a potential cycle. For any loop that is part of a cycle, if the POLCA card calculation results in only one card being recommended, you should start with two cards in that loop. For instance, if the flow of jobs in a particular loop is very low it could be that the calculation (even with a 10% safety margin) comes up with a number such as 0.7. In this case, rounding up to the next integer simply gives "1" as the recommended number of cards. If this loop is part of a cycle, it is recommended that you start with two cards in such a loop. The reason is that, without the use of a Cycle Card, simple probability calculations show that for a cycle of loops where each loop has only one card, putting two cards in each loop can reduce the probability of a gridlock by an order of magnitude or even more. But why should we worry about this when we have implemented a Cycle Card? Because even though with the Cycle Card in use a gridlock cannot occur, you could get close to a full cycle of stoppages before they are all released one by one, possibly with some delays. Adding a second card in any loop that has only one card will greatly reduce the occurrence of even this kind of situation.

In summary, if you decide that your situation does have the possibility for gridlock, here are the rules for use of a Cycle Card:

- Start with the Cycle Card being placed on the POLCA Board for any of the cells in the cycle. For the chosen cell, the card should be placed in the slot for the downstream cell in the cycle. (So, if you choose to start the card in Cell B, it must be placed in the B/G slot. If you choose to start it in Cell G, it must be placed in the G/A slot, and so on.)
- At a Decision Time, the Cycle Card can be used by the cell team just like a normal POLCA card for that slot. (There is no need for a special rule such as requiring the team to wait until no other cards are available in that slot; the team can just pick the Cycle Card and use it at any time, as indicated above.)
- When a job is completed and it has a Cycle Card on it, the cell team should review the two cards with the job, and decide which of these two situations are the case: (i) If the job has a regular upstream POLCA card, then the Cycle Card is acting as a downstream card; the upstream card goes to the upstream cell as usual, and the Cycle Card should go with the job to the downstream cell. (ii) If the job has a regular downstream POLCA card, then the Cycle Card came with the job from an upstream cell. In this case the job and the downstream POLCA card go to the downstream cell as usual, but the Cycle Card should be placed on the POLCA Board for *this* cell, in the slot for the cards for the next cell in the cycle.

That's it! The operation of the Cycle Card is also not that complicated.

Summary: POLCA Accommodates Exceptions Smoothly

All manufacturing companies have to deal with real-world events such as late arrivals of parts, machine failures, urgent and necessary schedule changes, and so on. This chapter has shown how these exceptional situations can be accommodated into your POLCA system. We have seen that in many cases POLCA simply adjusts its operation and does what we would expect of a control system, so nothing special is needed. In other cases, the addition of a simple mechanism (such as a Safety Card) helps to deal with the situation without adding a lot of complexity. For managers to fully commit to installing any manufacturing system, they need a level of confidence that the system will be able to deal with unusual situations. The intent of this chapter is to provide decision-makers with this confidence in regard to the POLCA system design and functioning.

Chapter 7

Launching POLCA

Now that you have verified the prerequisites for POLCA and designed the details of the system for your situation, it is time for the rubber to hit the road! This chapter will guide you through the steps for launching your POLCA system. Based on implementations at many companies of varying sizes, we have created a roadmap to help ensure a successful launch of POLCA in your factory.

The main steps in this roadmap are the following:

- Use of the POLCA Implementation Team and POLCA Champion.
- Conducting training for various groups.
- Creating a process for secondary activities.
- Scheduling review sessions and management updates.
- Planning how to roll out adjustments and corrections.
- Designing the checks to ensure POLCA rules are being followed.
- Tracking and debugging key metrics.

We now cover each of these points in detail.

Continue to Use the POLCA Implementation Team and the POLCA Champion

In Chapter 5 we explained that the POLCA system design and implementation needs to be carried out by a cross-functional team whose composition was described in that chapter, and also that, at the outset, management should designate one of the POLCA Implementation Team members as the POLCA Champion. The role of this person was also explained in Chapter 5. It is important that you continue to use the POLCA Implementation Team and POLCA Champion to support the launch of POLCA.

The first task of this team should be to create a timeline for all the activities needed for the launch, based on the items described in this chapter. The second task should be to establish baseline values for key metrics prior to the POLCA launch. For this, the team should review the goals that were established for the POLCA system (see Chapter 4) and decide how the progress toward these goals could best be measured using both hard data (numbers) and soft data (opinions, possibly based on surveys). Then the team should collect current values for these metrics, so that there is an established baseline against which to compare the progress after POLCA is in place. We will return to the issue of metrics at the end of this chapter with some additional pointers.

As a brief reminder about the responsibilities of the POLCA Champion, a key role is to serve as a central point of contact to whom questions regarding the operation of the POLCA system can be directed. Since the POLCA Implementation Team will consist of many people, it is helpful for everyone in the rest of the organization to know that if they have a question it can be addressed to a specific person, rather than to the team in general. Another important role of the POLCA Champion is to serve as the liaison between the POLCA Implementation Team and upper management. During the steps preceding the launch, as well as during the launch and initial operation of the system, questions may arise that require reevaluating or even rethinking existing policies. Examples are: changes that simplify the POLCA decision-making, such as reassigning a material-handling task to a different person; changing who is authorized to make a particular decision; redesigning how a particular exception is handled; and questions related to workload and staffing policies. When such issues arise, it is helpful to have one person who can be the advocate for the change and bring the issue to upper management for a decision.

If your system design includes some type of Safety Cards (Chapter 6) then the POLCA Implementation Team should also decide who will be in charge of these cards. This could be one person, such as a senior operations manager, or multiple people. For instance, if there are Safety Cards for different product lines or customers, each of the product line managers (or customer managers) could be in charge of their quota of cards.

Conduct Training for Everyone Impacted by the System

While companies understand the need for training prior to implementing a new system, they usually underestimate the extent of training and also don't involve enough people in the training. It is critical that you train not only the people directly involved in applying POLCA in their jobs, but also many other people who will be or might be affected by the new system. Below, we discuss the people that should be trained, and the various types of training to be conducted.

Training for the Core Group

First, we suggest the types of people that should be involved in the most extensive training and what should be included in the training. You can use this list to stimulate the thinking for your particular situation, and add to it as you see fit. We assume here that the POLCA Implementation Team members went through sufficient POLCA training prior to engaging in the system design. The most obvious next group of people to be trained includes all the operators that are in cells that will be part of POLCA loops. If you are initially implementing POLCA only in a portion of your shop floor, then you also need to train operators that will be in cells that are not on POLCA but will interface with cells that are on POLCA (see Chapter 5). If you have material handlers that are separate from the cell operators, then any such people who would move jobs that are on POLCA, or who would be involved in returning POLCA cards, need to be included as well. Next, you should review all the support personnel. Clearly, the relevant planners, schedulers, supervisors and managers should be included. In particular, if you will start your POLCA Chains with the Planning Cell as recommended in Chapter 5, then all the members of this cell need to be trained. Depending on your operation, you may feel that some customer service representatives need to be included too. Similarly, you should review if people in other parts of the organization might be impacted and should also be included. Finally, it is important to have key senior managers attend the training, so they can understand the need for some of the prerequisites and/or policy changes required for the success of POLCA.

As mentioned, some of these people would have been trained much earlier as part of the POLCA Implementation Team working on designing the POLCA system. However, that team only included a small subset of the people impacted by POLCA, and now you need to be sure to cover a wider swath consisting of the full set of people that will be affected.

For the core group of people mentioned above, you should anticipate a total of two to four days of training over a period of time. The first part of the training should involve both the general concepts of POLCA, and the motivation for implementing it—specifically, why it is needed for your company's manufacturing environment. The concepts of POLCA should include how it works (similar to Chapter 2) as well as its benefits (Chapter 3) so people can understand why it will be effective for your situation. You may find that it is useful to do some simple physical simulations for people to see how jobs and cards would move through the system. Appendix F has an example of one such game, and Appendix I on "Additional Resources" lists other POLCA games available around the world. You can use some of the simulation games already available, or you can make your own simulation if you feel you want to use your own terminology and manufacturing environment. Chapter 11 describes how the pharmaceutical company Patheon created its own simulation game and how it was used for both senior management and shop floor employees.

While the first part of the training above covers the general concepts of POLCA and how it works, for the next part of the training this core group of people needs to be educated on the procedures that have been designed for your implementation. Essentially this involves covering the specific details that you have decided on, based on the roadmap in Chapter 5, as well as the exception-handling procedures that you have constructed based on Chapter 6. As recommended in Chapter 5, you will find it helpful to create a flowchart that documents the rules and decision points, both for the flow of jobs as well as for cards. The teams might also find it helpful to have all the Decision Time triggers documented in a list for reference on each POLCA Board. Both these items—the flowchart and the list of triggers—are not only useful for the training, but serve as an ongoing reference for everyone throughout the launch and subsequent implementation. They are also valuable for training new employees and temporary workers. Figure 7.1 shows an example of a flowchart. This figure is intended to illustrate the types of details that should be in the chart, so some of the details are there as examples, not as concrete items that you need to copy for your flowchart. For instance, in the diamond box with the question "Do we have Authorized jobs delayed more than two days…" the chosen value of "two days" is just for illustration and you should decide what represents a significant delay in your case. Similarly, you should design the entire flowchart and

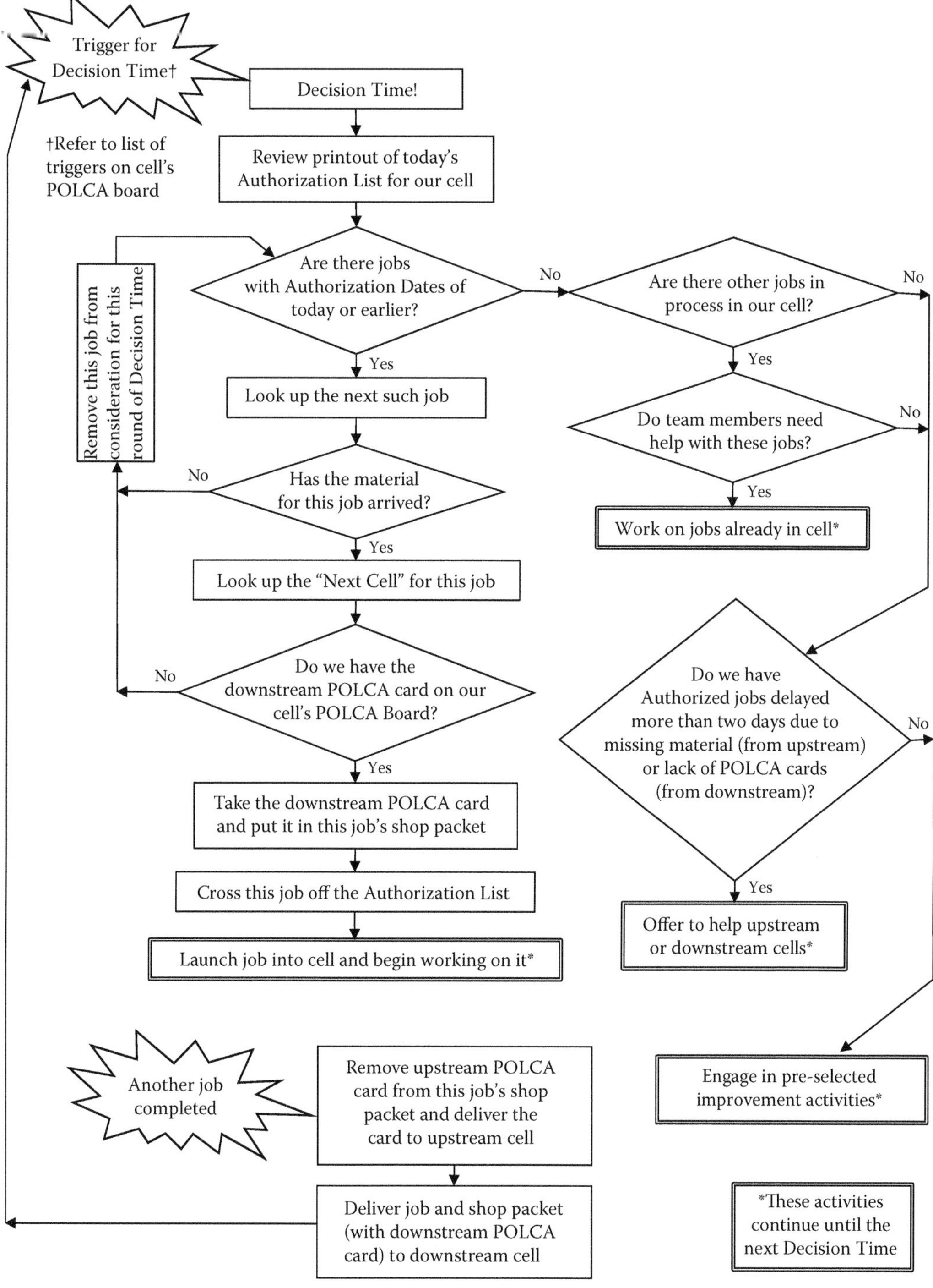

Figure 7.1 Example of flowchart to be used for training a cell team (this example assumes the Authorization List is in printed form and is printed once a day).

corresponding details to fit your situation. However, Figure 7.1 can serve as a model for your team in designing your flowchart.

Finally, prior to the actual launch, companies have found it extremely valuable to have a detailed physical simulation or "walk-through" with some make-believe jobs being carried through the shop floor. For this, you should pick one or more representative routings that will cover the key issues and decision points involved. Create a sample job packet for each such routing. Then the training group should walk through the complete decision logic and physical movements involved with each job packet. You will need to have sample POLCA cards made for these routings. Start with how the Planning Cell will release the job packet along with a POLCA card, and how the material arrives at the first step. Next, go over how the cell team uses the Authorization List and the Decision Time rules. It's not necessary to perform the manufacturing operations, but talk through what operations will be done at this cell and then walk the packet and material (or make-believe material), along with the appropriate POLCA card to the next cell. Then repeat the process at this next cell. As POLCA cards need to be returned to upstream cells at any of these steps, walk through the procedure for this (based on your decisions as explained in Chapter 5). Repeat this process until the end of the routing, being sure to explain any special procedures required in the last part of the POLCA chain.

It is possible that doing this physical walk-through for one routing will cover all the main points that people need to learn, or else you may feel that you must go over a few different routings to capture other situations. The POLCA Implementation Team should decide this based on the particulars of your system operation.

You will find that this physical simulation will not only help people to see how POLCA will work in your company and be more at ease about using the system, but it will also raise a number of issues that may not have been considered. In fact, it is a good idea for the Implementation Team to do a trial walk-through first, to uncover as many such issues ahead of time, and be more prepared for the walk-through with the larger group. (Tasks such as this should be considered when the Implementation Team is preparing the timeline for the launch.)

Training for the Rest of the Organization

While the previous section covered training for the core group of people, you should also plan on a shorter training session for more people in the organization, and possibly even the whole organization, depending on your company size and structure. There are several reasons for this. Even if people are not directly involved with the POLCA system, they will hear about it from their colleagues and co-workers, and will feel better if they have at least some knowledge about what it is. It's not good for people to feel that they've been excluded or, worse, intentionally kept in the dark. Also, keep in mind that the new system will impact the company's performance. It will change the way people approach decisions. The POLCA rules may hold up some jobs, and even if the performance overall is improving, some people may be frustrated or anxious about these delays. These and many similar reasons are why it is important for everyone to have a basic knowledge of POLCA.

We have found that a two-hour overview of the system and rules is sufficient to give people a first understanding of POLCA. Be sure to preface this overview with the reasons for implementing POLCA—a brief version of what you will have covered with the core group. Also, be sure to allow time for questions and discussion.

Based on the above descriptions, you can decide how far you want to extend across your organization for this overview session. For smaller companies, typically 100 people or less, we have recommended that the whole company attend such sessions (possibly in multiple smaller groups). For larger companies, you could confine the training to a division, or part of a division. However, don't make the mistake of limiting the training to people only in Operations. You must include people from all areas, including Marketing and Sales, Inside Sales and Order Entry, Engineering and R&D, Purchasing and Supply Chain, Materials Management, Manufacturing Engineering and Routing, Planning and Scheduling, Finance, and Human Resources. If some of these areas seem far removed from the topic of shop floor control, you should reflect on many of the discussions in the previous chapters to see why these functions need to be included. For instance, because there are times when cell teams cannot launch a job, Finance may need to rethink some of its metrics. An instance that involves both Finance and Human Resources is where temporary workers may need to be hired for spare capacity, as explained in Chapter 5. Purchasing people will have to understand the importance of limiting part shortages as well as the systems being put in place to deal with them and help to reduce them (see Chapter 6). Similar examples can be stated for all the other areas. Thus, it will be beneficial for people in these areas to have a basic understanding of POLCA.

Create a Standard Process for Secondary Activities Initiated by Decision Time Rules

When a cell team cannot launch a job based on the Decision Time rules, Step 4 of the rules requires that the team engage in secondary activities instead. As discussed in Chapter 3, this step provides a formal framework for improvement activities such as cross-training, setup reduction, and preventive maintenance, as well as for teams to offer assistance to other cells that are temporarily overloaded. It is best to assist your teams in creating a systematic process that will kick in when it's time for such secondary activities. Having such a process in place will preempt any bickering among the team members as to how to use this time.

The first step is for a cell team to develop a list of improvement activities that need to be addressed, along with an estimate of time that each activity would take; this list should be updated periodically. The second step is to prioritize this list and, again, these priorities should be revisited periodically. The result of these first two steps is that the team always has an up-to-date list ready to go. The third step is, when the team finds that no jobs can be launched and it has some time on its hands, it should first investigate roughly when it might be able to work on the next job. This can be done by (for instance) checking with a downstream team when a POLCA card might be returned and/or with an upstream team when a job might be completed. This step then gives the team an estimate of how much time is available. One option here is for the team to decide if upstream and downstream will be backed up for a while and if some team members should offer to help out in those cells (and thus help to alleviate the stoppage in their own cell). If the team (or part of the team) decides to engage in improvement activities, the final step is for the team to review its prioritized list along with the preceding time estimate and decide which improvement activity to begin. The result of this approach is that when production is stopped the team immediately has a direction to pursue, and such opportunities are not lost.

If management is concerned about the cell stoppages caused by POLCA and the resulting lower labor utilization metrics, one solution used by companies has been to have the employees record the time that they are working on secondary activities, such as improvement projects or helping in another cell. In the same way that a worker would log time to a specific job, he or she can log their time to another activity. In this way, there is no "idle" time of the employees: either someone is working on a production job, or they are working on a secondary activity. These logs will also give management an idea of the time and efforts spent in improvement activities. *In this way management can perceive such times as an investment, rather than a loss of labor productivity.*

Schedule Frequent Review Sessions and Management Updates

During the initial phases of the launch, it is beneficial to schedule frequent review sessions by the POLCA Implementation Team. This ensures that team members have already cleared their schedules for these sessions, rather than the POLCA Champion trying to put together an urgent meeting as issues arise. These reviews can be kept short, and are meant to address issues that may have gone unnoticed during the design of the system. For example, during the first few days of the launch, the team could schedule a half-hour meeting at the beginning or end of every day, or both. After a week or so, this could be scaled back to a 15-minute "stand-up meeting" at the beginning of each day, and one longer review meeting towards the end of the week. As the confidence in the implementation increases, a periodic review meeting might be all that is needed for example, once every couple of weeks.

To complement these POLCA Implementation Team meetings, you should also plan for and schedule frequent report-outs to upper management. The purpose is partly to keep management abreast of the progress, but also to pre-empt any surprises for upper management, such as possible short-term degrading of some performance metrics as the bugs are worked out of the implementation. In addition, these sessions can be used to address policy changes that might be needed, as discussed previously. Again, these report-outs can be short but frequent initially, and then scaled back to once a week or once every couple of weeks as the implementation gains traction.

Have a Plan for Rolling Out Adjustments and Corrections

As a follow-up to the points in the previous section, if you decide that some significant changes need to be implemented in your POLCA system, you also need a plan as to how these will be rolled out. For example, suppose you decide to change the way that cards are being returned, or to implement some new loops. How will you communicate the new method to everyone involved?

One way to institutionalize this process is to have a periodic (e.g., once a month) POLCA review meeting involving all the relevant shop floor employees and support functions. While companies may already have monthly meetings for employees, this should not be confused with those; you must earmark a meeting specifically to deal with POLCA matters, and the audience for this meeting might be selected differently in any case. You can use this periodic meeting to communicate and roll out system changes. It is also a good forum to solicit feedback on the system. After the POLCA system has been in place for a while and has matured sufficiently, you can reduce the frequency of this meeting or hold it only as needed when specific situations arise.

Design the Checks That Will Ensure POLCA Discipline Is Followed

Simply describing the POLCA rules to all involved is not enough—you need to ensure that the rules are followed! In case you are wondering why employees might not follow the rules, there are many reasons, and, most of all, remember that we are dealing with people in the system. In factories, a common reason for machine operators to change the recommended sequence of jobs is cherry-picking: choosing an easier job over a more difficult one. Another is ease of changeover. An employee might pick a job that requires minimal or no setup after the job just completed. Other reasons could involve personal schedules. For example, if a job has arrived at a cell but the Decision Time rules don't allow the cell team to start that job at this moment, employees might decide to work on it anyway because they want to be able to leave early or take some time off on the next day. And so on.

So, the Implementation Team needs to accept that these possibilities exist and think through two important points ahead of the launch. The first is, how will we check if the POLCA rules are being followed? Some companies are small enough that they feel the production manager or supervisors clearly know what is going on. Larger companies have instituted spontaneous audits where some members of the POLCA Implementation Team arrive at a cell and check if all the jobs in process are indeed Authorized and have the right POLCA cards attached to them. (Two types of POLCA audits will be described further in Chapter 9.)

The second point to decide is, if a POLCA rule has been broken, what are the consequences for the employee or the cell team? Again, companies have differed in their handling of this. Smaller companies have taken the softer approach that the errors could be simply due to misunderstanding and have used the occasion to explain and re-train the employee(s). Some larger companies with more wide-ranging employees have taken a harder line, even writing up a complaint about an employee if this is a repeated occurrence.

Regardless of which way you decide to go with these two points, it is important to think them through ahead of time and come up with your answers. This represents yet another reason why it is critical to include some representation from Human Resources (HR) in the POLCA Implementation Team as recommended in Chapter 5, and possibly include the whole HR group in the short overview training session described earlier.

Start Tracking and Debugging Key Metrics

As part of your POLCA launch you must determine the way that you will track key POLCA-related metrics, specifically, metrics that determine your progress toward the goals that were established for your POLCA implementation (Chapter 4). The core set of metrics that you should measure include WIP inventories, finished-goods inventories (if relevant to your type of manufacturing), lead times (you could use the MCT metric described in Appendix A, if your management has bought into this metric), total system throughput, and on-time delivery performance. Additional metrics should be designed to meet other goals that were set at the start. For example, if you felt that the tighter WIP control and communication between the cells would help to improve quality, then quality-related measures could be added into the basket of metrics.

There is an important issue to consider when implementing a lead time metric while using the POLCA system: lead time needs to be computed differently for individual cell teams versus for the POLCA system as a whole. The reason is, you need to correctly attribute the components of lead time for which a cell team or the POLCA system should be held responsible.

Let's start by considering this issue for a cell team. You want to make sure that the lead time (or MCT) metric for this cell only contains portions of time that the cell team is truly responsible for, and is able to impact with its actions. To understand the issue here, remember that when a job arrives at a cell the Decision Time rules could prevent the job from being launched. Therefore, the cell should not be penalized for the segment of time that the job is waiting due to POLCA rules.

However, in measuring the performance of the POLCA system as a whole, our metric would be designed differently. If a job has to wait at a cell because it is not yet Authorized or the right POLCA card is not available, then this is due to the operation of the POLCA system, and this waiting time *should* be included in the lead time metric. Once again, though, we need to be careful not to penalize the POLCA system for items that are outside the scope of the POLCA implementation.

Let's take lead time (or MCT) for a particular POLCA Chain as an example. We pick a given POLCA Chain because it makes sense to separate out the lead times for different routings—if you want to carefully track performance patterns then you don't want to mix the data for short and simple jobs with long, complex ones. So, for a given POLCA Chain the lead time clock for a job should only start ticking when both these conditions are satisfied: the job is Authorized at the *first* cell in the chain, *and* the material for this job is available. Clearly, if the material is not available at the first cell, this is an issue that is outside the POLCA system, and the lead time clock for this POLCA Chain should not start. However, the clock can start ticking regardless of the availability of a POLCA card! If the right POLCA card is not available and the job has to wait, this waiting time is definitely part of the POLCA system's design and operation, so it should be included in the lead time metric.

Another example is when there is a component part shortage and a Safety Card needs to be deployed. The job is delayed, but this delay should not be attributed to the cell nor to the POLCA system; the delay should be recorded in a separate category. If you have other such situations, you should use the preceding discussions as examples to help you decide whether to include, or not include, segments of lead time for a cell team or for a POLCA Chain.

Similarly, on-time delivery performance for the POLCA system should be measured at the end of the POLCA Chain, not at the shipping point. In order to fairly judge the POLCA system based on this data, you will need to compensate for whether material arrived at the first cell on time. You don't want POLCA to take the blame if the first cell didn't receive its material until much later than planned!

The preceding discussions highlight the point that you should think through some of your traditional metrics before applying them to measure the performance of POLCA. You want to make sure that you are fairly measuring the POLCA system, not coloring the data with problems that are outside the scope of POLCA. As you start implementing the metrics, shop floor employees may point out other situations that require adjusting the metrics; this is part of the debugging mentioned in the title to this section.

This chapter concludes the sections on designing and launching your POLCA system. If this and the preceding chapters appear somewhat daunting, in actual fact some of these steps and decisions can occur very quickly. The next chapter describes an inspiring case study where a company designed and launched POLCA in three days!

Chapter 8

How We Designed and Launched POLCA in Three Days

Guest author: Ananth Krishnamurthy

During workshops on planning and implementing POLCA, held at the Center for Quick Response Manufacturing (QRM), at the University of Wisconsin–Madison, I often get asked the questions: "How long does it take to implement POLCA?"; "Does a typical implementation take four to six months?"; and "Has anyone implemented POLCA in less time?" I had the good fortune of being part of a POLCA implementation that was done in three days! This seems truly impressive, and you might wonder if it could really happen. So, first, I am going to walk you through this POLCA implementation effort, and then I will summarize the key takeaways about reducing the time for a POLCA implementation so you can see that this target is achievable for other companies as well.

The Setting: A Factory in Canada

A few years ago, a leading manufacturer of electrical control systems approached the Center for QRM for guidance in implementing QRM across several of their facilities in the United States and Canada. The company's senior leadership was convinced that the focus on responsiveness would be a positive, unifying perspective for the entire business. This led to mutually beneficial collaborations that illustrated how QRM built on the company's existing foundations of implementing Lean principles but went further by enabling their facilities to provide custom solutions with still shorter lead times. After implementing cells in several of their facilities in the United States and Canada, the company realized that while the cells had improved local performance (*within* cells), the next big opportunity lay in improving the flow *across* cells. So, they initiated efforts to implement POLCA at some facilities in the U.S. Midwest. In parallel, one of their facilities in Canada also expressed interest in implementing POLCA and approached the QRM Center for assistance. I decided to visit their facility to provide training on POLCA and high-level guidance on POLCA implementation.

I visited their facility on a Wednesday and my plan was to take a return flight to Madison on Friday afternoon. My visit was hosted by Tracy, the Planning Department Manager. (Note: Since there has been a change in management at the company and many of the former managers are no longer there, all names have been changed for the purpose of this description.)

The agenda for the three days was as follows: on the first day, I was to provide an overview of POLCA, get a tour of the facility, and, along with a team from the company, collect preliminary information needed for implementing POLCA. On the remaining two days, I would work with their team to go over details of the potential POLCA implementation and provide as much guidance as I could.

Day One: POLCA Overview and Factory Tour

As explained, Day One started with an overview of POLCA. As I was getting ready to present the overview, Tracy said: "We have all been reading about POLCA and you should expect a lot of questions. Many of us think this would be a good fit for our facility, but I should warn you that some of the people in the group are not really convinced that POLCA is a good fit. I am hoping that you can convince them, and *we can implement POLCA before you leave*." I promised to do my best, but I deliberately did not comment on her goal of implementing POLCA before I left. I thought it was too ambitious, but did not want to dismiss her optimism outright. I figured that once they heard my overview and learned about examples of POLCA implementation at other facilities, they would realize that implementation might take a few weeks.

The group at my presentation consisted of around 15 people and represented a variety of functions (manufacturing, assembly, quality control, inventory planning, materials, customer support, and so on). It had a good mix of managers, supervisors, support functions, and lead workers. The group was very engaged during my presentation and, as warned, asked a lot of questions. At one of the breaks, Tracy commented: "The session seems to be going well, I think the skeptics are starting to believe that POLCA would be a good fit." I concluded my presentation with a roadmap for implementing POLCA (similar to that in Chapter 5) along with an overview of POLCA implementations at some other factories.

Tracy then reminded the group that since I was available till mid-afternoon on Friday, she wanted to make use of my time to implement POLCA before I left. The group seemed to nod in agreement. Again, although I was surprised at their optimism, I refrained from discouraging them. We spent the rest of the afternoon getting a tour of their facility and collecting some information about their manufacturing and assembly operations. During the tour, I observed that the facility made large electrical control unit assemblies. Each assembly was composed of smaller sub-assembly cabinets and enclosures that contained electrical components (such as switches, wire harnesses, breakers, printed circuit boards, and display panels). Most of the facility was organized into cells focused on fabrication, sub-assembly (SA), and final assembly (FA). However, they had excess inventory in many areas and it was evident that coordination between the different areas could be significantly improved. Following the tour, we had a report-out to the senior leadership (Vice Presidents, Directors, and Plant Manager). We informed the leadership that the group had collectively decided to pursue POLCA and we would try to make as much progress on the implementation before I left. We committed to updating the leadership every day.

Day Two: Checking the Prerequisites and Some Initial Design

Early the next day, the group convened and started with the pre-POLCA assessment, including checking the prerequisites for POLCA (see Chapter 4). First, we needed to identify areas for potential POLCA implementation. Almost unanimously, the group proposed implementing POLCA between the SA cells and FA cells. There was significant inventory in and between these cells, and shipping commitments from FA were being compromised daily, creating constant pressure to expedite orders within the FA cells. These expediting requests snowballed into changes to their supplying cells, resulting in more confusion and delays as well as more upstream inventory—jobs that had been moved aside in order to work on other orders.

I felt that first we needed to make sure it wasn't simply a capacity problem. So, based on a question from me about the overall capacities of each of the cells, Marge, the senior planner, went back to her office briefly and extracted some data for the past few months. She did a quick estimate of the total production hours spent on making subassemblies versus the hours required to make the units required for final assembly. The calculation revealed that, for every month, the SA cells did have the capacity to meet the FA cell needs. So, it seemed that it was not because of lack of capacity, but rather, because of changing priorities that schedules were in a constant flux and shipments were getting affected. This convinced everyone that implementing POLCA in these areas would provide the required discipline and prioritization needed to make better use of available capacity and improve shipments.

Checking the Remaining Prerequisites

Next, we walked through the remaining list of POLCA prerequisites. We observed that both the SA and FA areas were already organized into cells, and these did meet the criteria of POLCA-enabling cells (Chapter 4). Further, on a weekly basis, the facility was using MRP-generated Dispatch Lists for capacity planning in the various SA and FA cells (with hours of work and schedules), so there was already reasonable rough-cut capacity planning in place. (However, changes in priorities meant that these schedules were tweaked often.) From a POLCA prerequisite point of view, we felt comfortable about the level of planning. However, the Dispatch Lists needed to be modified to create Authorization Lists (as described in Chapter 4). Currently, the Dispatch Lists only had the shipping dates after final assembly. We would need to add Authorization Dates for each cell based on the individual lead times of the cells. We asked Jeff, who managed planning and IT systems, if this modification was possible. "We already have the routing information and planning lead times in the system," replied Jeff. "I think we should be able to write a script to modify our Dispatch Lists. Let me see if I can get one of my staff to get working on this right away." He stepped out of the room to follow up on this, and a few minutes later he returned and informed us that the changes would be done either by the end of the day or by the morning of the following day. The wheels were in motion!

As part of setting the goals for the POLCA implementation, we set targets for work-in-process (WIP) and lead time reduction. At this point we were satisfied that all the prerequisites would be in place, so we moved on to the other items needed to implement POLCA at the facility.

Initial Design: The POLCA Loops

The next step for our group was to identify the POLCA loops. Using POLCA between the SA and FA cells was an obvious starting point. However, there was some discussion about whether all the SA and FA cells should be linked using POLCA or if we should restrict the initial implementation to only a few of these cells. After some discussion, we felt that we would plan based on the assumption that all the SA and FA cells would be linked, and only eliminate some of the cells from the scope of implementation if we ran into major roadblocks. We then discussed whether the areas upstream of SA (fabrication, machining, and the warehouse supplying parts to the SA cells) and downstream of FA (testing, packing, and shipping areas that were customers of the FA cells) needed to be included in the scope of the POLCA system. Fortunately, supervisors from these areas were part of our group. Their opinion was that the bulk of the inventory and expediting issues were due to coordination issues between the SA and FA cells. Although the upstream and downstream areas might benefit from being in the POLCA process, they suggested that we start with implementing POLCA only between the SA and FA areas and expand the implementation into the other areas later. The team agreed that doing so would be a reasonable strategy.

Deciding on the Quantum

Next, we had to determine the quantum corresponding to a POLCA card. The SA areas were making cabinets and enclosures that were assembled into larger units in the FA cells. What would be a good unit for the quantum? Should a quantum correspond to an individual cabinet? Or should it correspond to an assembly order? Or should it be something else? We decided to walk through the shop floor to get a better understanding of the possibilities. As our group of about 15 walked through the areas, we were approached by a few shop floor workers who were curious to know more about what the team members were doing huddled in the conference room, skipping their regular duties for the day. Someone said, "We are all working on POLCA. We are going to be doing this tomorrow as well." I did not think that response was well understood. I had the feeling that they were imagining our group playing polka music and practicing dancing steps, while the rest of the workers were trying to get shipments out!

Our walking tour helped us visualize a few options for the quantum. The first option was to set the quantum to be one cabinet. However, if we were to attach POLCA cards to every cabinet, we would be dealing with a very large number of POLCA cards. An alternative was to set the quantum equal to a fixed number of cabinets, say, five cabinets. Since sub-assemblies occasionally needed as many as 12 cabinets, and it was also common to see orders that needed six to eight cabinets, we would need to ensure that workers used the right number of POLCA cards and understood the rules for this (see Chapter 5). So, this would add a bit more complexity to the POLCA operation for the workers. A more relevant consideration was that some of these cabinets were small and others were large (housing more components and requiring more time to assemble). Since the POLCA quantum is an indication of capacity, clearly the quantum should consider the difference in assembly times needed for small and large cabinets. Hearing this discussion, one of the cell supervisors mentioned that they use carts to move the cabinets from station to station. Each cart held anywhere from 3 to 12 cabinets, depending on whether they were small or large cabinets. He wondered if we could set the quantum to be one cart of cabinets. Although in terms of the number of cabinets our quantum would represent a number that varied significantly, in terms of the amount of work, each cart was a more steady measure: assembling the 12 small cabinets on a cart took roughly the same time as assembling three large cabinets on a cart. Furthermore, it would be easy to attach a POLCA card to each cart. The supervisors all agreed. Bingo! We had found a unit of capacity that was easy to visualize, manage, and robust to the mix of products flowing between various SA and FA cells. And so the quantum decision was made: one POLCA card would correspond to one cart of cabinets.

Designing the POLCA Card

The final task of the day was to design the POLCA card. We discussed a few options, and then, since the carts were made of steel, the idea came up to make a POLCA card with a thin sheet of magnetic backing (like a refrigerator magnet). This way the card could be stuck to the side of the cart and would not easily be lost or misplaced. Marge, the senior planner, sketched out a rough design on the white board. The group provided inputs on the information to be displayed on the POLCA card, the card layout, and other design details. Once the design was approved, Marge volunteered to finalize it in a MS Word document. To close out a day of hard work, Tracy decided to order in pizza for everyone, as a mini-celebration of our accomplishments. One of the cell supervisors volunteered to pick up the pizza and offered to buy the materials needed to create the POLCA cards on his way back. Not long after, we were all eating pizza and sticking POLCA cards on the magnetic sheets! I felt we had made an amazing amount of progress for the second day. I was looking forward to the final day.

Day Three: Final Design Decisions

The group was back in action early on my third day. The next item on the POLCA implementation roadmap was to calculate the number of POLCA cards. I explained the formula for calculating the number of cards (see Chapter 5). The group saw that we needed lead times for each cell and the flow rate of jobs between the various cells. Fortunately, we had the cell supervisors in the room, along with Marge, the previously mentioned senior planner. We referred to the work order lists that she had printed earlier for the capacity calculations. As explained in Chapter 5, we had to convert the jobs from cabinets to quanta, using the knowledge of how many cabinets of a given type would fit on a cart. After doing this we computed the number of POLCA cards that would flow between various SA and FA cells during one planning period. The supervisors chimed in with their opinions on the calculations (either validating the numbers or suggesting modifications because of exceptions related to work orders). We used lead time estimates for the various cells from the planning system, and before long we were looking at a table that listed all the types of POLCA cards and the number of cards in each loop. Although we felt we had been careful in the approach and our calculations, the question still arose from the group: if we wanted to have some degree of comfort in these numbers before jumping into the implementation, how could we validate our calculations?

With some brainstorming, we realized that since each POLCA card was to be attached to a cart, we could simply go out to the factory and see how many carts were in use on the floor. So, without further delay, the group fanned out around the factory and started to count the number of carts in use and take pictures (just in case we needed to count again, we could refer to the pictures). One of the workers, who envied the fun that we seemed to be having, asked "When is the next POLCA session? I would like to be part of it!"

We came back to the meeting room and compared the calculated number of POLCA cards with the cart count. Our estimate of the number of POLCA cards was much lower! We felt we had hit a road block. Did we make errors in our calculations? We reviewed them again; there were no obvious errors. Then it dawned on us that we should not be surprised at this discrepancy. We knew that we had excess WIP in the system. Since we were confident with the data we used in the calculations, the number of POLCA cards we computed was providing us an estimate of the WIP (in terms of carts) needed to support the production targets. The difference between the shop floor observation and the POLCA card calculation was the WIP reduction opportunity, and in fact corresponded well with the targets we had set on the previous day. We felt relieved that the pieces of the puzzle were fitting together.

Defining the Decision-Time Rules and Exceptions

We moved on to determine the details of the decision-making process for POLCA implementation. We refined the basic Decision Time rules (described in Chapter 2) to include specifics that would be useful for workers and supervisors using the system. It was really helpful to have supervisors, lead workers, planners, and managers during this discussion, because they brought up several situations. It was useful to walk through each scenario and confirm how the POLCA system would work, and whether we needed to identify any exceptions to the basic rules. After an hour of discussion of several potential scenarios that might call for exceptions, we stared down our list of rules for these exceptions. We realized that there were very few additional rules that were needed. Consistent with the point made in several places in Chapter 6, we saw that, in most situations, merely following the POLCA rules would lead to the right action. That was also encouraging to the group.

However, we did need to identify additional rules to deal with unexpected part shortages in the SA and FA cells. We decided to use Safety Cards (described in Chapter 6) to deal with these shortage instances. We estimated the number of safety cards and documented the process for their use. At that point, one of the supervisors jumped in with the remark, "I hate to bring this up, but John's expedite requests are going to mess up our well-thought POLCA plan. How are we going to deal with that?" That statement silenced everyone in the room; you could have heard a pin drop. Everyone had a defeated look on their face. Tracy acknowledged, "Yes, John is going to want to expedite something every day and that would mess up our POLCA implementation, wouldn't it?" Who is John, I was wondering, when they explained that he was the Plant Manager, and when customers called him personally with expedite requests on their late orders it was hard for him to say no. Such expedites would violate the rules of POLCA. We pondered over potential solutions, and came up with an idea. "Let's give John a limited number of special cards. John can use them to expedite orders." We came up with the rules of how these cards would be deployed. Finally, we decided to limit the number of these special cards to three, so that he could only have three expedited orders in the system at one time. This would bound the disturbance to the POLCA system, but still keep John happy by giving him some ability to make his customers feel good.

Here, I would like to add a historical note to this story. The POLCA implementation that I am describing was one of the earliest applications of POLCA in industry. What we did on that day, in terms of designing the special cards for John, was to create the basis for what later became known as *Bullet Cards* (see Chapter 6)—an important addition to the POLCA toolbox for other companies to use in their implementations.

Decision to Launch

During the afternoon report-out, the leadership was impressed with how far we had come in the planning process. In fact, they were anxious to see POLCA in action, helping to reduce their inventory and improve their delivery performance. Toward the end of the update, John commented "Great work! This is impressive. I am assuming I can still expedite orders when I want to, right?" Tracy jumped in with: "We thought about that too, John! You will get three special cards that you can use to expedite. Be sure to use them wisely." After the laughter died down, we explained the concept. John was concerned that he could not expedite more than three at a time. We told him that based on experience with other implementations, once POLCA was working, he would not even need three cards! While at first still dubious about this, John soon realized that when it came to following the POLCA discipline, he had to lead by example. He agreed to the rules.

There were just a couple of hours left before I had to leave for the airport to fly back to Madison, Wisconsin. As I looked at the list of steps needed to implement POLCA, I could see that we had made significant progress. We then talked about the timeline to launch POLCA. Before anyone else could suggest a launch date, the supervisors of SA and FA suggested, "Why not roll this out Monday morning? We all agree that this will make our operation more efficient, so the sooner we implement, the better. We're all together in this room right now, decision-makers and key stakeholders; if we don't do it right away, we will keep putting it off for a better time." Everyone agreed. We made Marge the POLCA Champion, with duties similar to those listed in Chapters 5 and 7. The materials manager was appointed Safety Card Owner. And John, of course, had his special cards, the precursors to the Bullet Cards.

Tracy informed us that throughout the three days she had been typing notes based on our discussions and that her notes could be used for training other workers. I was pleasantly surprised to hear that. To be honest, I had noticed her typing periodically during our discussions, but I thought she was responding to urgent emails. The lead workers in the group immediately volunteered to train their colleagues. They informed us that they had been talking to their colleagues during our discussions and most of them were already on board. They offered to do some initial training before the end of the shift that day. The group also decided that this team would meet every morning for 30 minutes for the next two weeks to discuss any implementation issues. Subsequently, the team would meet once a week to review the progress. It looked like we had accomplished everything we needed to launch the POLCA implementation on Monday morning!

As I left for the airport and was saying my final goodbyes to Tracy and the team I confessed, "I thought you were joking when you said you wanted to implement POLCA in three days. Until today, all the POLCA implementations done by the QRM Center have taken several weeks of planning. This is going to be a record in terms of the fastest implementation. Do let me know how this goes on Monday." Indeed, we received word that the company did launch POLCA on Monday morning. There you have it—POLCA implemented in three days!

Postscript: Keys to Success and a Challenge to Others

During my journey back to Madison, I reflected on the achievement and realized that it was not a coincidence. This impressive progress was a result of significant foresight and deliberate planning by Tracy. In my mind, there were two key factors that contributed to the success. First, Tracy had anticipated many of the challenges. Even though she did not know the answers to these challenges, she ensured that the team she assembled included everyone that would be impacted by the POLCA implementation as well as people with the expertise to do any supporting analysis needed along the way. She was confident that this group would come up with a solution every time we came upon an obstacle. When I reflected on the two days, I could not agree more. On several occasions, we had hit roadblocks and had collectively found a way to overcome them. Second, Tracy used my limited availability to create a sense of urgency for the implementation. She had the entire team dedicated to POLCA implementation for three days and personally ensured that we stayed focused on the task at hand. She knew that POLCA was the right solution for the facility, but she gave the team time to come to that conclusion themselves. After getting everyone on board, she leveraged the excitement and my presence to get issues resolved and an implementation rolled out in three days.

A few weeks later, the QRM Center received a note from Tracy thanking us for the help and support. She reported that POLCA was running successfully. The shop floor had embraced POLCA. There were a few minor issues during the initial days of implementation, but these were easily overcome. The initial results had been very encouraging—30% reduction in WIP. An unexpected result was a 10% increase in shipments, attained without adding any people, equipment, or overtime! (This underscores the point made earlier in the book: a lot of people's time and capacity goes into rescheduling and expediting jobs, and when you reduce the need for this, they can focus on getting jobs out, instead of moving things around.) Finally, everyone was happy with the results since they exceeded the target we had set for ourselves initially, and, in particular, John was also happy and wasn't complaining about not having enough special cards! Tracy's note concluded with a final comment: "Just wanted to remind all of you that Canada now holds the record for POLCA implementation." There you go! That is an open challenge to all of you around the world planning to implement POLCA. If you beat this benchmark, let us know, and the trophy will move from Canada into your hands.

About the Author

Ananth Krishnamurthy is a professor of Industrial and Systems Engineering at the University of Wisconsin–Madison, where he has also been Director of the Center for Quick Response Manufacturing (QRM) since

2008. Ananth has an M.S. in Manufacturing Systems Engineering and a Ph.D. in Industrial Engineering, both from the University of Wisconsin–Madison, and an undergraduate degree in Mechanical Engineering from the Indian Institute of Technology in Mumbai, India. Ananth has conducted numerous QRM projects with many types of companies, and has been a frequent speaker at QRM workshops and conferences. Holding training events in the United States and Europe has earned him worldwide recognition as an expert in the implementation of POLCA. Ananth has consulted for leading companies including Alcoa, Ingersoll, John Deere, Johnson Controls, Rockwell Automation, and Trek Bicycle. He has authored numerous publications in international journals, and is a Senior Member of several professional organizations. Ananth served as an assistant professor at Rensselaer Polytechnic Institute (RPI) from 2002 to 2007, and was named winner of RPI's Excellence in Education Award for 2007.

Chapter 9

Sustaining POLCA: Post-Implementation Activities

After you have launched POLCA and it appears to be working smoothly in the short term, you need to ensure that there are ongoing activities to support the long-term success of the system. Too often, companies find that improvement programs are backed by a lot of enthusiasm in their early stages, but then the energy level subsides and people resort to their prior habits and business as usual. It is important that you anticipate this and put mechanisms in place to sustain the POLCA system and its positive performance.

Create a Steering Committee and Stay the Course

The first advice for sustaining POLCA in the long-term is to be resolute and disciplined about the system and the rules. Managers or planners are tempted to work around the system to meet some short-term requirements, but this usually results in a vicious cycle of breaking the rules and undermining the system. The resulting message to the shop floor personnel is that these rules don't matter, and so eventually they start ignoring them as well, leading to a slippery slope ending in failure of the implementation. This is documented by the experience of Alexandria Industries (Chapter 10), where the authors clearly advise at the end of the chapter that you should "Stay tenacious and disciplined." As stated in that case study, their company got extremely busy and ran out of capacity. In the authors' words, "As the chaos grew, we found ourselves trying to out-think the system … We thought that we could better control the job flow manually. It is easy to predict what happened next. On-time delivery diminished and WIP increased … we were expediting more and moving WIP around because it was in the way of what we needed to work on next. Looking back, it is clear that reverting to processes that didn't work previously is not the answer! Hence our advice is that you should stick with the POLCA system and rules regardless of business conditions."

Therefore, as one of the first activities to help sustain POLCA, you should institute a process to review and handle unusual circumstances or extreme business conditions without changing the basic POLCA rules. An effective way of institutionalizing this process is to create a *POLCA Steering Committee* that includes the POLCA Champion and representatives from a few functional areas as well senior management. You could see this POLCA Steering Committee as a logical evolution of the POLCA Implementation Team. The latter was created to justify, design, and then implement your POLCA system. Now, the overall goal of this POLCA Steering Committee should be to support and sustain the POLCA implementation in the long term; dealing with urgent issues that arise can be among its shorter-term tasks. For continuity, and in consideration of their knowledge of POLCA, you could consider having several people from the POLCA Implementation Team continue to serve on the POLCA Steering Committee. However, you should also consider some new members based on the different role of this committee. For instance, some additions could be from senior management—people that have the standing to review and change policy decisions in areas impacted by POLCA.

The POLCA Champion should be empowered to call a meeting of this committee if an urgent situation arises, such as unexpectedly high demand, unusually short lead time requests by key customers, or unanticipated reduction in capacity due to machine or employee problems. It is important to note that you can in fact deal with most of these situations while still working within the POLCA system rules. Here are examples of techniques you can use to handle unusual demand or unexpected short-lead-time orders. You can:

- Add cards to some loops for a limited period, and then remove them at the end of that period.
- Increase the planned lead time for some routings. (See Chapter 5 for a discussion on the choice between this and the previous option.)

- Change the due dates of a few jobs. When the Authorization List is regenerated (daily, or at the start of each shift, or even more frequently) the Authorization Dates of all remaining steps for those jobs will be recalculated and these jobs will be expedited or held back based on the changes you made. Note that the Decision Time rules will still be followed by all the cell teams for those remaining operations.
- Add temporary capacity to some areas. Chapter 5 gives examples of how companies plan for reserve capacity that can be kicked in at short notice.
- Outsource some operations. For example, if you are experiencing unusually high requirements for turning of large-diameter parts and you only have a couple of Lathes that can do this, consider sending parts to a subcontractor for this operation (and possibly one or two preceding or following operations). For this outsourcing to be feasible in a short lead time, it requires that you already have a plan for this possibility ahead of time, along with pre-approved subcontractors for various types of operations.

The main point to note with all the above solutions is that there is no need to tamper with the POLCA system; the shop floor operators will keep working to the standard Decision Time rules. For example, if a customer needs a spare part urgently and this order needs to be expedited, while at the same time management knows that another order is just going into safety stock at a distribution center, then it can decide to allocate capacity to the urgent order first. This can be accomplished by adjusting the due dates for these jobs as explained above, and the people on the shop floor might not even know that this has been done—they will simply go by the latest Decision Time rules as the jobs flow through the shop floor.

To be able to use the above solutions effectively, the Steering Committee should conduct some up-front brainstorming about the possible techniques it might use, and then it should get pre-approval from management for some of the decisions that may have a cost impact, such as adding temporary capacity, or outsourcing an operation. This approval can be within pre-specified boundaries—for example, "You can deploy up to 20% overtime in Cells X and Y, for up to two weeks in a given quarter," or "You can subcontract up to 20 hours of Lathe work in a month."

The main aim of requiring the POLCA Champion to call a POLCA Steering Committee meeting is to ensure that these mechanisms aren't being misused or overused. For example, the approach of changing due dates for jobs seems simple enough to implement, and managers might be tempted to do this regularly. This would just lead the way back to the pre-POLCA situation of constantly changing priorities. So, the committee could act as a check on such practices; simply requiring a committee meeting and then also needing the committee's approval would help to filter out unneeded usage of this technique.

Over time, the POLCA Steering Committee can enhance the techniques it is using to deal with exceptional situations. It could learn how to better utilize some of the approaches, or come up with additional techniques based on suggestions from employees or committee members. This learning will also help with sustaining the POLCA implementation in the long term, an issue that is discussed at the end of this chapter.

The POLCA Steering Committee should also monitor the methods in place for training new employees and temporary workers on the basics of POLCA, and ensure that these methods are consistently used for such additions to the workforce. It should also engage in regular communications and updates for employees on the performance of the POLCA system. This leads us to the next section.

Track the Key Metrics and Celebrate Successes

You must keep tracking the key metrics that were set in place at the POLCA launch (see Chapter 7). It is best if some of the high-level metrics are published regularly for the whole factory to see, preferably in easy-to-read visual charts.

You should also continuously evaluate the fairness of the metrics. You don't want to penalize teams for performance degradations that are due to events which are out of their span of control: Chapter 7 gave several examples of such situations. The employee feedback mechanisms discussed in the next section can assist you in monitoring the fairness of the current metrics.

Plan to recognize and celebrate when pre-specified targets are achieved. For instance, if you have historically struggled to get your on-time performance better than 70%, you could set targets of 80%, 90%, and 95% as goals and, as each is reached, arrange for some formal recognition from management along with a small celebration for employees. Don't feel that these celebrations need to be large events. The literature shows that even small amounts of recognition go a long way toward making employees happier. For instance, a pizza party or cookout for employees and their families at a neighboring park gives enough of a message to everyone that management recognizes and appreciates what has been achieved. In addition, such events help to reenergize and spur people on to achieving the next goal.

Use Surveys and Feedback Forms to Assess the Qualitative Benefits

As has been stressed throughout this book, a good part of the success of POLCA comes from the fact that it is a people-oriented system. So, its success depends heavily on the buy-in from your workforce. Hence, you shouldn't judge POLCA based only on "hard" performance data, you should also look for some "soft" data based on people's opinions and impressions.

A survey form is an excellent instrument to gather such soft data. You should work with your Human Resources (HR) department to design such a survey, and ask employees to fill it in periodically, for example, one month after the start of POLCA, then three months later, and six months later. Typical questions that companies have included in their surveys are in three categories:

1. *Employee understanding and effectiveness of training.* Do employees understand why the company opted to implement POLCA? Do they feel that they understand all the workings of the POLCA system? Was the training adequate? Did they get sufficient opportunities to have their questions and concerns addressed?
2. *System performance.* In the employees' eyes, have there been improvements in performance? For example, do they have less WIP and more space in their cells? Are there fewer hot jobs? Less frequent schedule changes?
3. *Morale and satisfaction.* Did employees feel they were sufficiently involved in the implementation process? Has POLCA improved the teamwork in their cell? Is the communication with upstream and downstream operations improved? Do they feel less stress in their jobs? Is their overall morale higher than before?

Even though we have used the term "soft" for such data, you can use a widely accepted method to turn the data into quantitative metrics that you can track over time. The method involves using an ordinal scale for the answers. For example, in response to a question about whether a particular problem is better or worse than before, employees could be asked to rate their answer from 1 to 5, as follows:

1	2	3	4	5
Much Worse	Worse	No Change	Better	Much Better

Alternatively, the question could be turned into a statement such as: "I have to deal with fewer schedule changes every day." The employee could then be asked to select one of these 5 responses:

1	2	3	4	5
Completely Disagree	Mostly Disagree	No Opinion	Mostly Agree	Completely Agree

Both these approaches enable some statistical analysis since the data from surveying a number of employees can then be summarized using a histogram or median for each question. Note that for such ordinal answers, the *mean* is *not* a correct summary statistic, only the *median* or *mode* may be used; the median is arguably a better statistic than the mode, as it is based on the whole sample of answers. By looking at the histograms and medians for the answers over several time periods, you can see if you are making progress on a particular issue.

The final part of your questionnaire should allow employees to write open-ended answers. Typical questions companies have used for this portion include the following: How has POLCA affected your job on a daily basis? What improvements have you noticed in other areas outside of yours? Has POLCA affected how you deal with other areas? Do you have any suggestions for the POLCA Steering Committee on how to improve the system?

These answers should be collated in categories and reviewed by the POLCA Steering Committee. If HR is not represented on the POLCA Steering Committee then a representative from HR should also be involved in reviewing these comments as well as the statistics from the ordinal questions. Based on the collection of answers, the POLCA Steering Committee can decide if it needs to craft a response for employees and, if so, this can be included as part of the regular communications and updates by the committee that were mentioned earlier.

A last word of advice. It is best if you make the surveys anonymous—and make sure employees trust that they will indeed be kept anonymous! This will enable the truest data collection.

Conduct Regular POLCA Audits

To maintain the integrity of POLCA you need to plan audits of the system. You should consider two types of audits: (i) periodic audits, to check on the status of cards, and (ii) random audits, to ensure the POLCA rules are being followed.

Periodic Audits

Let's begin by discussing the periodic audits. Such audits have several purposes: to examine whether the number of cards in a given loop needs to be reevaluated; to see if the procedures in a given loop need to be improved; and to check whether any cards are missing. This audit involves taking a snapshot of each POLCA loop in turn and recording the location of the cards in that loop. This data can be summarized for all the loops in a simple visual chart. Let's go over an example to illustrate the details of a typical audit and the results.

Suppose a company has decided that, as part of the POLCA card handling procedures, the moving of a job and associated POLCA card from an originating cell to a destination cell will be done by material-handling operators, but the returning of a card from the destination cell to the originating cell will be everyone's job, as explained in Chapter 5. In this situation, we decide to collect statistics on the location of each POLCA card based on the following seven possibilities (the abbreviations on the left side of each row correspond to the headings in Figure 9.1):

Orig. Board:	On originating cell's POLCA Board
Orig. Cell:	In process in originating cell
Moving:	Job and card being moved to destination cell
Dest. Board:	On destination cell's POLCA Board
Dest. Cell:	In process in destination cell
Returning:	Card being returned to originating cell
Missing:	Did not find card during this audit

Note that these categories are meant as examples; you can either combine or further divide categories to suit your situation. For example, if the moving of the job and card will be done by the team in the originating cell, you could include the "Moving" category in with the "Orig. Cell" since the move can be considered an extension of the process in the cell.

Charles Casings Company – POLCA Card Audit Results

Audit date: 4/4/2018

Loop		Card Serial No.	Card Location							Total
			Orig. Board	Orig. Cell	Moving	Dest. Board	Dest. Cell	Returning	Missing	
A	B	001		1						
A	B	002	1							
A	B	003						1		
A	B	004							1	
A	B	005		1						
A	B	006					1			
A	B	007		1						
A	B	008					1			
A	B	009						1		
A	B	010						1		
Total			1	3	0	0	2	3	1	10
%			10.0	30.0	0.0	0.0	20.0	30.0	10.0	100.0
B	G	001			1					
B	G	002						1		
B	G	003	1							
B	G	004	1							
B	G	005				1				
B	G	006						1		
Total			2	0	1	1	0	2	0	6
%			33.3	0.0	16.7	16.7	0.0	33.3	0.0	100.0

Figure 9.1 Spreadsheet used to summarize data from a POLCA audit.

The audit team should then record the location of each card, by serial number, in a sheet similar to the one shown in Figure 9.1. The figure shows as examples, data for two loops: the A/B loop and the B/G loop. Entering the data into a spreadsheet, with a "1" for the location of each card, allows you to easily perform statistical analyses on the locations of cards as also shown in Figure 9.1 in the last two rows for each loop. An immediate observation from the "%" row for both loops is that, during this audit, around a third of the cards in each loop are in the process of being returned. Let's say that during the POLCA card calculation you expected the return time to be relatively short so it did not contribute much to the number of cards in the loop, then this audit shows that in actuality there is a significant reduction in the total number of available cards for cells to use in production. This might just be a temporary phenomenon, so you could decide to monitor this situation over the next two weeks or so. If this continues to be the case, then it means you need to look into the process of returning cards and see what can be done to make this happen more quickly. The POLCA audit helps provide such clues that you can follow up on as needed—without the audit, this situation might not have been noticed for a while.

In some cases, you may need several audits to get data that is statistically valid. Again, having the audit numbers in a spreadsheet enables easy analysis over several samples. As you conduct more audits, you can view the tables side-by-side and also summarize the data using averages in order to detect any patterns. (Note that if you decide to go with an electronic POLCA system, such statistics can be continuously collected and easily summarized by the system without need for manual audits.) For example, if it seems that for a given loop cards are mostly in use and there are no free cards on the originating cell's POLCA Board, you may consider increasing the number of cards. Conversely, if there are always several cards of a given type available on a POLCA Board, you can remove a card and tighten up the system.

The nature of the periodic audit is such that you do not need to stress about doing the whole audit at once and requiring a large team for this. You can audit one loop at a time and complete the audit over a few days. The aim is to get a snapshot of each loop and so any random moment in time is okay for such a snapshot. We emphasize the use of the word "random" here, because you don't want to always use a fixed time like the start of a shift or the end of a break, because that might lead to some bias due to starting conditions.

Since POLCA cards have serial numbers, as part of this process the audit can help to determine if all cards are accounted for, or if a particular card is missing. If a card is missing, don't replace it right away—our experience has been that missing cards have a way of turning up unexpectedly! You should wait for a period of time (such as a week or two) and check again to see if it has surfaced. In the meantime, if you feel that a particular loop is starving for cards—say it only had four cards to begin with, then a missing card would reduce the number of cards by 25%—you can deploy a Safety Card. The advantage of using a Safety Card is that it is always removed at the end of the loop, and requires an active decision in order to be used again. This will remind you to check for the missing card, and also force you to reexamine whether the extra card is still needed. If the missing card still has not appeared after a given period, such as two weeks, companies typically "retire" the serial number of that card and issue a card with a new serial number. This also ensures that if the card does show up at a future time, you don't end up with two cards with the same serial number!

In terms of the frequency of the audits, you should start with more frequent audits and then lengthen the period between audits as your faith in the POLCA system increases. Initially, companies have used weekly audits until they felt they could move to monthly, and then quarterly. Since demand can change significantly over quarters, it is advisable to continue with conducting the audits at least quarterly.

Random Audits

The purpose of random or "surprise" audits is to ensure that the POLCA rules are being followed. In some cases, there might simply be a misunderstanding or misinterpretation of the rules. But it could be that employees are intentionally ignoring some of the rules. Chapter 7 explained several reasons why employees might want to bend the rules, such as a desire for cherry-picking their next job. In smaller companies where everyone can see what is going on, management might feel more confident that employees are keeping to the rules and the random audits may not be necessary. Larger companies have found that the random audits are needed to reinforce the discipline needed for POLCA to succeed.

The random audit should be used to check at least these three items for each cell that is being audited:

- For every job that is currently launched into the cell: is it Authorized, and does it have the right downstream POLCA card attached?
- Is the quantum rule being correctly implemented for jobs that are in the cell or jobs that are waiting?
- Are there jobs on the Authorization List that should have been launched before some of the jobs that are already in the cell?

In addition, since the audit team will be at the cell gathering data, it is an easy addition to their task to get data on the location of each POLCA card, as explained in the preceding section on "Periodic Audits," and to use this data for analysis and insights as also previously explained.

Unlike the periodic audit, the random audit does need to be conducted all at once, or at least, at the same time for a number of loops. The reason is, if employees are bending the rules, you don't want to give them advance warning so they can correct the situation. Therefore, the way to conduct this audit is to have a team of people fan out across the shop floor and hit all the targeted loops and cells at the same time.

Coupled with the random audit is your decision about how to deal with infractions. If you find some teams have broken the POLCA rules, what are you going to do about it? This issue was also discussed in Chapter 7 and you should think it through. If there will be consequences for breaking the rules then employees need to know about these ahead of time. Once again, this underscores the importance of involving HR personnel in the POLCA implementation process.

Use POLCA to Drive Long-Term Improvements

Another mechanism that will help support the POLCA implementation is to make clear the connection between POLCA and various continuous improvement activities in your company. Use the data as well as anecdotal experiences generated through the POLCA operations to motivate and support improvement activities. As examples of this:

- The Safety Card mechanism used to deal with occasional component part shortages can also be used to gather data on the root causes for these shortages and their impact on the system (see Chapter 6). Analysis of these data can point to the most critical areas to tackle for reduction of the shortages.
- Management notices that a particular team often complains about waiting for cards from a downstream cell. However, according to the current data on workload, capacity, and planned lead times for the downstream cell, there should not be a problem with the number of cards available. This can result in an investigation of the data in the planning system as well as the operation of the cell: perhaps jobs are more complex than estimated; or there is more absenteeism in the cell than management was aware of; or a given machine breaks down more often than the planners knew; and so on. Whatever the root cause or causes, this will give some directions for improvement activities.
- Employees in a particular POLCA Chain that is currently using the quantum rules point out that if batch sizes can be reduced by around 50% they could use a single POLCA card to deal with each work order. This would simplify the operations, including the decision-making, material-handling, and control of an order that is split into multiple units. This provides a concrete target for improvement activities: what needs to be done in order to enable the smaller batch sizes to be used without overloading the system? For example, do some of the setups need to be reduced? Could some of the operations be done differently or performed on a different machine?

As such instances arise and improvements are made, the POLCA Champion or Steering Committee should be sure to document these occurrences as well as the results—such as quality improvements or shortage reductions—that ensue. The formal recognition and celebration events suggested in an earlier section will help to underscore these achievements.

Over time, this will demonstrate to management as well the employees that POLCA is supporting long-term continuous improvement at your company. The message this conveys is that the fabric of the future of your company is interwoven with the successful operation of POLCA, and this is a powerful way to maintain support for and sustain the POLCA implementation.

This concludes the second part of this book, which has provided a systematic guide to implementing POLCA. Part III of the book serves to demonstrate that POLCA is not just a theoretical concept; it has been implemented in factories around the world with impressive results.

INDUSTRY CASE STUDIES

Introduction

A major goal of this book is to demonstrate that POLCA is not just a theoretical concept; it has been implemented in factories around the world with impressive results. Part III of this book documents six such implementations, with case studies from the U.S., Canada, the Netherlands, Belgium, Poland, and Germany. The companies involved range in size from a large multinational to a factory with only a dozen employees! Industrial applications span from the more common metalworking operations such as extrusion, metal-cutting, and fabrication, to more unusual factories such as glass and even pharmaceuticals. It was important to us that you, the reader, should experience these case studies from the eyes of the people who were involved. Thus, the six case studies in Part III are all written by guest authors—senior managers, employees, and consultants at the companies that implemented POLCA—who share their personal journeys, insights, and of course results. This also lends credibility to the results presented, as they are directly supported by the writings of these guest authors.

Chapter 10

From the United States: Using POLCA to Eliminate Material Flow Chaos in an Aluminum Extrusion Operation

Guest authors: Jeff Cypher and Todd Carlson

Located in Alexandria, Minnesota, Alexandria Industries (AI) started its operations in 1966 by providing a local boat manufacturer with lineal aluminum extrusions. Shortly after that, another Minnesota company needed precision-machined components. AI stepped up to the challenge, which resulted in AI being one of the first aluminum extrusion companies to have machining as part of its core capabilities. AI continued to develop a vertically integrated structure to align with its customers' needs, including growth through strategic acquisitions based on customer requirements. In addition to its aluminum extrusion operations, AI now has capabilities of precision machining of ferrous and non-ferrous products, heatsinks, plastic injection, and foam-molding components, as well as finishing, high-level assemblies, and welding services (Figure 10.1). At the same time, AI has expanded its market far beyond the original marine products and now serves numerous other industry segments including electronics, medical, firearms, recreation, lighting, and solar and renewable energy.

AI currently has ISO-certified manufacturing locations in Minnesota, Indiana, and Texas producing high-quality, short-lead-time engineered products, with around 600 employees in total. These operations include

Figure 10.1 Examples of aluminum extrusions (left) and a heat sink enclosed housing (right).

four extrusion presses, with three located in Alexandria, Minnesota and one located in Indianapolis, Indiana. AI has many high production saws, several types of bending machines and stretch-forming equipment. Automation takes a prevalent role at AI, for example, through specialty machines that can punch, drill, tap, and cut parts in one operation. CNC Machining is accomplished at two of AI's factories in Alexandria, Minnesota and in Carrollton, Texas, with a total of around 80 CNC machines, 30% of which are robotically tended. It is common for aluminum products to have finishing operations such as anodizing and painting, and all of AI's finishing applications are completed by subcontractors at their facilities outside of AI. The typical process sequence within AI involves extruding, artificially aging, cutting, machining and/or fabrication, finishing (done outside AI), and finally packing. Since AI's strategic focus is on make-to-order business, there are numerous variations of the "typical" process, leading to complexity of the job flows: in fact, AI currently has over 3,500 active extrusion dies resulting in more than 10,000 finished end items.

The Start of Our POLCA Journey

In the late 1990s, lead times in the extrusion industry were typically between 8 and 12 weeks, and AI's lead time was in this range as well. This was an accepted industry standard, so AI felt that there was no reason to change. With the telecommunications collapse of the early 2000s, AI was looking for something that would strategically differentiate it from the competition. In 2002, Tom Schabel, president of AI, identified Quick Response Manufacturing (QRM) as a competitive strategy for AI because of QRM's strengths in dealing with high-mix, low-volume and custom (HMLVC) production, which was AI's core business.

The authors, along with a core team of people, were involved from the start with AI's initiative to implement QRM, with Jeff Cypher acting as the QRM Steering Committee Champion. The original co-champions of the POLCA Implementation Team were Jason Bachman and Todd Carlson, and other members of that team were Brian Larson and Kurt Norling, both in manufacturing leadership. The rest of this article will take the reader through our personal journey.

In early 2004, a couple of years after starting QRM, we had several cells in place, both in manufacturing and the office, but we still had a material flow problem between the cells as well as between workcenters that were not in cells. We were a typical manufacturing facility utilizing an MRP (Push) scheduling system. Most of our product was made to order; however, to meet very short lead-time requirements for a few customers, we did make some product to forecast. Again, in line with a typical manufacturing facility, high work-in-process (WIP) and expediting were commonplace. In spite of our best attempts at expediting, we were still experiencing poor on-time delivery performance, resulting in customer dissatisfaction.

At this point, a number of us recalled reading the chapter on POLCA during our study of Rajan Suri's book, *Quick Response Manufacturing*. A group of us re-read the chapter, but we still didn't understand how POLCA would work for us, and the specifics of how we could implement it at AI. Fortunately, we all had a strong gut feeling that we had an environment that would benefit from application of POLCA, and we were also lucky to have Jason Bachman in the team, who tenaciously continued pushing us to pursue the concept. So, we decided that we needed some expert help to assist us with our first steps. We approached the Center for Quick Response Manufacturing at the University of Wisconsin–Madison for assistance. Two of the staff from the Center came to Alexandria to conduct a training session that included both the theory of POLCA as well as a hands-on POLCA simulation game where we "made" a high variety of products in a make-believe factory. Seeing the principles explained in slides, and experiencing the way POLCA worked for the pseudo-factory, drove the details home for us. After that training session, we knew that POLCA would work at AI.

But there were many remaining obstacles, and we still had a lot of work to do! One of the biggest issues was overcoming fear. Many people wanted to wait because we simply didn't have the time to go through the changes—we were already running late and it seemed like we never had enough capacity. This point was raised repeatedly: "What if we implement POLCA and it doesn't work? We'll be even worse off than now!"

To overcome some of the fear, our team toured a facility that was using POLCA. The collaboration was mutual: each company invested in the effort to visit the other with a team of people. This was not trivial considering the companies are located 500 miles apart. The factory that we visited had not implemented QRM principles but was using POLCA, and it had greatly reduced their WIP and improved their material flow. The product flows in both factories were similar, and after seeing POLCA work for them, the AI team felt confident that POLCA would work for us as well. The team at the other company stressed that we should stop hesitating and "Just Do It!"

We also gained some useful tips from visiting the other factory, including ideas for card layout, color-coding of cells, and use of Bullet Cards.

The other point we had to keep reminding ourselves, to overcome the fear of change, was that our current Push system obviously wasn't working! We had experimented with a Pull system, but we soon realized that it wasn't right for our business either. (Appendix C explains further why both Push and Pull approaches were not working in our HMLVC environment.) It was clear that we needed to try something different.

Experimenting with a Small-Scale POLCA Implementation

As a final way to overcome the fears of change or failure, the team decided to try POLCA in a small, mostly self-contained area of our operation. If it didn't work, this would limit the damage. Specifically, we chose products going from one of our extrusion presses, the Loewy Press, to our Centerless Grinding cell. This seemed like a logical test case because this cell serviced one customer's needs and we were performing poorly. We didn't understand the reasons for poor performance because the product mix was relatively low and the customer had a consistent ordering pattern. In addition, our capacity matched demand, we had few quality issues, and it was a relatively simple manufacturing process. Therefore, it was clear that our current system was not working even in this simple situation.

This first POLCA loop took about six months to complete. Frankly, it didn't need to take that long at all, but our fear of failing prolonged the process. There was a lot of investigating, education, and hesitation. As one instance of this, to calculate the number of POLCA cards in the loop, we used the formula from the above-mentioned POLCA chapter in Suri's book. By the rules of what we had plugged into our MRP system, the formula recommended a number of cards that appeared to be way too many, and this caused some concern with the team members. So we went back to the drawing board and reevaluated the days of queue time and move time that had been put into our routings. We found that there was a lot of padding in these times, which had been put in place as a safety margin to allow for unexpected problems, errors, schedule changes, and so on. We decided that POLCA would alleviate many of these issues and so this safety margin was way too high. After reducing the days of queue and move time to only what was necessary for normal operations, the new information was plugged back into the formula. The result was a significantly lower number of cards.

We were more comfortable with that and we used this as the starting point for the number of cards in the loop.

The team then looked at the product structure, which included the bill of material (BOM) and routing structure. The product going through this cell had five different shapes, each with two lengths and three colors, resulting in 30 part numbers. The BOMs and routings were designed to have pieces of each shape and length stocked (i.e., 10 part numbers of semi-finished stock) so that we could "quickly" produce a finished part in any color. After evaluating this process, it wasn't as effective as we thought: a lot of MRP transactions were involved in tracking material going in and out of inventory, and despite the stocking attempts we still often didn't have the right semi-finished goods!

Therefore, the first thing we did before starting POLCA was to change the product structure so that each of the 30 part numbers had its own BOM and routing from raw material (extrusion) to customer's part (colored anodized). We also observed that due to historical policies based on economizing on changeover times at the extrusion press, the extrusion batch sizes were set at 960 pieces, while customer delivery quantities of the finished parts were 240 pieces, which again forced us to store the remaining pieces and be involved in inventory transactions. To ensure that a POLCA card would be associated with a specific order from start to finish, we decided to change the extrusion batch quantity to 240 to match the customer order size.

To summarize, we made three significant changes to the BOM and routings:

1. The queue times and move times in our routings were reduced to reflect the actual processes. Specifically, all "padding" was removed.
2. Stocking of intermediate (semi-finished) product was eliminated.
3. The extrusion batch size was changed to reflect the customer order quantity for finished product.

It was pointed out to us that the third item would increase the standard cost of the product. This was yet another potential obstacle, because from an accounting standpoint, these changes to support POLCA didn't make sense. However, from our QRM training and book reading, we understood that this was an example of the situation where standard costing doesn't account for the resulting reductions in cost through the reductions in WIP, stocked inventory, floor space, and expediting, and nor did it account for the market value of improved on-time delivery performance. We discussed these tradeoffs with the team members and AI's management, and because of top management's commitment to QRM and support for this type of thinking beyond standard costing, we were able to forge ahead with POLCA.

The points above highlight the fact that when you are designing your POLCA system, you shouldn't simply instrument the current operation with POLCA. You should take a step back and think through what changes would make sense to better support POLCA operation.

System Design and Training

It was now time for the POLCA implementation team to develop the system and complete the training. We trained the leadership in the Loewy Press area, which was the first cell of the POLCA Loop, all of the personnel and leadership of the Centerless Grinding Cell (the second cell of the Loop), and the office cell involved with this product. Since the Loewy Press supplied product to other cells that would not be on POLCA, we explained that those jobs would be run based on the normal MRP dispatch list, attempting to keep as close to the MRP-assigned start dates, but the jobs on POLCA needed to obey the full Decision Time rules (Authorization Date plus POLCA card). Since we already had an MRP system with start date calculations, we used this to create the Authorization Dates as well. However, in order to have some safety margin as we started up the POLCA system, we decided that the Authorization Date for jobs on POLCA would be three days earlier than the MRP calculated start date. For example, if the MRP start date for a given job at extrusion was November 10, the job would actually be Authorized from November 7. Since there was just one simple loop involved we decided to go with the Release-and-Flow version of POLCA, so that the downstream cell did not need to use Authorization Dates.

We then designed and printed the POLCA cards, including four Bullet Cards (Figures 10.2 and 10.3). Next, we spent some time in both these production areas so that everyone could see the tools and discipline required for POLCA. This included:

- Where the free POLCA cards would be placed, on a POLCA Board at the first cell (Figure 10.4);
- How a POLCA card would be incorporated into the shop packet once it was associated with a launched job, and where the shop packet along

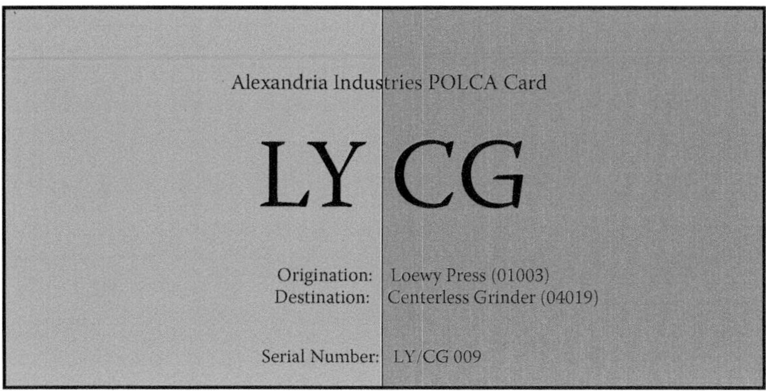

Figure 10.2 POLCA card for the first loop.

Figure 10.3 One of the bullet cards (QROC refers to the Quick Response Office Cell responsible for a set of products).

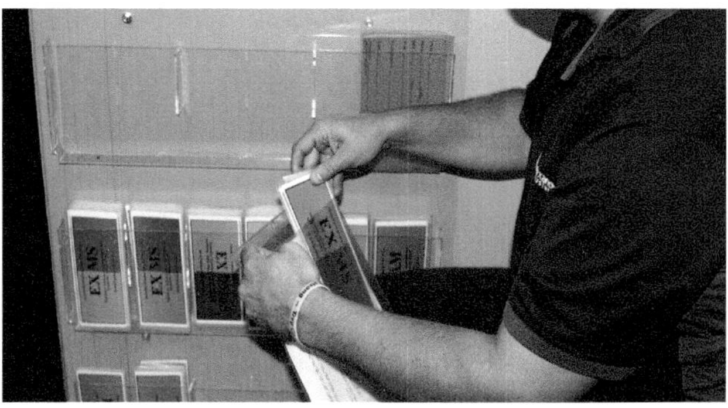

Figure 10.4 POLCA Board for placing the cards (this picture is for the current situation with more POLCA loops in place, but the original board was similar to this).

with the card would be placed on the baskets used to carry the product through the production areas (Figure 10.5);

- Marking where the WIP would be placed as it progressed through the two cells;
- How the POLCA card would be returned: we decided that the team in the second cell would be responsible for returning the card to the upstream cell as soon as they were finished with the job. (This worked fine for us and we have continued with this rule for all POLCA loops implemented subsequently.)

Figure 10.5 POLCA cards placed in shop packets, which are then placed in the baskets used for transporting material.

Following the training and the specification of these details on the shop floor, the team and the shop personnel felt confident about how the system would work. It was finally time to implement!

System Startup and Early Results

At the startup, we still had inventory of semi-finished products, so we had to institute a short-term policy to work through this stock. What we did might be helpful to other readers going through a similar startup. We started by giving the full set of POLCA cards to the scheduler. When the scheduler saw that an order needed material that wasn't already in stock he would issue a POLCA card to enable production at the upstream cell. On the other hand, if the material was already in semi-finished stock, he would place a POLCA card on the basket of material and have it delivered to the downstream cell as if it had arrived from upstream. In this way, we bled down the stock and then moved to fully make-to-order production, following the full POLCA rules for these products.

The transition from Push to POLCA was relatively smooth, and the speed of the transition was astonishing. It didn't take long for us to see that POLCA was going to work in our environment. Floor space started to visibly clear up as WIP went down, and our delivery performance started improving. Amazingly, in less than two months this first POLCA loop had already achieved the following results:

- Floor space occupied by WIP in this area shrank from 250 square feet to 63 square feet: a 75% reduction!
- The value of WIP in this loop went from $27,720 down to $6,930, also a 75% reduction.
- Team leaders and managers noticed that expediting efforts were reduced dramatically.

Traditionally, manufacturing companies have stuffed themselves with inventory, thinking that this would improve their ability to serve customers. When our first POLCA results were tallied, what was amazing to all of us was that we had *less* inventory, and yet we had a *much higher* service level to our clients! These results were unambiguous enough to convince us that we should implement POLCA for the rest of the shop floor.

Extending POLCA to the Whole Factory

After the success with the initial POLCA loop, we went through the same process for the next four loops. Thanks to our learning from the first loop, implementation time for these loops was cut down to about eight weeks. After these loops were implemented, most shop employees were either part of a POLCA loop or certainly aware of POLCA and what we were accomplishing with it, and were prepared for additional loops to be put in place. As a result, the next set of loops—around 30—were implemented very quickly. The final set of loops is shown in Figure 10.6. Some of the boxes in the figure represent cells, while others are standalone machines. Our experience demonstrates that you can use POLCA with or without cells, or with a combination of cells and individual workcenters. POLCA has worked well for us since we instrumented the entire factory, and, as explained at the end of this chapter, the results have continued to pour in.

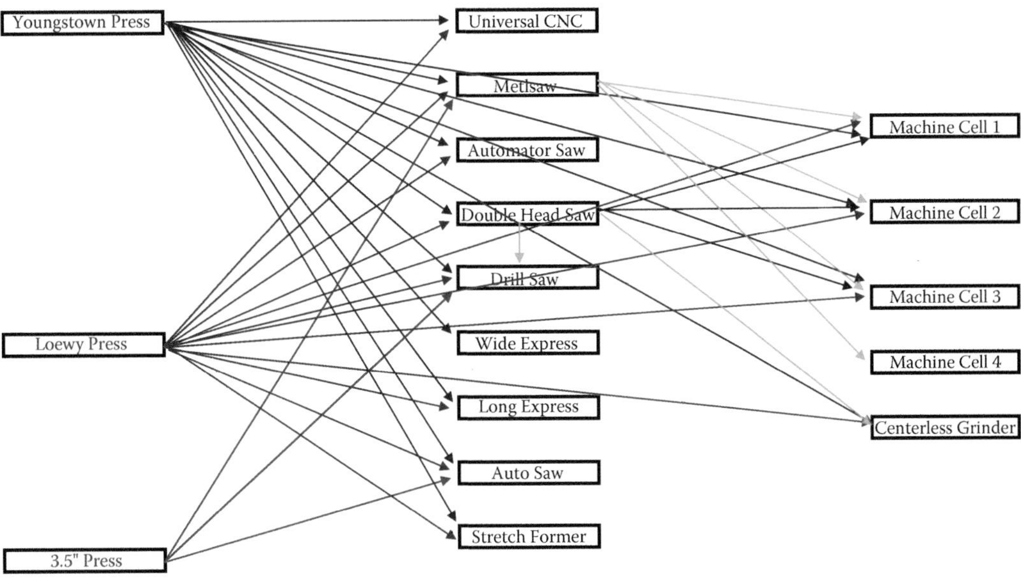

Figure 10.6 Full set of POLCA loops at the factory in Alexandria. To keep the figure less crowded, arrows are used for each loop. Each arrow represents a POLCA loop with the direction indicating the upstream and downstream cells.

What We Learned, and Advice to Other Practitioners

We were among the first companies in the world to implement POLCA, and we only visited one other company before our implementation, so we had to learn many things by ourselves along the way. We would like to share some of these lessons to help others succeed in their POLCA implementations, and to potentially shave some time off the implementation process.

- As mentioned earlier, don't just instrument your current operation with POLCA. Think about how POLCA works and whether you need to make some changes first, to better support the system's operation. Thinking through how POLCA works thus gives you a chance to step back and reexamine your current operating norms.
- When you are putting in a POLCA loop, don't sweat about whether the number of cards in the loop is perfect: after implementation of a loop, it is easy to tell if the number of cards is correct, and also easy to adjust this number. We found that the formula for calculating the number of cards was a very good starting point, as long as you check and remove any unnecessary padding in the planned lead times (as explained earlier). You can confidently go with this number and start. If you are constantly starving the downstream cell, add a card or two. If you consistently have too many cards—witnessed by the fact that the needed card is always available to the cell—and the downstream cell seems to be overloaded and accumulating a lot of WIP, remove cards one at a time until it appears that the upstream cell is occasionally being shut down to let the downstream cell catch up. In our case, the formula for number of cards worked well enough that in some loops we did not need to adjust the number of cards at all. In other loops, we adjusted the number of cards very little. Hence our final advice on this point is that you don't need to spend a lot of time worrying if you have the correct number of cards in each loop. Just get going, and then it's easy to fine tune!
- Also, don't stress on deciding on the number of Bullet Cards. As a reminder, the purpose of a Bullet Card is to take care of unplanned or unusual circumstances when an order needs to be expedited without having to wait on the normal POLCA decision rules. This will typically happen when there is an urgent unanticipated customer requirement and management feels it is important to service this customer without delay. As explained in Chapter 6, the Bullet Card allows a job to flow quickly through the whole shop floor. We had four office cells, each servicing

orders for a segment of customers, and we issued one Bullet Card to each of the office cells. This meant that if a particular office cell used its Bullet Card for an order, it could not release another "bullet" order until this card was returned to the office cell after the previous bullet order had been completed. The idea was to impose some restraints on using this card, and for an office cell to think carefully before it "played its card." In the beginning, the office cells thought that four Bullet Cards were too little; each cell felt that having just one Bullet Card wouldn't come close to being enough for its needs. However, the office personnel had not anticipated the performance improvements that result with POLCA, which dramatically enhanced our ability to service customer needs. In fact, after POLCA had been in place in the whole factory for about six months, we realized that even just two Bullet Cards, shared by all four office cells, would have sufficed! However, implementing four cards was the right thing to do, as it took the politics out of the decisions, with each cell controlling its own card.

- We recommend that you implement an audit plan for the number of cards. This is similar to the auditing needed for a 5S or ISO system. Manufacturing involves a lot of materials and movements, and cards have a way of disappearing among all this. As explained in Chapters 5 and 9, cards need to have serial numbers, and the audit needs to identify the location of all cards. Not only will the audit pinpoint missing cards, but over time statistical analysis will show if loops have frequently unused cards (too many cards) or almost never have unused cards (too few cards) and this can help fine-tune the numbers as well. As with any auditing system, once confidence is gained, the frequency of audits can be reduced. We started with weekly audits. After four weeks, we switched this to monthly. Then after the first quarter, we extended this to quarterly audits, which is where it remains now.
- Be prepared to adjust your scheduling process: POLCA *will* change how you schedule your manufacturing. Your frequency of scheduling may change (for example, every shift versus every day, or every other day versus every week). You may find that you have less discussion between the office and manufacturing. The manufacturing communication will be different between portions of the organization. Regardless of how it changes for your particular situation, it *will* change, but for the better. So, be prepared, be flexible, and embrace it!
- Don't feel like you have to implement overlapping POLCA loops to be deemed successful. At AI, we found great value even in single POLCA loops. Our manufacturing leadership grasped the concept of single

POLCA loops very easily and their ownership was apparent very early in the process.

- Stay tenacious and disciplined. In 2015, AI got extremely busy and we ran out of capacity. As the chaos grew, we found ourselves trying to outthink the system. We stopped using POLCA in a disciplined fashion. We thought that we could better control the job flow manually. It is easy to predict what happened next. On-time delivery diminished and WIP increased. Along with that came other sins. We found that we were expediting more and moving WIP around because it was in the way of what we needed to work on next. Compounding these challenges was the hiring of many new employees and the changing of manufacturing leadership. As these happened, we did a poor job of training the newcomers on POLCA principles and maintaining focus. Looking back, it is clear that reverting to processes that didn't work previously is not the answer! Hence our advice is that you should stick with the POLCA system and rules regardless of business conditions.
- POLCA should be part of your orientation for new employees, particularly for people involved in production areas. We have developed a 45-minute simulation that shows how POLCA works and the advantages of utilizing it at AI.

Impact of POLCA on Our Organization

Implementing POLCA had more profound and far-reaching consequences than we had expected. It fundamentally transformed our manufacturing methodology. The changes from implementing POLCA resulted in a better understanding and deeper penetration of QRM principles. It proved the value of batch-size reduction and queue-time removal, both of which had previously been difficult to do on faith alone. In combination, the cutting of batch sizes, removal of queue time, and reduction of stocking levels was a scary journey. However, the result was that it broke down standard costing barriers and promoted QRM activities, resulting in the intended gains. In 2002, before starting our QRM journey, our sales averaged $274 per square foot of manufacturing floor space. In 2012 this number had grown to $434 per square foot—a 60% increase! Most of this increase came from a reduction in WIP—which can directly be attributed to POLCA—and in place of the nonproductive WIP we could put more productive resources on the shop floor, significantly increasing our output in the same space.

Also, as compared with the 8–12 week lead times for extrusions mentioned at the start of this chapter, our extrusion-only lead time is now just one week. And in addition, our on-time delivery is almost perfect. The anecdote that follows helps to drive home the impact of all these improvements.

A few months before we implemented POLCA, the Plant Manager grabbed one of the office cell members by the shoulder one day and led her through a particular production area while asking, "What am I supposed to do with all of this material?" The press would run large batches so that this area could work over the weekend. We would then send an entire truckload to an outside supplier on Monday morning, and Purchasing would call us and ask why we plugged up the supplier. The supplier would then work hard on expediting this big order and send all the material to the customer. However, four different parts need to be assembled together at the customer's site for them to have one finished unit. After the customer received product from our supplier, he would call us and ask why we sent a truckload of material of one of the parts, because he could not do any assembly without the other three!

After we put this product line on POLCA, the effort that we put in to satisfy this customer is virtually nonexistent. The customer orders material, Manufacturing produces it, we ship it in the right quantities and mix, and the Plant Manager hasn't asked that office cell member about these products since that eventful day before we started POLCA.

About the Authors

Jeff Cypher is Director of Business Integration at Alexandria Industries. He has over 20 years of experience in business systems and materials management, and has actively participated in Alexandria Industries' QRM journey. Alexandria Industries has four divisions and Jeff is assisting in the integration of the Business Systems and implementing QRM principles.

Todd Carlson is the Corporate Capacity and CNC Robotics Manager at Alexandria Industries. He has been responsible for implementing automation in the machine shop, managing the operations of the shop, and also actively engaged in Alexandria Industries' QRM journey for over 15 years. Todd is assisting in bringing automation to each of the four divisions of Alexandria Industries, and balancing the capacities and workloads at each location. Todd has a B.S. in Manufacturing Engineering and an M.B.A., both from St. Cloud State University.

Chapter 11

From Canada: Applying POLCA in a Pharmaceutical Environment

Guest authors: Justin Bos and Susan Ferris

Patheon is a global pharmaceuticals contract development and manufacturing organization (CDMO). Patheon offers a simplified, end-to-end supply chain solution for pharmaceutical and biopharmaceutical companies of all sizes. The Toronto Regional Operations site specializes in high-potency oral solid dose development and commercial manufacturing.

A typical process flow in an oral solid dose plant has five stages: Dispensing (where the materials are measured to the exact amounts required); Granulation (where the materials are combined to create a uniform blend); Compression (where the blend is formed into tablets); Coating (where the tablet is coated with a liquid solution); and Packaging (where tablets are packed in bottles or blisters). Each of these manufacturing stages is completed on a different workcenter, so each order will run on several pieces of equipment before it is ready for shipment. We have eight granulators, eight compression machines, five coaters, and seven packaging lines, and in principle, an order can be routed from a workcenter to any of the workcenters in the next stage, resulting in a complex collection of possible routings, as shown in Figure 11.1.

Since Patheon does not manufacture its own products, it operates make-to-order plants, managing orders from multiple clients that share Patheon's equipment capacity. To further complicate matters, each product must be

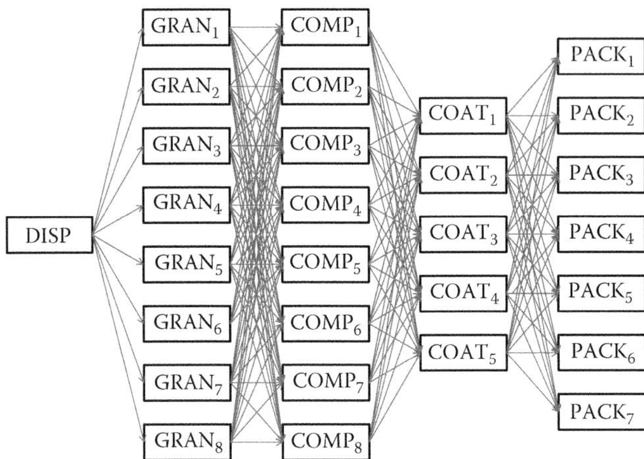

Figure 11.1 Schematic of all the workcenters and possible routings.

validated on specific equipment to ensure repeatability. This creates rigid manufacturing paths that cannot be easily changed, and, when combined with the large number of routings along with volatile forecasts, makes for a challenging planning environment.

Why Did We Pursue POLCA?

In 2011, prior to POLCA, the plant was finitely scheduled using the MRP portion of our ERP system. This schedule required almost constant revision to keep accurate when there was any deviation from it. In addition, gaps in the schedule were filled to keep machines utilized and it wasn't long before work-in-process (WIP) started to pile up in front of bottlenecks. *In our plants, WIP poses an extra threat: it must be consumed quickly or else it requires testing to confirm that it is still safe to use.*

More WIP in the system also meant that our production lead time—as measured from the start of the first stage (Dispensing) to the end of the final stage (Packaging)—was starting to creep up to over 20 days. This resulted in more schedule changes in an effort to avoid stocking out the market of our clients' life-saving drugs. Those schedule changes often forced us to run in less optimal sequences, further compounding the problem.

We had been focusing our Operational Excellence efforts on constrained resources, trying to maximize Overall Equipment Effectiveness (OEE). We use the traditional definition of OEE, which takes the total of the standard run-time hours for products made on a given workcenter, divided by the total available hours on that workcenter, both for a given period. We noticed that from our focus on OEE, the individual workcenter results were promising, but we weren't seeing systematic success. Bottlenecks would shift depending on client order patterns and our throughput had flatlined, despite our hard work. We understood that we had to change the way our plant was planned. We needed to stop trying to *push* product through and instead begin to *pull* it.

A traditional Pull system, such as Kanban, was not suitable in our high-mix, low-volume operations (see Appendix C for more on this point). We started researching more recently introduced systems, including CONWIP (see Appendix C) and POLCA. After a preliminary analysis of both these systems, we concluded that for our environment POLCA was the best option for two main reasons. To understand these reasons, note that CONWIP assigns a loop to each particular routing from the start of the first operation all the way through the last operation for that routing. So, the first reason for our choice was that, as mentioned, our routings can go from any workcenter at one stage to any of the workcenters at the next stage. Referring to Figure 11.1, this means that there are 8 × 8 × 5 × 7 = 2,240 possible routings in our factory, requiring 2,240 different CONWIP loops to be managed with cards. In contrast, POLCA would require at most 147 loops—or more than an order of magnitude fewer loops. (Each workcenter needs a loop to all its downstream workcenters. From Figure 11.1 this results in 8 + 64 + 40 + 35 = 147 loops.) Second, the CONWIP loop signal comes only from the end of the last step back to the start of the first step. With our lead times of around 20 days, we felt that this would be a highly delayed signal, and we wanted more frequent signals plus tighter coupling between each stage, and POLCA offered both of these capabilities.

The POLCA Kaizen Workshop

POLCA began the same way many of our Operational Excellence projects do—with a five-day kaizen workshop. A small team was assembled with representation from Production, Planning and the Operational Excellence department. The first day was spent getting the entire team up to speed with the principles of POLCA and working through a white paper, "How to Plan and Implement POLCA" by R. Suri and A. Krishnamurthy, available from the website of the Center for Quick Response Manufacturing. Our team felt that POLCA would be able to address the scheduling and WIP issues we had been experiencing, but the team still had one reservation. A change of this magnitude would pose a substantial behavioral challenge for both our management and our front-line employees at the site. We had to come out of the workshop with a robust change management plan.

Deciding on All the Fundamentals of the POLCA System

We started by designing a spreadsheet that would become a working document for all our developments throughout the week. Product forecast and routing information were loaded into the spreadsheet and we began to identify the POLCA loops. As already mentioned, our plant was so complex that by the end of the exercise most workcenters were connected to all workcenters in the following stage. We went with the recommended standard design for the POLCA cards. Specifically, each resource was given a unique color that is used on the cards and on the POLCA Board. Figure 11.2 shows the generic layout of our cards. As shown in the figure, we followed the advice to assign a unique serial number to each card, to enable auditing of the cards. We also laminated the cards to help them survive a factory floor environment!

To calculate the number of cards, we chose a three-month planning period because we have a 90-day firm order window. We used the standard formula (see Chapter 5), but then minor adjustments were made to the number of cards. For example, more WIP (i.e., POLCA cards) was permitted before the final manufacturing stages because we wanted a larger buffer in the event

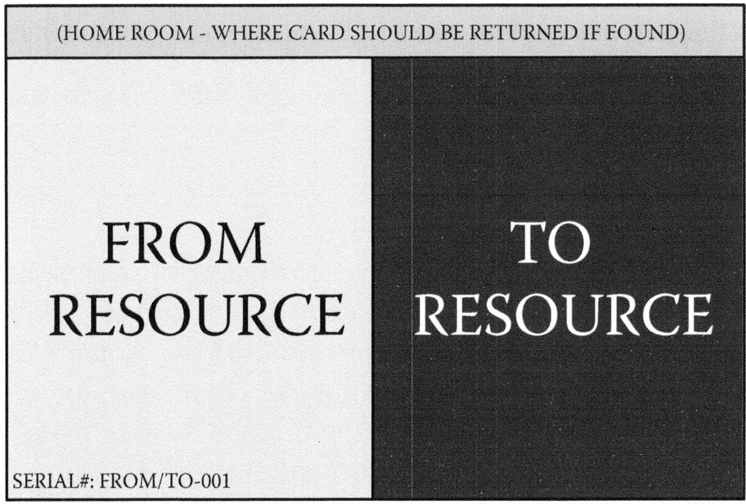

Figure 11.2 Design of the POLCA card.

of an upstream manufacturing disruption. Since our product mix is constantly changing, we also decided that we would we recalculate the POLCA card requirements every three months based on our clients' upcoming forecasts, and that minor adjustments would be made on the floor to add/remove cards as required based on observing the system operation for that quarter.

The team created a tabletop game to simulate the manufacturing steps and the operation of POLCA in our factory. We tested different card levels to understand how it would change the flow of the batches. There was a lot of discussion around the quantum, i.e., what amount of work a POLCA card should represent (see Chapter 5). Our products have significant variation in operation times at the machines, so initially we were inclined to make a POLCA card represent a unit of time (e.g., four hours). However, this would require our front-line employees to collect multiple POLCA cards based on routing times before having the green light to start a particular batch. After running simulations, and ultimately prioritizing simplicity over accuracy, we decided to have a POLCA card represent a single batch. This also made sense in terms of quality and integrity of product, as each batch is earmarked for one (single) customer.

Finally, we decided that the Release-and-Flow version (RF-POLCA) would work fine for us. While individual routings may differ, all batches have the same number of routing steps, and as already mentioned, once a batch is started it is important to keep it moving for quality reasons. So, we felt we didn't need to re-sequence jobs downstream, but simply to control for capacity constraints and WIP buildup in those downstream areas. Therefore, the Authorization Date for each batch is assigned only for the first stage. This date is calculated by our ERP system via standard MRP logic that uses the planned lead time for our start-to-finish production operation.

Rules to Break Rules

POLCA was not necessarily designed with the pharmaceutical industry in mind. There are specific challenges that we face in our manufacturing operations, and we needed to come up with creative ways to overcome them while still working within the general rules of POLCA. The POLCA system design rules do recognize that you need a way to handle real-world exceptions, and there are techniques such as Safety Cards explained in the design rules (see Chapter 6). Using these ideas and after some additional brainstorming, our team came up with the idea of three special types of cards: Campaign Cards, Quality Deviation Cards, and Bullet Batch Cards. Each is a temporary card that is requested by Production and destroyed after a single use.

Campaign Cards are issued when Production wants to execute an extra batch of the same product without waiting for a capacity signal (POLCA card) from the relevant downstream resource. Since we use the same equipment to manufacture multiple products, changeovers from one product to another can be time-consuming. The Campaign Card authorizes controlled overproduction of a product where normally we would pause and wait for a card to return. During this pause we would need to change over and begin making another product based on a different POLCA card, and this would cause some loss of capacity due to the changeover time.

Quality Deviation Cards are issued when a batch has a quality hold since it can sometimes take days to disposition it as accepted or rejected. During that time, we want to authorize upstream resources to continue making product. So, this temporary card is issued and kept with the batch on hold, while the original POLCA card is sent back to the previous process to signal capacity. (Chapter 6 further explains the need for such a process.) If the batch is ultimately accepted, there is one additional batch ahead of the downstream resource, but the WIP level will return to normal once the card is destroyed after use.

Both the Campaign Card and the Quality Deviation Card are identical to the normal POLCA card, but we put a yellow border around the outside with large words such as "CAMPAIGN CARD" on the top to highlight that it is a different card. These cards are not created beforehand, but they are printed and issued when required (and not laminated like the other cards). Both these types of cards can be issued by Production Management.

The final custom card, the Bullet Batch Card, is used for extreme situations when the execution of a batch must be expedited (e.g., there are clients on-site observing a batch from start to finish, or there is a market stock-out). A Bullet Batch Card enables that batch to move as quickly as possible through the system, jumping any queued batches in WIP. This is extremely disruptive to the batches already in the system and their lead time will increase, which is why it reserved for the most critical of circumstances. A Bullet Batch Card can only be issued by senior leadership: it requires both the Production Director and the Supply Chain Director to sign off on its use. This card does allow us to be flexible and respond to our clients' needs when absolutely necessary. However, because it is also disruptive to the overall operation and to other batches, funneling the decision through two senior people ensures sufficient scrutiny that it is in the best overall interest of our factory. Bullet Batch Cards are pre-made and laminated, and, in fact, we have *only two* of these cards! Interestingly, this has not been a problem. As described later in this chapter, with the POLCA system in place, better overall control, and much shorter lead times, we have found that the Bullet Batch Cards are rarely used—they have been issued only two or three times per quarter.

Senior Leadership Plays the POLCA Game

As mentioned above, coming from an ERP/MRP environment, implementing a system like POLCA involved considerable mindset challenges. It was a foreign concept for most of us, even for those who had Lean training. While the whitepaper was read by the whole team, some members of the team struggled to understand the mechanics of POLCA until they were involved with running scenarios through the game we had developed. So we knew that in order to get other people on board, we needed to have them experience the game as well. Thus, once we were confident that we wanted to pursue POLCA, the team concentrated its efforts on honing the game.

The five-day kaizen ended with a report-out to senior leadership at the site. As has been stressed in other sections of this book, implementing POLCA requires top management buy-in and commitment, so readers may find it useful to see the level and breadth of management that attended our report-out. Attendance included our site General Manager and direct reports from several business units. These people included the Senior Director of Pharmaceutical Development Services, Director of Supply Chain, Director of Production, Director of Operational Excellence, and Director of Business Management. Also in attendance were all five people who were directly involved in the kaizen activity.

Our findings and recommendations were summarized in a slide presentation, but we also had them play the game to demonstrate how POLCA would work for our factory. The first round was played with our existing ERP/MRP conditions (i.e., a finite schedule, maximizing the OEE of individual machines). During the round, we simulated machine breakdowns and client escalations to underscore how disruptive those things could be to the system. The second round was played using POLCA. With the POLCA cards and rules in place, managers could see that the WIP was more balanced, so that even when there was a breakdown or client escalation, the system experienced little disruption. At the end of each round three metrics were measured: the amount of idle time at each machine, the amount of WIP in the system and the total batches produced (throughput). Comparing those numbers made it clear that not only did POLCA work in theory, but it would work for us. After playing the game, senior leadership was convinced and they sponsored an implementation of POLCA for the whole factory.

Training, Training, Training

It took 12 weeks to completely implement POLCA. Most of the design was done during the workshop, but the majority of those 12 weeks were spent training nearly every business unit at the site. To truly appreciate the magnitude of this training effort, you should know that we held more than 20 training sessions that were attended by over 250 people in production and another 80 in support roles! This may sound like a lot of organizational time spent on training, but we knew we could not overdo the communication for this magnitude of change. In each session, overview slides were presented and then the participants played the POLCA game. There were a lot of questions and concerns, and each was addressed before moving forward. For example, people from Planning were reluctant to let go of their traditional tasks of scheduling workcenters, and to focus on more high-level capacity planning. Yet, on the other hand, staff from Production were excited because they were constantly dealing with changing plans and welcomed the ability to have more clear visibility into their destiny. Such discussions helped to bring the teams to a mutual understanding. We needed to have complete buy-in from all these people to not only understand but support the initiative.

Training the front-line Production staff was more tactical and got into the finer details. Where would the cards be stored? What would the cards look like? What would you do if you found a card on the floor? How could we audit the quantity of cards? What would happen if we didn't use cards to authorize production? Every single Production employee attended one of these training sessions, sometimes more than once. These were the people who would make POLCA a great success or a great failure. It was absolutely critical that they understood not only what to do but why it was so important to follow its rules.

The POLCA Boards and POLCA Dashboard

With the training complete and the POLCA cards laminated, we had to figure out how to store and manage all the cards. In our highly regulated environment, every executed batch has controlled documents called a "batch record" that travels in a transparent plastic briefcase. This offered the perfect vehicle to transport cards when they were in use. We labeled and hung steel seven-pocket wall racks to be our POLCA Boards, and this is where batch records waiting to be executed on a specific workcenter queue up (Figure 11.3). Unused cards for downstream resources are stored in the bottom pocket, so before starting a batch, Production staff can confirm that the correct downstream card is available.

After producing the batch, the card from the previous stage is returned to its home POLCA Board. Again, readers may find our decision on this process helpful. We first thought about having designated "runners" for returning cards. But then we realized there was enough traffic throughout the day that we could just set up a mailbox system and make returning POLCA cards everybody's job. (This is one of the options recommended in Chapter 5.) So, if someone was walking past a POLCA Board and noticed cards in the mailbox (the box labeled "POLCA CARD HOLDER" on the top right of Figure 11.3), they would pick them up and take them to the other POLCA Board if they were going that way. At a minimum, this would happen every two hours anyway with our scheduled breaks. This procedure has worked well for us, and it has also helped to drive ownership of the system to everyone in the factory.

Instead of having a POLCA Board at each workcenter, we have four locations throughout the plant where these POLCA Boards are placed for nearby equipment. This is because we have a lot of workcenters, which are divided into zones with leadership oversight for each area, and we aligned the POLCA

Figure 11.3 POLCA Board showing the pockets for the batch records, unused POLCA cards (in the bottom pockets), and POLCA cards that need to be returned (top right holder).

Boards with those zones to enable decision-making for a zone to be done in one place. In addition, our WIP is too large to be stored directly in front of resources where it would be visible, so having these POLCA Boards clearly displays how much WIP is actually waiting for each resource. This also allows front-line employees to easily identify bottlenecks and understand priorities.

Taking that idea one step further, we developed a large plant-wide "POLCA Dashboard" where batches are represented by 2" × 3.5" magnets (Figure 11.4). Prior to implementing this board, managers and support departments would refer to a printed Master Production Schedule to determine what to do next and when things were expected to happen. This printout was harder to absorb as it was not really visual, but the real downside was that our schedule was very volatile so the printout was obsolete almost as soon as it was printed! The POLCA Dashboard offers a live picture of where all the batches are, and can be continuously updated. It has become a central location where all departments come to discuss strategy and execution sequence based on current events. If a machine breaks down, we discuss it at this board and look for batches that can be started or expedited to avoid the resource with the breakdown. The POLCA cards and decision rules would naturally create the same response, but this large visual board offers a macro view of what the cards are doing and supports buy-in from employees and supervisors about why changes are occurring.

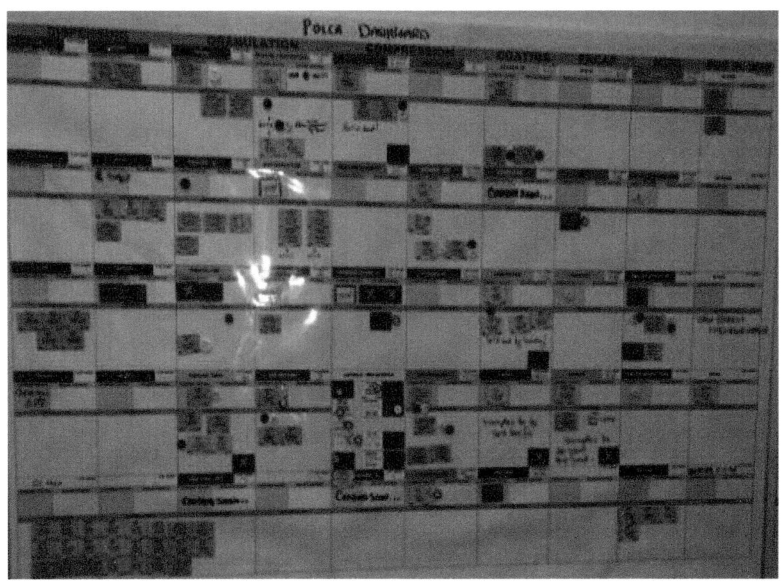

Figure 11.4 Plant-wide POLCA Dashboard.

Not everyone embraced the POLCA Dashboard at the beginning. Some considered it cumbersome to manually move magnets from resource to resource. However, it wasn't long before the benefits began to outweigh the effort. A proud moment for the project team was when employees started improving the Dashboard and using it to track additional information to make their work easier; like proud parents, that's when we knew the Dashboard had been accepted. A few months after the POLCA launch we wanted to permanently capture some of the improvements, so we redesigned the Dashboard. As we were taking down the old Dashboard, there were concerned looks from people coming to the Dashboard for information. We had to assure them that we weren't taking it away, but trying to make it better.

Life with POLCA

Within weeks of our launching POLCA the product flow through the plant had already improved. Within three months our metrics were showing sustained improvements:

- Production lead time was reduced from over 20 days to around 10 days (a 50% reduction).
- On average, we find we are using only about 80% of the total calculated number of POLCA cards in the system. When combined with the 50% lead time reduction, there is a substantial reduction in WIP; a stark contrast from having difficulty moving through the plant because WIP rooms were crammed full and hallways were storing the overflow.
- Maybe most impressively, and most impactful to our business, the throughput increased 20%, as measured by the number of batches produced. (Although individual batches vary significantly in their demands for equipment, over a longer period this aggregate number shows a clear trend.) Finally, we saw a systematic improvement throughout the entire plant (rather than individual resources), as measured by another aggregate metric, kg produced/FTE (FTE stands for Full Time Equivalent, a common metric for headcount in organizations).
- Escalations still happened but they became less disruptive to the shop floor. In the past, shop floor staff would be pestered to expedite certain batches, but at the same time they would get conflicting direction causing frustration as well as more changeovers. With POLCA in place, most of the time the shop floor people are unaware that a batch has even

been escalated, because a 10-day shop floor lead time offers the flexibility for sales and planning to respond to clients' needs without changing priorities of batches already in production.
- Production staff, who often knew best, were empowered to make real-time scheduling decisions rather than blindly following a schedule that had been created days ago by someone in an office.

Pursuing a project like POLCA was a risk at the time. It truly changed core business practices all the way back from how we procured materials (to ensure flexibility), how we scheduled batches, and how we executed them. Yet we were able to successfully apply the concepts and adjust them to our unique situation.

A change of this magnitude would not have been adopted without the constant training and communication. Anyone looking to implement POLCA should not underestimate those aspects of the project. Our plant was such that we could not pilot the concepts of POLCA in a small area. We do, however, recommend that others start this way if possible. It would have been much easier to launch and control if the scope were smaller.

POLCA has changed the way Patheon runs its business. The improvement in production lead time has made us much more flexible and agile (words not typically associated with our highly regulated industry), allowing us to respond quickly to our clients' ever-changing needs. POLCA has enabled continued growth at our site and now that it is so embedded in our culture, it's hard to imagine ever going back to finite schedules and WIP in our corridors!

About the Authors

Susan Ferris joined Patheon in 2002. Having worked in various positions, she currently holds the position of production director. After completing a Black Belt program within Patheon, she was assigned a project to improve the overall lead time and throughput in Production, which led her and the rest of the team to POLCA. Susan holds a Bachelor of Science degree with Honors in Neuroscience from Brock University in St. Catharines, Ontario.

Justin Bos joined Patheon in 2010 and currently holds the position of Manager, Operations Strategy. He was the Operational Excellence Black Belt that supported the POLCA implementation at the Toronto site. Justin studied Mechanical Engineering Design at Conestoga College in Kitchener, Ontario.

Chapter 12

From the Netherlands: Creating Flow and Improving Delivery Performance at a Custom Hinge Manufacturer through QRM and POLCA

Guest Authors: Godfried Kaanen and Robert Peters

SECTION I: THE JOURNEY TO POLCA, by Godfried Kaanen

BOSCH Hinges is a high-mix, low-volume manufacturer of custom-engineered hinges (bespoke hinges) in steel and stainless steel. The company specializes in small orders ranging from a single prototype to 2,500 pieces. BOSCH Hinges serves customers in many industries, including equipment manufacturers, land transportation (both rail and vehicle industries), aviation, and the maritime sector. Short delivery times and a very high delivery reliability are key attributes of the company. As the owner of BOSCH Hinges, in Section I of this chapter I will lead you through my personal journey to POLCA.

I Get a Rude Wake-Up Call

In 2004, we had started to embrace the concepts of Lean and were determined to succeed on a never-ending journey of continuous improvements. The battle against high volumes, low prices, and competition from low-cost countries left us only one opening: strive for maximum productivity of our staff and seek the niche position of high-mix, low-volume and custom (HMLVC) producer. Serving our clients with the best quality hinges, engineered and produced in short lead times and in exact quantities, however small, became our goal. It seemed that we had hit upon the right strategy. Customers were glad that we accepted smaller orders without negotiating for larger quantities, and they were satisfied with the quality of our products. Our performance as a company seemed to be improving, both in the eyes of our customers as well as in our financials, and we felt we were on a good course. *But then in the winter of 2006, I received a loud and unexpected wake-up call.*

I had an annual meeting with the purchasing department of an important German customer in the rail sector, an international corporation that we supplied almost every week. Even though the orders came weekly, they were widely different, with each order consisting of numerous types of hinges in varying quantities, and with options specific for this customer. During the meeting, which involved three purchasing managers from the customer, I felt like I was being waterboarded. Although we had been supplying this customer for more than 15 years, nothing seemed to be right in their eyes: our delivery times were poor and our delivery reliability even worse. Thanks to our quality, which was never an issue, we didn't completely fail their criteria, but their overall vendor rating report for us was nothing to be proud of: they gave us a "C" grade. In summary, their decree to me was: "Cut your delivery times from four weeks to two, and improve your reliability to above 95% on-time." They also warned me that if our overall grade did not improve to an "A" within 12 months our position as vendor would be at risk.

With that load on my shoulders, I drove 160 miles back to my company. On the way back, I had time to reflect on how we had arrived at this point.

Our Initial Improvement Efforts

We started our efforts in 2004 by taking courses in Lean and also arranging for in-house training from a consulting company. Applying toolbox elements of Lean led to improved productivity. We reduced stocks of raw materials to a strategic minimum. We stopped making finished goods to stock, and started making only to specific customer orders. We eliminated in-between stocks of semi-finished goods, and greatly reduced our work-in-process (WIP). Times for equipment setup and tooling changeovers were shortened. Lead times started going down.

It seemed that we were on the right path. But even as we continued down this path, customers kept pressuring us for ever-shorter and more reliable delivery times. This was also starting to take its toll on our staff. When we first started our Lean efforts, our employees were enthusiastic. However, they were also hesitant to change and, especially as we felt more pressure from the market and tried to keep up, the continuous changes felt like a burden for many of the employees. In evaluating the situation, I realized that we were struggling with two major issues: our poor performance on delivery reliability, and the expertise of our staff.

The Pastry Shop Needs a Breakthrough

For a better understanding of these points, let me explain our situation at the time. Every year we released around 1,600 production orders onto our shop floor, and these orders covered over 400 individual designs. And to make things more difficult, order sizes ranged from a single prototype hinge to 2,500 pieces.

The choice of materials that we work with is also large: mild plain steel in two grades, stainless steel in grades 304 and 316, pre-galvanized steel, and Zincor. Material thicknesses range from 0.8 to 5 mm. In total, we stock more than 50 items as raw material in sheet metal. The number of manufacturing operations needed to produce a custom-engineered rolled hinge is also large: laser cutting or die-nibbling of male and female cut-out flat parts, pre-rolling, rolling, calibrating, axial and radial corrections, bending, folding, countersinking, threading, galvanizing, powder coating, production of the pin, assembly, packing, and shipping. Even before all these, you have to add the processes of design engineering, manufacturing engineering, and routing preparation for every order. Every hinge we produce is a customized mix of the above (see Figure 12.1 for examples). With all this to be done, you can see that having a staff of just 25 people on the shop floor plus some temporary workers put a great strain on the required skills of the employees.

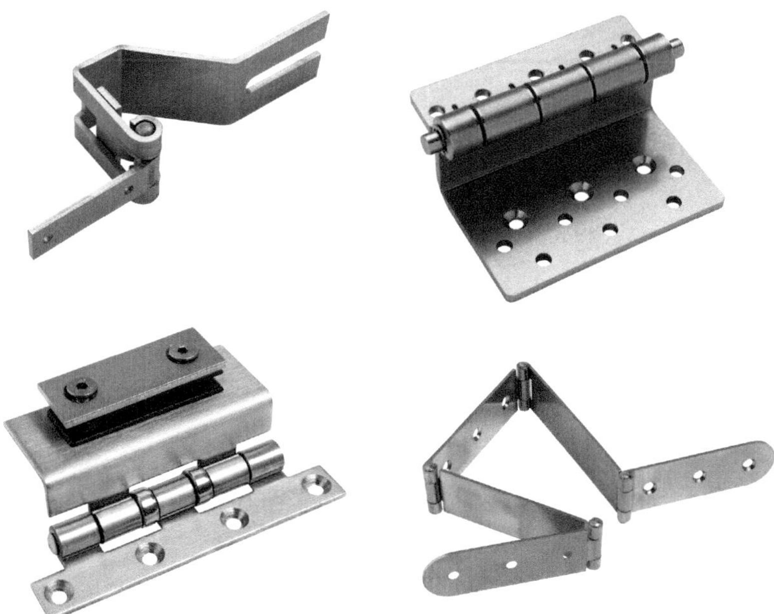

Figure 12.1 Examples of custom hinges.

This is how complex a small job shop operation can be. We are a classic example of HMLVC production. I often compare our company with a gourmet pastry shop: an enormous variety, every item freshly baked in small batches, and everything must be ready every day just in time for the demanding customers!

In reflecting on our journey, it became clear to me that, despite our Lean efforts, production planning and control were still troublesome: too much to plan, too many details to take care of, and too much dependency on the skills of our individual employees. So even though our company had become leaner, we still faced the phantom of late deliveries. And it wasn't going to get any easier. The pressure from our customers for even shorter delivery times was growing.

As I drove those 160 miles back home, I realized that fine-tuning our Lean efforts wasn't going to do it: we needed a breakthrough!

Thus it was that, in the spring of 2007, I decided to initiate a whole new and major improvement project, which we called "Getting a grip on production planning." The keys that we were looking for in this project were how to preserve operational flexibility while maintaining reliable delivery, and how to cope with all this in our HMLVC production environment without stress for our employees.

Our First Step: The Rhineland Model

We have always been inspired by theoretical models, so we started researching possible approaches, and found one that seemed suited to our situation. It was the Rhineland model described by Jaap Peters and Mathieu Weggeman in *The Rhineland Way* (Business Contact, 2013). Two elements particularly interested us—a cell-based structure and greater employee responsibility—because these could potentially address both our challenges.

To determine a cell-based structure, we analyzed the routings for four months of production orders—around 600 orders in total. We identified every single operation for each production order and marked it in an enormous spreadsheet, ending up with a spreadsheet that was 600 lines deep and 50 columns wide—no wonder that we had constant planning problems! Using this data, we defined cells as clusters of coherent operations and gave each cell a name. This led us to identify nine work cells on the shop floor. Drawing the cells in a factory layout on paper gave us an immediate overview. We realized that the planner was not to blame for orders being late without this overview.

We also realized that our staff had been assigned highly diverse individual jobs. In our existing process, an employee had to manage an individual job from end to end. It appeared much too complicated for the less-skilled employees. Not everyone knew the characteristics of each equipment and mistakes were made in setting up the tooling. Worst of all, employees would bump into each other at their next production step, creating queues and low productivity. Orders ending up being late, and as employees couldn't do much about this, morale was low. Instead, we felt that assigning employees to a particular cell (their "home" position) would limit the variety of the work to be done by each of them, but would raise the quality of their work.

Next, we created a buffer zone for every cell: an identified space where WIP that needed to be handled at that cell was to be placed. (To this day, when I visit companies I see few examples of shop floors where clear areas are defined for placement of WIP during the production process.)

As we moved forward with these ideas, we found that creating nine cells on the shop floor provided a good overview for the production manager and planner but it still did not control the production flow. The consultant helping with our Lean implementation had told us that you need to pull orders rather than push them. We had learned about Kanban and the Pull system. But with our diverse and customized orders, we didn't see how to apply this system.

At this point, we were trying to solve a puzzle that involved several issues:

- How do we prevent our upstream cells from pushing work to the downstream cells and creating an overflow of WIP in the buffers?
- How do our upstream cells know specifically what to make at a given moment to supply the right items needed by the downstream cells?
- How do we quickly communicate and deal with day-to-day capacity limitations of cells due to unanticipated issues, such as jobs that take longer than expected, employee sick leave, last-minute holiday requests, and breakdown of equipment or tooling?

In short, even after the cells were in place, we needed some form of shop floor control, but it had to build on the cellular structure we had created.

We Are Introduced to POLCA

The answer to our puzzle came from a junior consultant, Jacob Pieffers of the consulting firm, who along with his advisor Professor Jan Riezebos had written a masters thesis, "POLCA als innovatief materiaal-beheersingssysteem" (POLCA as an innovative material control system) (University of Groningen, the Netherlands, 2006). For his thesis, he had taken a course at the Center for Quick Response Manufacturing (QRM) at the University of Wisconsin–Madison. Jacob explained the concepts of QRM to us, and also POLCA as a control mechanism as part of QRM, specifically designed for the control and communication between cells. There was instant recognition for us that this applied to our situation. Jacob showed us Rajan Suri's book, *Quick Response Manufacturing* (Productivity Press, 1998). I immediately ordered the book.

By this time, it was the summer of 2007. My wife and I had rented a cottage in Rügen, a beautiful island in northeast Germany. Rügen is the German version of Cape Cod. In the 1920s and 1930s it was a resort for rich industrialists of the time. Later, from the 1960s to the 1980s, leaders of the communist party of East Germany would spend their holidays here, hidden away from their people. I was really intrigued by Suri's book, and decided to spend the holiday reading the book. But there is a lot to see in Rügen, so I had to negotiate a deal with my wife, who stated categorically: "In the morning you can read your book, then in the afternoon we will cycle around the island, and in the evening you will take me to a restaurant of my choice."

I didn't dare refuse! So, every morning I read some more of the book. Chapter after chapter hit home with me; I felt as though "this book is about my company!"

In particular, POLCA as a shop floor balancing and control mechanism to complement the cells fitted in perfectly with our company's needs. My decision to go for POLCA was made.

Next came the implementation phase, which took a few months. I realized that we needed to do several things to improve the cellular structure based on QRM thinking, and to create the foundation for POLCA, so I decided to engage the consulting firm for a second assignment. With their help and revived enthusiasm of our own staff we rearranged the shop floor to create an improved layout for the defined cells. We placed the cells as much as possible in a horseshoe layout based on overall flow patterns, and gave each cell a name based on a color (Figure 12.2). We created clearly marked buffer zones for every cell for WIP to be placed. This first set of tasks was completed between August and December 2007.

Next, we designed the POLCA cards and boards, and a board was placed in each buffer zone. Production order sheets needed to be revised for each order to specify the cell sequence. This workflow had to be specified in our ERP system. It was a period of three very intense months.

We weren't worried about achieving perfection. We felt like eager Boy Scouts on an adventure and moved rapidly along! Obviously, mistakes will be made in such a situation, but we were ready to adjust in real-time. The first POLCA board is an example. After Rajan Suri visited us in 2008 and made some suggestions, we quickly made a second version

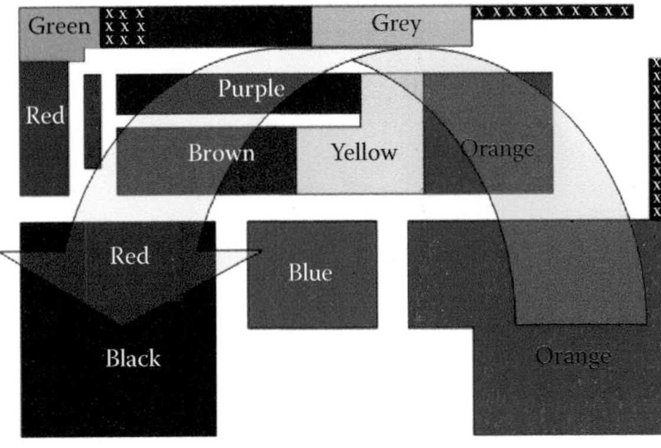

Figure 12.2 Improved cell layout with defined colors and overall horseshoe flow.

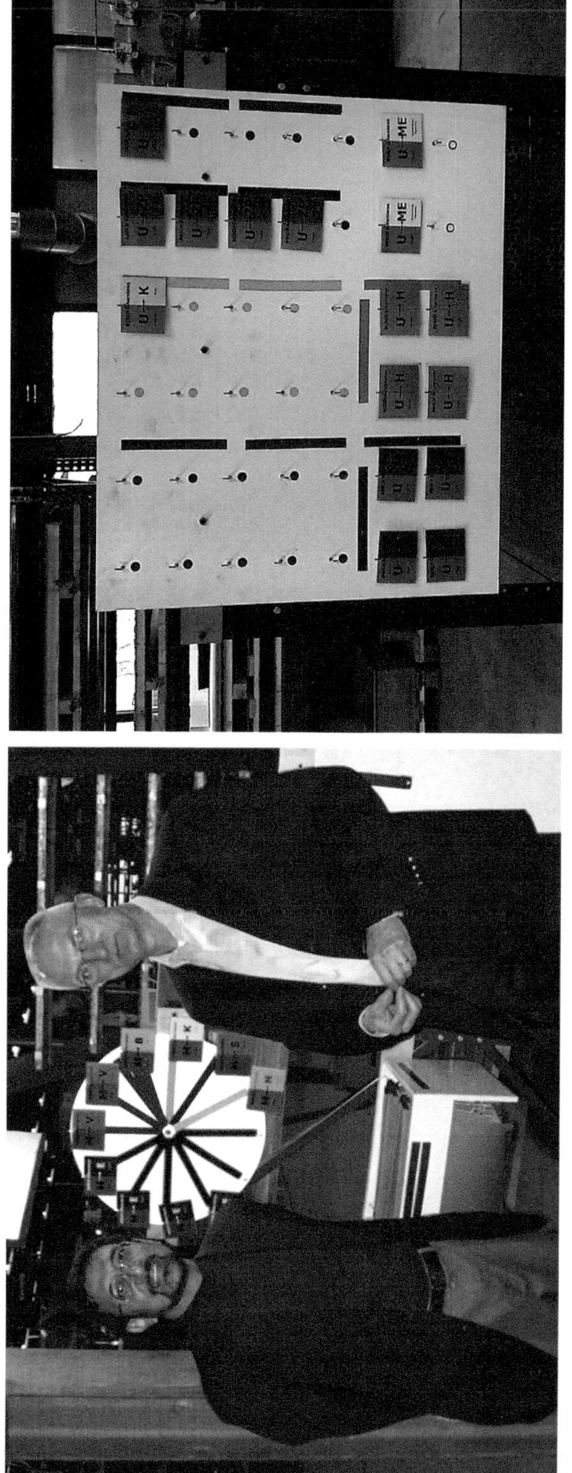

Figure 12.3 Examples of POLCA Boards. The first attempt is on the left, pictured with Rajan Suri and Godfried Kaanen. The improved board after Suri's feedback is on the right.

(see both versions in Figure 12.3). This happened several times with many of our choices.

We felt that our staff needed training in the details of POLCA and how the system would work. Our operators are used to "hands-on" work and we felt that a physical simulation would be better than an overview of the theory. Such a simulation didn't exist so we had it developed by the consulting firm. We were so determined that the principles of QRM and POLCA as job shop control system were right for our situation that every expense was taken. The consultants developed a POLCA Game, which included a physical simulation (see Appendix F). We used this game to train our staff on working with POLCA.

Implementing the Color-Coded Visual System

Our next step was creating a visual system for communicating the order flow to the shop floor people. Each job has to follow a specific routing related to the customer order. At first, we glued the cell sequence on each order sheet with colored stickers (Figure 12.4). For over a year our production manager spent up to two hours every day reading order details and gluing stickers in the right sequence on every order sheet. A boring and repetitive job! But knowing the value of the visual signals to the workers kept him going.

It was clear that this was exactly the kind of job that should be automated using information technology. So, we asked our ERP vendor if they could print the colors on the order sheets along with the routings. Their answer was that that it couldn't be done! "Do you realize what that means in terms of programming work?" they said. "And what it will cost you in terms of the number of full-color pages printed in a year?" To which we replied, "Yes, we realize it very well, and we are determined to buy a color printer that can handle this volume of printing. Whatever the cost, the printer will pay back very soon. We will save so much time of the production manager attaching stickers every day. And do you realize the value of visual communication? Our employees need the visual cues to attach the right POLCA card to the right job. If you make this improvement in your system, you will see that other companies will follow." The discussion felt like David against Goliath, but we won (see Figure 12.4). The add-on to our ERP system paid back soon and kept paying back. We couldn't have maintained our POLCA operations without it.

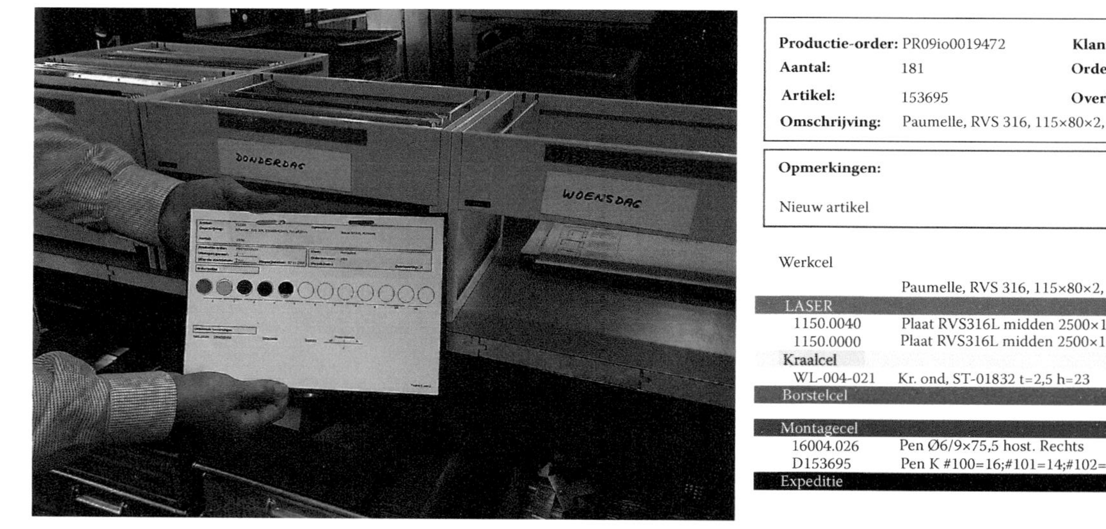

Productie-order: PR09io0019472 **Klant:** Van keulen Mobielbouw
Aantal: 181 **Ordernummer:** 6529
Artikel: 153695 **Overlevering:** JA **Verpak.instr.:** IW 000
Omschrijving: Paumelle, RVS 316, 115×80×2, 5mm, Hostoform pen

Opmerkingen:

Nieuw artikel

Werkcel		Locatie:	Tekening	Productietijd: Machine	Arbeid
	Paumelle, RVS 316, 115×80×2, 5mm, Hostoform p		5		
LASER				2,52	
1150.0040	Plaat RVS316L midden 2500×1250×2.5				
1150.0000	Plaat RVS316L midden 2500×1250×1.0				
Kraalcel					3,11
WL-004-021	Kr. ond, ST-01832 t=2,5 h=23	Geel 15-D	105		
Borstelcel					3,02
Montagecel					1,76
16004.026	Pen Ø6/9×75,5 host. Rechts				
D153695	Pen K #100=16;#101=14;#102=14:#103=9				
Expeditie					
			Total:	2,52	7,89

Figure 12.4 Putting Visual Signals into the Routing Sheets. The left side shows how we started by attaching colored stickers. Then we worked with our ERP vendor to incorporate the colors in the routing printouts, as shown on the right side.

We kept the POLCA cards simple, as specified in the theory, with the two cell colors and abbreviations of the cell names, and we also added an arrow to clarify the direction of flow (Figure 12.5). In the figure, you will notice that although the POLCA principles recommend putting serial numbers on the cards, we decided we didn't need them. We had a small factory and the production manager would periodically check on the total number of cards. The initial number of cards per loop was calculated by the consulting team. We soon found that, in practice, we could improve on these numbers by making judgment calls from observing the

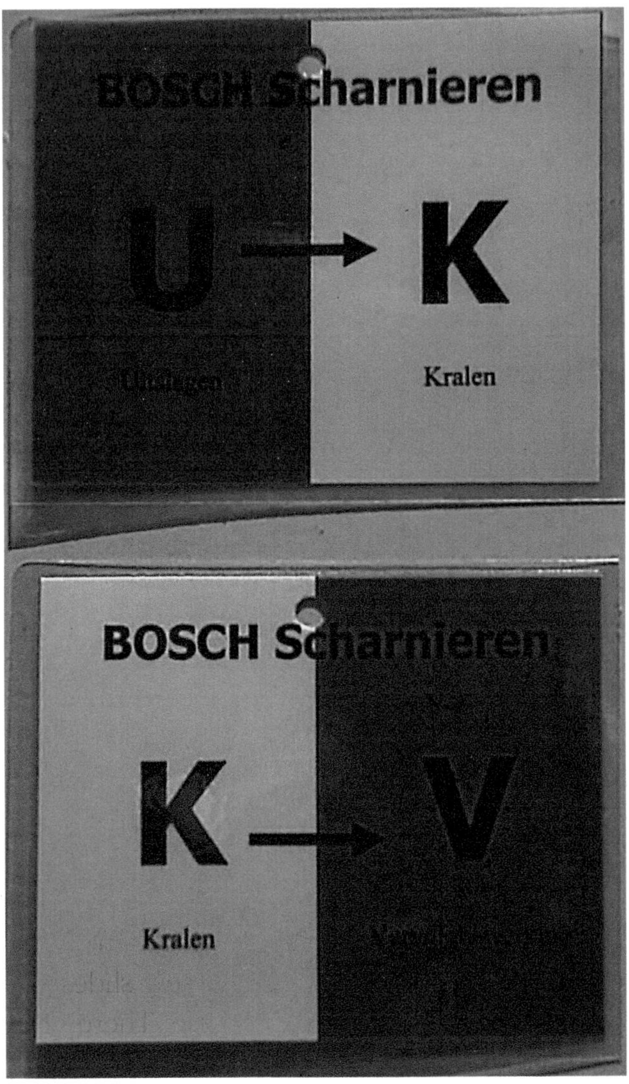

Figure 12.5 Examples of the first POLCA cards.

WIP as occupancies changed during the year, and thus tighten up the system. In fact, we started with 90 cards and this was reduced to around 65 cards.

Initial Results from POLCA

With the introduction of cells and POLCA, the whole role of planning changed: the entire job of actively planning orders in detailed steps by our planner disappeared. He got a different job within the company. Planning became a part-time responsibility of our production manager, who released the orders for the shop floor. Let's remember that the A of POLCA stands for Authorization. When we planned our POLCA implementation, we had decided on the "Release-and-Flow" option or RF-POLCA (see Chapter 5), and thus for each job we only needed the Authorization Date for the release of that job to the shop floor. (We later upgraded our system to the standard POLCA; see the second section of this chapter.) As mentioned, we gave the responsibility of setting this Authorization Date to the production manager, who functioned as the Planning Cell described in Chapter 5. He would sort the orders based on due date and then for each job, based on his experience and the estimated total touch time, he would back-schedule the job from its due date and set the Authorization Date of when to release the job to the first cell in its routing.

In keeping with the RF-POLCA rules, once a job was released, Authorization Dates were not needed for subsequent operations. Instead, each cell had a basket to keep the orders in sequence. As orders arrived at cells, their production orders were arranged in the cell's basket based on a FIFO (first-in first-out) sequence, as required by the RF-POLCA rules. (The only person allowed to change the sequence for any cell was the production manager.) So, when a cell team had capacity to launch the next job into its cell, it would look in this basket for the next job in the FIFO sequence, check the color of its destination cell, and with the visual aid of the colored sticker, it would look to see if the right POLCA card was available to launch the job.

After we made the initial changes of creating the cells and implementing POLCA we saw rapid improvements. Before POLCA, our lead times ranged as high as six weeks and we could sometimes be an additional two weeks late in delivering beyond that. Without putting pressure on employees, our lead times dropped to around four weeks in the first year, along with high reliability of delivery. At the time of writing this chapter, the lead times have been further reduced to two to three weeks, with very high reliability. With our employees working in teams in the cells, we saw that their productivity rose immediately. This resulted in less overtime and temporary workers. Equally important, fellowship and communication between the cells has created a team spirit and made a tremendous difference.

At this stage, perhaps you are thinking, "What about the German customer you talked about at the beginning?" Oh yes, they are still there. Our relationship is better than ever. We did indeed cut our delivery times to two weeks, and for over six years we have received an A+ vendor rating.

SECTION II: FROM PHYSICAL POLCA TO DIGITAL POLCA, by Robert Peters

After the POLCA system had been operating for over a year, we observed where our employees were spending time on repetitive tasks, and decided to evolve to a more automated system while still keeping to the POLCA principles. In reviewing the activities of the production manager, one of the first priorities for automation was to create a daily Authorization List. We created a computer program to do the back-scheduling based on the routing and estimated lead times at cells, and thus calculate the Authorization Date for each job. With this list created automatically each day, the production manager didn't need to spend time on the piles of order sheets. This was the first step in what later became the PROPOS shop floor control system described in the rest of this chapter. This also meant that the production manager could spend more time in dealing with strategic improvements and occasionally making minor adjustments if urgent customer orders were received or other unexpected events occurred.

Shortly thereafter, the idea arose to provide each cell with a screen that showed the current situation at each cell based on the routing of each order. Figure 12.6 shows an example of a screen at the Welding Cell. Each line on this screen represents a production order, and the screen is separated into three sections: the "In process" section shows orders that have already been launched and are in process in the cell; the "In buffer" section displays orders that have arrived in the cell's buffer and have not yet been launched; and the "Expected" section displays orders that have been released to the shop floor and are on their way to this cell, but have not yet arrived in the buffer. This last section allows the cell team to anticipate which jobs might arrive soon and helps them plan their capacity and workload in the near term.

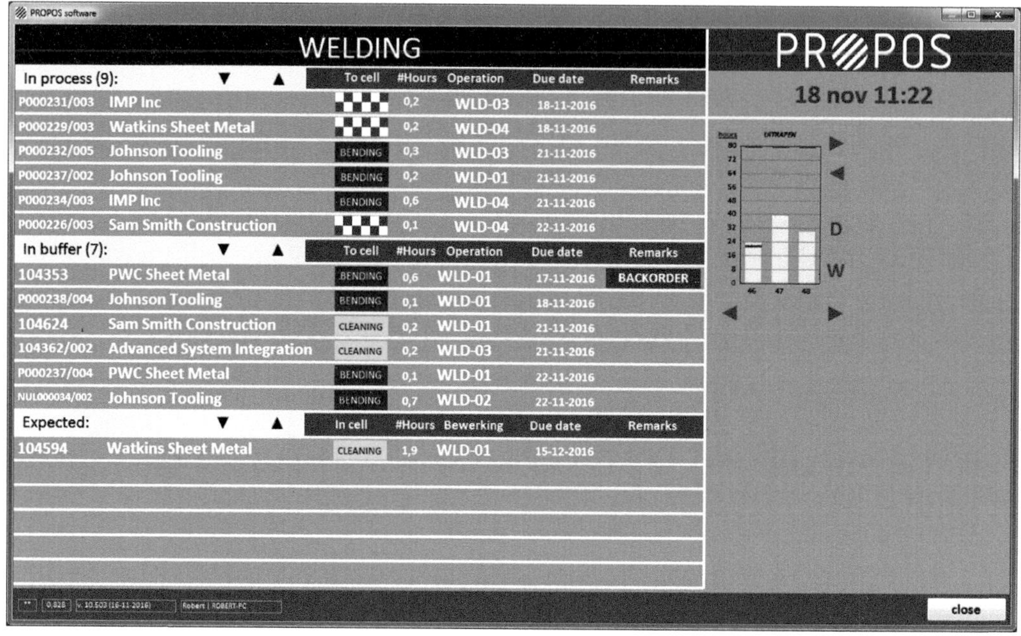

Figure 12.6 Example of a PROPOS screen at the Welding Cell showing the status of orders in or destined for the cell.

In this first stage of implementation the PROPOS screens supported the initial order release as well as the sequence at each cell, and we were able to eliminate the FIFO baskets at the cells. Further, the sequence at each cell was now based on back-scheduling from the order due date, and thus we upgraded from RF-POLCA to the standard POLCA with Authorization Dates now being used at each cell. Finally, to support our QRM strategy, the PROPOS system also included the measurement of MCT, including the breakdown into white space and gray space (see Appendix A for more details on MCT and white/gray space).

By the summer of 2010, we had a situation where the PROPOS system specified the work for the cells on the shop floor, but there was also the POLCA system giving people additional information. We realized that people had to look at two different systems (Figure 12.7). First, they had to look at the PROPOS screen for the first Authorized production order (the top line in the "In buffer" section of the screen) and note the downstream cell for that order. Next, they had to check their POLCA board to see if a POLCA card was available for that POLCA loop. If so, they would launch the order in PROPOS and remove the corresponding POLCA card from the board and place it with the material. If no POLCA card was available, they had to look

Figure 12.7 When the PROPOS screens were first implemented, the cells still used physical POLCA cards. Here we see a PROPOS screen at a cell, as well as the cards on the cell's POLCA board.

at the next Authorized order on the PROPOS screen and check if a POLCA card was available for that order. And so on.

So, although the systems complemented each other and were easy to use, we thought we could make things easier for the people in the cells by unifying the two into one digital system. In addition, there were two practical issues that would be solved by designing a unified system. The first practical issue that we ran into was that a lot of POLCA cards were flowing between a few pairs of cells that were not in the same building. As a result, people had to walk a lot (sometimes through the rain) to return the POLCA cards, causing people to hold back the free POLCA cards and return all of them once a day, which both delayed the POLCA signals for the upstream cells and created lumpy demand for those cells when a batch of cards was returned. The second practical issue was that sometimes POLCA cards got lost so they needed to be counted and replaced regularly. Counting, re-creating and replacing physical POLCA cards in a busy production environment asks for a lot of discipline. When in the course of 2012 the QRM cells Brown and Yellow were merged and when, at the same time, new cells were created—things that happen in a dynamic company—it took some effort to recalculate and create the new cards for the new routes. We felt that all this could be solved by incorporating a digital version of POLCA into the existing PROPOS system.

In early 2012, we started the development of Digital POLCA. The PROPOS system already contained information about all the production orders and their routings. Based on that data, we let PROPOS automatically identify all occurring POLCA loops in the factory. Next, for each POLCA loop, the number of POLCA cards was calculated using the standard formula (Chapter 5). By the summer of 2012 we had implemented the Digital POLCA system throughout the shop floor and people were working with an integrated system for both the production order information and POLCA signals.

Figure 12.8 shows an example of the first Digital POLCA window that we added to the PROPOS screens. This electronic display replaced the physical POLCA board. The figure shows the Digital POLCA board for the cell Yellow.

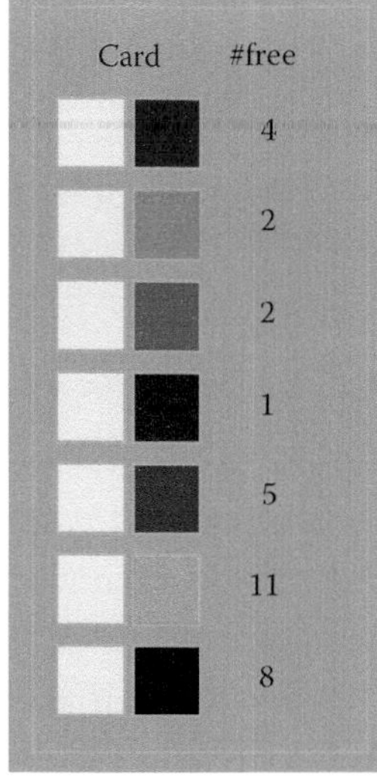

Figure 12.8 Example of the first Digital POLCA window added to the PROPOS screens. This is the electronic version of the POLCA board at cell Yellow. It shows the POLCA cards to seven destination cells, along with the number of cards available for each destination.

78150 Production order 343-B BLUE

Figure 12.9 This is a line in the PROPOS screen for a production order going from cell Yellow to cell Blue. The traffic light on the right is red, indicating that there are no Yellow/Blue POLCA cards available and the order is blocked at this moment.

There are seven POLCA loops from the Yellow cell to downstream cells, and you can see the electronic images for the corresponding seven POLCA cards. Next to each image you see the number of cards of that type that are currently available. For instance, for the Yellow/Blue loop, there are four unused POLCA cards at this moment. When the cell team launches a job using the PROPOS screens, the Digital POLCA system automatically reduces the number of cards by one (for the type of card used by the job).

On the PROPOS screen, the line listing a production order also contains a traffic light symbol (Figure 12.9). If no cards are available for a particular POLCA route, the PROPOS screen displays a red traffic light for all orders with that route. It also blocks those production orders from being accidentally launched. So, the cell team will only be allowed to launch the first production order without a red traffic light.

One more feature that we added to the digital system was the ability to include temporary POLCA cards along the lines of the Safety Cards and Bullet Cards described in Chapter 6. In some unusual circumstances, as described in that chapter, management may decide to add a few extra POLCA cards. In that case, there had to be a system to ensure that these cards were taken out at some later time. PROPOS takes care of this automatically. You can add temporary POLCA cards for a certain POLCA loop and at the same time provide an expiration date, after which the cards automatically disappear from the loop.

Implementing the Quantum through Load-Based POLCA

News of our first Digital POLCA implementation and its success quickly spread through our region, and we received inquiries from several companies that wanted to implement this system. As we rolled out the PROPOS system at other companies and gained more experience, we added a Load-Based POLCA option to PROPOS in 2013. This option electronically implements the idea of the quantum discussed in Chapter 5, and is ideal for situations where there is a large variation in workload between production orders. If you choose to implement the Load-Based POLCA option, then instead of a production order using one POLCA card, it now uses "x" cards, where x is the workload in hours for this job at the downstream cell. Also, x does not need to be an integer, as required with physical cards; it can be any number, including fractions.

We can explain the implementation using the PROPOS screen shown in Figure 12.10 for the Welding cell. Looking at the "In buffer" section (see the enlarged portion in the lower half of the figure), the first Authorized order is blocked by the system, as shown by the Red traffic light on that line. The reason is, it needs 0.6 POLCA cards of type Purple/Blue, but from the 45 cards in the system, there are currently none available. The second order is also going to Blue, so it too is blocked for now. So, if the cell team has capacity, the system will allow it to launch the third order, which is destined for Light Blue. This one needs 0.2 cards of type Purple/Light Blue, which is possible because there are 38 available.

The preceding discussion highlights one more advantage of an automated system: it would be difficult and time-consuming for people to calculate and

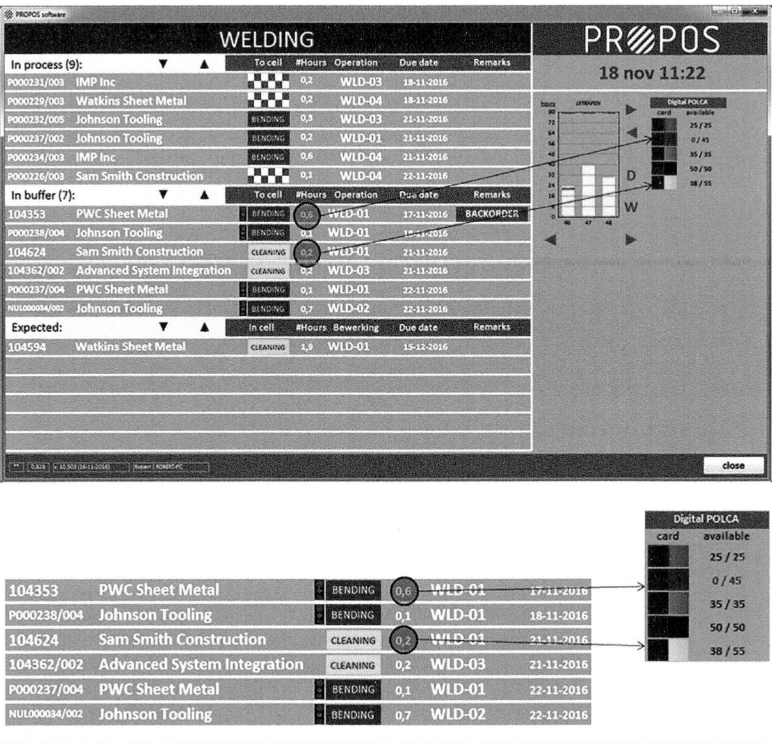

Figure 12.10 Upper picture shows a PROPOS screen for the Load-Based POLCA option. Lower picture enlarges the "In buffer" portion of the screen so you can see the number of POLCA cards needed and the resulting red lights.

manage Load-Based POLCA at this fine level of detail. Along the same lines, the automated system has enabled us to add specific features requested by customers based on their unique situation. In one case, we added some additional logic to help reduce the idle time at highly constrained resources; this is described in the Provan case study in Chapter 13. In another situation, we added flexible POLCA cards where the system can move cards between upstream loops that are all destined for the same downstream cell. The total number of cards for the downstream cell remains the same, thus still constraining WIP and congestion, but the cards get moved between loops when demand varies a lot at the upstream cells. In all such modifications, our aim has still been to keep to the main goals of POLCA and also aim for simplicity, not making the computerized system too complex, and thus retaining the buy-in from the shop floor people.

Impact of PROPOS and POLCA on the Company's Operations

As can be expected from any computer-based system, it is easy to look up the status of orders at any time to see where they are in the shop. The difference with PROPOS, incorporating the POLCA logic and operating rules, is that the shop floor performance is very reliable, and management can have confidence in meeting the due dates in the system. Godfried Kaanen (first author) had a recent personal experience with this. He was just about to leave at the end of the day before the Easter holiday weekend, when the phone rang. He picked it up and it was an important customer wanting to be sure his order would arrive on time the following week. In the past, Godfried would have had to walk into the factory, try to find the order, or maybe even chase down his production manager who had already left for the holiday. Next, he would have had to check how many steps the order still had to go through and estimate if the order would be ready on time. With PROPOS, Godfried could immediately see which step the order was in, that the system predicted it would be shipped on Tuesday, and that the customer would receive it on Thursday, as promised. Godfried could confidently answer the customer, knowing that due date performance is now very reliable.

Beyond some of these functional enhancements from the system, there have been broader impacts on the organization as a whole. Prior to implementing the full functionality in PROPOS, there was a lot of staff time involved in planning and scheduling. As explained earlier in the chapter, the company had a full-time planner, plus the production manager spent a lot of time calculating release dates. In addition, if customers changed delivery dates for orders already on the shop floor, the production manager had to manually reprioritize the orders in various FIFO baskets at the cells. And finally, the managers had to keep an eye on the progress of jobs to detect any jobs that were in danger of being late.

With the POLCA logic incorporated in the PROPOS system, combined with the other functionalities described above, several benefits have been realized. First, the planner's job was no longer needed and he could be productively employed in another capacity. Adding the equivalent of one person's capacity is quite significant for a small company! Second, changes in customer request dates get automatically reflected in the revised Authorization Dates in the PROPOS screens, and this along with the POLCA rules helps to keep jobs on track. And third, the PROPOS screens clearly highlight jobs that are behind schedule (if they should have been launched already, but are not yet in the cell) so shop floor personnel can see if they need to take some corrective action, without requiring a manager to constantly be on top of this situation. The shop floor workers also like having the information clearly on the screen. In the past, they would often check with the production manager before starting a job, just in case something else had come up. Now they can see clearly what they need to do and don't have to wait for a second opinion and reassurance. All these preceding points have given the workers a higher feeling of ownership and pride in their jobs. Equally important is that the production manager, instead of spending time on day-to-day micromanagement, can now spend much more time on strategic improvements with cumulating benefits for the company.

As a final conclusion for both sections of this chapter, at BOSCH Hinges we have seen significant quantitative as well as qualitative benefits from the combination of the POLCA system rules and the PROPOS digital system's implementation features.

About the Authors

Godfried Kaanen is an entrepreneur in the SME (small- and medium-sized enterprises) metal industry in the Netherlands. He believes that the industry and in particular SMEs are an indispensable pillar to any economy. His motive is the pursuit of operational excellence. He combines his private business at BOSCH Hinges with the chairmanship of *Koninklijke Metaalunie*, the Dutch employers' organization for SMEs in the metal industry. The more than 13,000 affiliated members employ approximately 150,000 people and jointly represent a turnover of more than 22 billion Euros. Godfried is also known as one of the ambassadors for QRM (Quick Response Manufacturing) in Europe.

Robert Peters is co-owner of PROPOS software. He has over 18 years of experience in business (sales, marketing, and project management in different sectors) and over 12 years of experience in developing and implementing software solutions at innovative companies to help them improve their results. Robert has a B.ICT degree from the HAN University of Applied Sciences in Arnhem, the Netherlands.

Chapter 13

From Belgium: Metalworking Subcontractor Reduces Lead Times by 85% Using QRM and POLCA

Guest authors: Pascal Pollet and Ben Proesmans

Provan was founded in 1998 by two young entrepreneurs, Ben Proesmans and Luc Vanhees, with the dream of establishing a thriving metalworking subcontractor. Thanks to the right technology and the necessary know-how, Provan has emerged as a trend-setting metalworking subcontractor and supplier of metal products. Provan offers its customers a total solution for welded structures, laser and sheet-metal work, profile machining, and assembly. The company now employs around 80 employees and serves a variety of industries including medical, automotive, agriculture, heating, and machinery.

Motivation for Rethinking the Shop Floor System

For the first few years of its operation, Provan had used its ERP system together with Lean techniques like Kanban to manage its order flow. However, management noticed that Provan's customers were getting more demanding: the variation of parts ordered had grown, and batch sizes in the orders had shrunk considerably. At the same time, customers wanted more flexibility and shorter delivery times. As a result, it became more and more difficult to satisfy customers using the material management systems that were in use.

The issues with the existing systems became more pronounced during 2012 in the production of stoves for a major customer. The stoves were comprised of 130 parts that were produced in several steps. These steps mainly involved laser cutting, bending, and threading. Then these 130 parts were welded or fastened together and assembled into one stove.

The parts were stored between the successive fabrication and assembly steps in a warehouse that occupied over 600 square meters (around 6,500 square feet). The stoves were produced in batches of 60, and the lead time from the initial laser cutting operation until the final inspection took about four weeks.

The customer was pleased with Provan's performance, and asked the company if it could supply three variations of the stove product. This meant that three types of stoves needed to be welded and assembled, and 130 components per stove—often with different routings—needed to be manufactured. So, while management was pleased with the increase in the customer's orders, it also feared that this would almost triple the needed warehouse space to store the components. In addition, they were concerned that the increased complexity could lengthen their lead times, which would also add to the WIP and storage space for semi-finished parts. For a small company, this extra space would add considerable expense.

Implementing the Stove Cell

As luck would have it, at about this time Provan was introduced to Quick Response Manufacturing (QRM) at a workshop given by Rajan Suri in Belgium. Management felt that the QRM approach could help attenuate the warehousing problem and the company decided to create a QRM Cell for the production of stoves.

In creating such a stove cell, a major issue that needed to be resolved was the organization of the complex work flow within the cell. Remember, three types of stoves needed to be fabricated and 130 components per type needed to be produced, often with different routings within the cell. How could the operators manage these material flows without needing to resort to a miniature version of an MRP system within the cell, with all its associated complexity and problems?

Introducing Color-Coded Visual Work Management

After six months of discussions and tinkering with ideas, an elegant solution was developed that would be simple and yet avoid the typical problems. This solution got the employees to become familiar with visual management and using color-coding to signal tasks and priorities, and prepared them for the companywide POLCA implementation later.

The solution involved the formation of a cell that would undertake all the processing steps except for laser cutting. Laser cutting was kept separate from the cell, as this expensive machine was shared among many different products and clients. Buying an additional laser cutting machine for the cell couldn't be reasonably justified and wasn't necessary to accomplish the goals. Other workstations such as bending, rolling, threading, grinding, and welding were moved and positioned in a U-shaped cell (Figure 13.1). A color code was marked on every workstation in the cell to support the visualization of the workflow. Once the cell was in place, the order steering system, described next, was used to manage the flow of work.

When parts for a batch of stoves arrive from the laser station, they are put on a set of carts in a standardized way as follows. First, a numbered metal

Figure 13.1 Stove cell. Note the material on carts in the middle and work stations on the outside of the cell.

flag with a specific color is attached to an empty cart. The color on the numbered flag corresponds to the stove type. So, with three types of stoves, three different colors are used. The number on the flag indicates which parts should be put on which cart. Multiple parts are grouped together on a cart based on their routing within the cell. Then on every cart, a row of additional colored flags is added. These flags indicate the routing the cart has to follow within the cell. The first color in the flag row corresponds with the color of the first workstation in the routing. The second color corresponds with the color of the second workstation, and so on. In summary, two types of colored flags are used to guide the material flow: the colored numbered flag indicates the product type, while the colors on the (unnumbered) flags specify the routing (Figure 13.2).

When this initial step of coding the parts with the flags is completed, the carts are placed in the middle of the stove cell. The cell is now ready to process the parts. After a cart is processed at a workstation, the corresponding colored flag of the workstation is removed by the operator and the cart is put back in the middle of the cell. The cart is now available for processing at its next workstation. With this system, operators can easily spot which parts they have to process without consulting a computer system or printed shop orders. They just have to look at the colored flags to know which carts are available to be processed next at any given workstation.

Figure 13.2 Details of the flag system on a material cart: A red numbered flag, and three flags (green, yellow, grey) indicating the routing.

The production control system in the stove cell is not a POLCA system. Nevertheless, it shares several important characteristics with POLCA that help to explain its success:

- As in POLCA, the work in process (WIP) in the cell is strictly limited by the finite and small number of available carts. Limiting the WIP has several advantages for the cell. Material piling up can lead to all kinds of waste (searching for material, larger walking distances, damage, and so on), and so all these wastes are avoided. Also, limiting the WIP in turn limits the lead time of jobs in the cell.
- It is a highly visual system that is easy to understand and use for the operators.
- The order information (the status indicated by the flags) is always up to date. So, the production control system doesn't rely on outdated information, which used to be the situation.

As a result of the simple rules, the workflow is completely self-steering. Employees are in control of the process, which means they feel more involved. Detailed planning of the separate workstations by a planner, and supervisory tasks like shifting workers between the work stations, have become unnecessary. Interestingly, the quality of the products has also improved significantly: scrap and complaints have been reduced by 60%.

The switch to a single production cell meant that material movement on the shop floor was also reduced and stoves could be made in batches of 15. The lead time per batch decreased by a massive 85%, from around four weeks to just three days. The packing process in the cell was also observed to be twice as fast because the products were directly on the right pallet and weren't stored in the warehouse waiting to be picked.

The success of the stove cell and its lead time and near-perfect delivery performance resulted in more products being ordered by the customer, and Provan now makes eight models of stoves. Even so, the on-time delivery performance is still 100% even with the many more models, and the promised delivery time to the customer has been shortened from five weeks to three weeks. And, of course, one of the best results for Provan's management has been that instead of tripling the warehouse space (or more), the stocks of parts have actually been eliminated, freeing up 600 square meters of space for Provan to expand its production operations.

Expanding to the Rest of the Factory with POLCA

After this initial success, Provan looked for further opportunities to implement QRM cells. However, this proved much harder. After analyzing its product mix and routings, Provan concluded that there were limited opportunities to form additional cells focused on clear product families. In addition to the stoves cell, an assembly cell for medical treatment couches was put in place, and a third cell was dedicated to prototyping work. These three cells represented about 50% of the work volume of the company. For the remaining 50% of the work, since Provan served as a subcontractor for a wide variety of customers and products, its environment represented a typical job shop. The volume and work content of jobs at Provan was rapidly varying, which made it difficult to dedicate work to particular cells. So, after the success of the first three cells, the company was left with the complex challenge of managing the workflow for the rest of its jobs.

For these remaining high-variety jobs, Provan suffered from several problems on its shop floor: information at the work centers was often already outdated even by the time the production orders were distributed to the shop floor; planning and re-planning activities formed a heavy burden for the company; supervisors had to run around the shop floor to expedite orders; and priorities were shifting continuously. Finally, despite all the follow-up and rescheduling, the necessary parts to start an order were often missing. As a result, the order information was not trusted by the operators.

To solve these issues, Provan undertook several steps. First, the planning was simplified by organizing the shop floor into manufacturing cells in which similar operations were grouped together. In the past, planning and order routing was done at the detailed level of individual machines. Now, the planning is done at a higher level of cells, and operators in the cells are given more responsibility so that they are able to handle the work themselves in a more flexible way. (For more insights into this approach, also see Appendix C.) Second, the high variety of jobs can be accommodated by routing jobs through different cells based on their needs.

Third, Provan decided to implement a POLCA system to control the work flow between the cells. As the team at Provan started looking into POLCA, it had several concerns about implementing the system with physical cards, and, after some deliberation, the team chose to implement an electronic version of POLCA. Several reasons motivated the option of an electronic solution:

- After analyzing the typical variety of orders going through the shop, it was calculated that a large number of different routings were possible between the cells. This would result in a huge number of possible POLCA loops and cards and it was felt it would be difficult to maintain and update all of these.

- The work content of the subcontracted jobs had large variations: some orders might take only a few minutes, while others represented several hours or even days of work. This meant that to smooth out the work flow to enable POLCA to work effectively, Provan would need to implement a quantum rule (see Chapter 5). However, with the large variation in jobs, as well as new and custom jobs that frequently arrived, someone would have to analyze the quantum for each job and the operators would need to deal with varying numbers of POLCA cards attached to each order. This would complicate the implementation for both the planners and the shop floor operators.
- The team felt that with the large number of cards and loops, physically returning the cards to the previous cell would be seen by the operators as a waste of their time, which would undermine their acceptance of the system.

After considering various electronic options, Provan chose to implement a system called *Digital POLCA* from PROPOS software. The system includes one screen for each cell on the shop floor (Figure 13.3), which displays the information on jobs at or destined for that cell, and a central screen for the

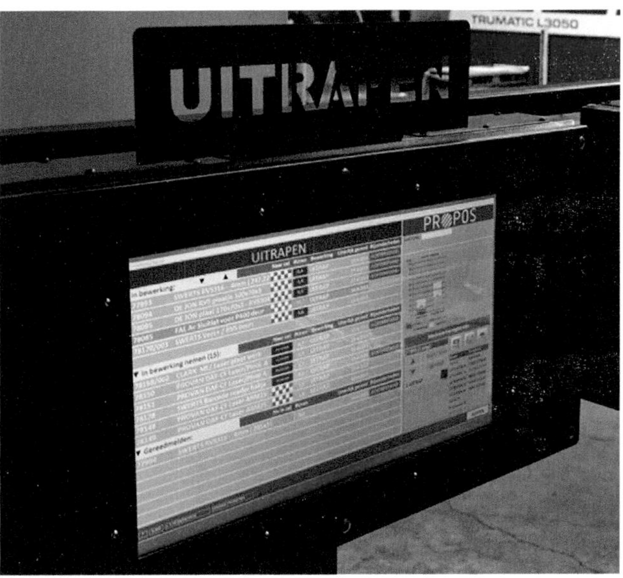

Figure 13.3 PROPOS screen at one of the cells.

supervisor and planner with more detailed information (see Chapter 12 for the background on PROPOS as well as more details of the system). Implementing an electronic version also offered the advantage that the system could be customized for Provan's needs, as described next.

Customizing the System for Better Use of Scarce Resources

In addition to the POLCA functionality, one straightforward customization of the PROPOS system at Provan was to use the system for recording the time spent on each job at each cell. However, a more elaborate customization proved beneficial for Provan's unique situation.

Lead times at Provan were already quite short because of the flexibility of its resources and management's belief in planning for spare capacity. This meant that with the operation of the POLCA control rules, there would typically be very low WIP levels in the factory. At times, some of the cells were standing idle waiting for orders. Of course, this can happen naturally when there are simply no orders left for these cells, and if you plan for spare capacity this is bound to happen occasionally. However, at other times management noticed that orders that would use the downstream idling cells had been released to the shop floor, but these orders were delayed at upstream cells that happened to be overloaded at this point of time because of the unplanned variability of the incoming work. Again, in some cases this might be acceptable, and is just part of the dynamics of high variety operations, but this situation was particularly concerning to Provan's management when the idling cells were staffed by welders. Because welders are hard to get in the labor market in Belgium, it is not easy to have extra capacity in welding. At the same time, Provan's expertise in welding gave it an edge in getting orders. So, any idle time at the welding stations was seen by management as a lost opportunity.

These factors motivated Provan to ask PROPOS if they could come up with a solution for this situation. The aim was to have a solution that did not involve complex shop floor scheduling algorithms because of the drawbacks of such systems (see Chapter 3), and also kept to the main framework of the POLCA rules.

As is the case in a typical POLCA implementation for high-variety jobs, upstream cells service multiple downstream stations. At certain points in time, some of the downstream cells were idle or almost idle, while others still had sufficient orders to keep them busy for a while. This situation offered the possibility to prioritize orders at upstream stations that were destined to go to stations that were soon to experience a shortage of work. In fact, POLCA offered an easy way to find the stations with a low work load! A low work load at a downstream station corresponds to a large number of available POLCA cards at the upstream station. So, the algorithm just had to prioritize orders based on how many POLCA cards of the downstream stations were available. This was still done under the constraint of respecting the authorization dates. In other words, if multiple jobs at an upstream station were authorized, then the system would pick the job based on this prioritization. This added an elementary layer of logic that was simple enough for everyone on the shop floor to understand and buy into, while maintaining the simplicity and clarity of the POLCA system.

The enhancements provided by PROPOS have been received well at Provan. The Digital POLCA system now improves the flow of work in both these situations: when there are a lot of orders at Provan, the POLCA logic limits the work at every station, prevents excessive WIP, and also ensures that upstream cells work on jobs that will be needed by downstream stations, using the normal POLCA logic. However, when there is lower demand, the added logic helps to ensure that critical cells such as welding do not unnecessarily run out of work.

Results: The People Side

The first benefit that Provan noticed after the POLCA implementation was the increased transparency on the shop floor. In the past, only the planner and the supervisors were (more or less) aware of what was going on. If a problem arose, they had to be consulted, which unavoidably led to additional delays. With the transparency of the system, everybody had access to the same, real-time, and thus up-to-date information. Over time, everyone found that this information was typically accurate, so then the operators started to trust the system, and this provided a huge benefit of buy-in from the shop floor.

Now, if an order is in danger of being late, a warning appears automatically on the upstream production screens, and since the shop floor personnel now trust this information, they can start to think how they can resolve this situation. The system further supports this by displaying the reasons for delay (such as waiting for parts from suppliers).

The system also had a major impact on the activities of the supervisors. The supervisors had serious doubts before the implementation: they objected that the system might not be as smart as an experienced supervisor. They also feared that their jobs would be even harder with two added duties: trying to follow the system, and at the same time trying to correct its deficiencies. However, with the system now in place for a while, these same supervisors actively endorse the system and are glad to have its support! They have experienced that the system has liberated them from many of their cumbersome tasks. Instead, these supervisors can now focus their energy on coaching and developing their people, cross-training, and other improvement activities, and leave the logistics issues to the PROPOS system. This is a better use of the experienced resources represented by the supervisors, and also reinforces Provan's competitive advantage through ongoing personnel development and other continuous improvements.

Results: The Commercial Side

Provan got an early signal during this journey that it was on the right track when a major customer re-sourced some of their products from a low-cost country (LCC) to Provan. This is particularly impressive when you realize that Belgium has among the highest labor costs in the world after you factor in all the benefits and taxes in addition to the wages. In fact, the customer calculated that because of Provan's extremely short lead times and high delivery reliability, they could agree to a price from Provan that was 11% higher than the price from the LCC: Provan's lead times and delivery record provided the customer with substantial savings in indirect costs such as inventory, warehousing space and personnel, rush freight charges, rescheduling, and so on. This was a classic win-win situation with lower total cost of ownership for the customer while at the same time supporting a reasonable profit margin for Provan.

Thanks in large part to its successful implementation of QRM and POLCA, in 2015 Provan received the prestigious "Factory of the Future" award. This distinction is awarded by the two Belgian industry federations Agoria and Sirris to companies that are best prepared for the future. This award typically places these companies among the global leaders in their sector. Some of the most striking results that were mentioned by the award jury included the significant reduction in lead times, and a tripling of the "added value" of the company (a metric used in Belgium that measures the difference between the annual revenue and the purchase costs) over a five-year time span.

About the Authors

Pascal Pollet is a principal engineer at Sirris, the collective center of the Belgian technology industry, where he started doing research on POLCA in 2005. Pascal has been pioneering the introduction of Quick Response Manufacturing (QRM) in Belgian industry and has helped many companies to radically shorten their lead times with QRM. Pascal obtained a Masters degree in Electrical Engineering from Ghent University in 1995, and is a member of the European QRM Institute.

Ben Proesmans is the co-founder and owner of the Belgian company Provan (www.provan.be), a fast-growing metal-working supplier. Ben is one of the QRM pioneers in Belgian industry, and is also affiliated with the European QRM Institute as an Ambassador. His company was recently awarded the prestigious Factory of the Future Award as a recognition for its pioneering work.

Chapter 14

From Poland: Dancing the POLCA in a Glass Factory

Guest author: Karol Bąk

Szklo Sp. z o.o. is a Polish company that has been specializing in the treatment of glass and mirrors since 1991. The company is active in the Polish as well as international markets in both business-to-business (B2B) and business-to-customer (B2C) segments. Typical products include items for the household and furniture industry such as table tops, shelves, mirrors, and lighting fixtures (Figure 14.1). Szklo receives around 500 orders a week from customers that range from small shops to large companies. All orders are executed from scratch because of the large variety and the fact that over 60% of the orders are unique. Promised delivery times range from two days to three weeks. However, in recent years Szklo has experienced that the market is expecting smaller order quantities coupled with shorter lead times. To add to this, the important B2B customers are increasingly changing their orders even after submission. Management at Szklo realized that their manufacturing system, as originally organized in their factory, was not able to meet these changing market conditions. Some of the issues they were facing included excessive inventory, frequent schedule changes, constant expediting of hot jobs, and high use of overtime (all of these are explained below).

Since the company was struggling with short-term planning and they knew about my expertise in this area, they asked me to help them select and implement an appropriate planning software package. As a consultant, I had already engaged in many projects involving Lean and Quick Response Manufacturing (QRM), including several training engagements on these

Figure 14.1 Example of glass shelf.

topics. After reviewing the company's situation, I felt it was an ideal situation where POLCA would be an appropriate solution instead of some complex planning software, and management agreed to explore the implementation of POLCA.

The Factory and Process Flows

Szklo's factory is located in Lidzbark Warminski, and employs 100 employees in a production area of around 3,000 square meters (about 32,000 sq. ft.).

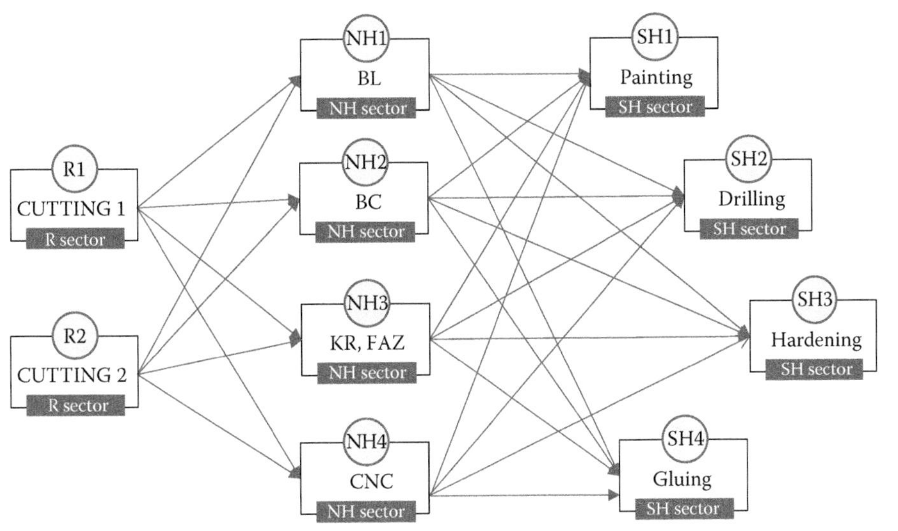

Figure 14.2 Complexity of multiple flows through the factory.

The manufacturing facilities include numerically controlled machines for processes such as cutting, grinding, chamfering, engraving, milling, and drilling. Additional processes include painting, laminating, gluing, screen printing, and hardening. Due to an increase in orders but because of their capacity constraints, the company had turned to traditional efficiency methods such as running larger batch sizes to reduce the occurrence of setups. However, this had resulted in large buffer stocks between operations, longer lead times, and hence more expediting and rush orders, as explained below.

Figure 14.2 provides an overview of the processes in the factory. There are three main types of manufacturing processes: cutting (R1 and R2), machining (NH1 through NH4), and finishing (SH1 through SH4): all these are described in more detail below. As an explanation for the abbreviations for these three areas, R comes from *Rozkrój*, the Polish word for cutting, NH is from *Nowa Hala*, the new production hall, while SH is for *Stara Hala*, the old production hall. The figure illustrates the complexity of the possible flows through the factory, depending on the particulars of each order. Now we describe each of the main processes in detail, as this will be needed to understand many of the problems that the company was experiencing.

The Cutting Operations and Resulting Inventory

The cutting was done by starting with a large pane of glass. Because of the specialized nature of the materials, and their high cost, the emphasis was on optimizing the nesting of parts being cut from one pane. So, the nesting planners looked at a large number of orders to nest parts and minimize wasted glass. The result was a large buffer filled with orders waiting in front of the machine for their turn in the nesting. Next, the supervisor of Cutting would search through this buffer to find orders from future dates that required similar material and setups, and he would execute them all together to minimize changeovers. Also, in order to further optimize the nesting, the planners used a database of frequently ordered products to fill in the nests and build these parts to stock, rather than to particular customer orders. Both these aspects—nesting far in advance, and cutting parts for stock—meant that there was a large amount of work-in-process (WIP) after the Cutting operation.

The next result was that the Cutting operators frequently complained about the lack of free pallets and storage space. Therefore, often, makeshift wooden pallets were prepared to temporarily store the WIP. To make things worse, because of lack of space some of the parts had to be transported outside the factory and stored in the open air. Ironically, although the whole idea of the nesting was to economize on material costs, the multiple movements along with external storage of the parts resulted in quality defects such as abrasions, cracks and scratches, thus creating rework and scrap!

A final issue was that because of the changes that customers were making to orders already in progress, the Cutting supervisor was engaged in frequent changes of schedule and related fire-fighting.

The Machining Department and Sequencing Problems

The Machining department has three linear grinding machines and a group of CNC machines. Again, in attempts to minimize the frequency of changeovers, the Machining supervisor tried to batch together orders that could be machined with minimal changeover times. However, the operators of the grinding machines would complain that the sequences being used in the Cutting department prevented similar types of orders from arriving at their machines, so they would make frequent trips to Cutting to negotiate with its supervisor to get suitable sets of parts that they could run together on their grinding machines.

In addition to this, there were frequent priority changes due to quality problems and scrap (as discussed above), or because of customer specification changes. All these resulted in orders being interrupted and waiting in queue, again adding to the WIP in the factory.

Finishing Processes

The final stages of production involve finishing processes including painting, hardening, screen printing, gluing, or drilling. The highest degree of nervousness in the schedule was experienced in this area. The workers in these finishing processes stated that the arrival of material in this area felt like a big lottery! Some days there was so much material that it was hard to move around the shop floor, yet on other days their machines were waiting for jobs to arrive because of the resequencing at earlier processes. On such days, the supervisors of these areas spent a lot of time visiting upstream processes to find jobs that could be expedited and brought to the Finishing department. Yet, on the days that the finishing processes were busy, they were often working on orders that had been changed by the customer, but the information had not caught up with the jobs on the shop floor!

One more issue that impacted all the departments was that management believed in planning the factory at even more than 100% of available production capacity. This, combined with the various problems mentioned above, meant that a large amount of overtime was needed, with Saturdays and Sundays being used in an attempt to catch up with all the backlogs.

Initial Design of the POLCA System

In 2011, I had the opportunity to educate the management about Quick Response Manufacturing (QRM), along with its production control system, POLCA. Management felt that this would fit their needs and help to resolve their problems, and they proceeded to investigate QRM and POLCA with my assistance.

The first step was for key people in the company to be trained in QRM, which included the concept of cells. Then, prior to moving to the POLCA implementation, they went through the process of checking the prerequisites. One of the next steps was to check the possibility of forming cells before implementing POLCA. However, they soon decided that, because of the large size of the machines, and the fact that each machine was actually handled by two to four operators, in most cases they could treat each machine as a cell.

Initial POLCA Loops, Quantum, and Number of Cards

With this decision as a starting point, the project team decided on all the loops as shown in Figure 14.3. This resulted in a total of 24 loops. For clarity

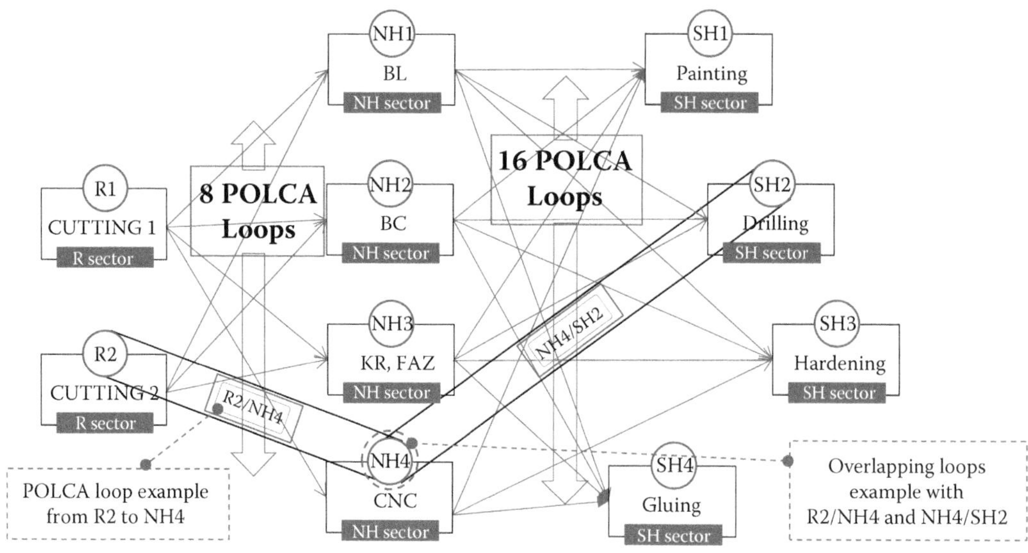

Figure 14.3 Initial set of POLCA loops. Each arrow represents a loop. For example, the loops R2/NH4 and NH4/SH2 are shown explicitly.

of representation, each arrow in this figure represents a POLCA loop, with the loops R2/NH4 and NH4/SH2 shown in detail, along with their associated cards.

To decide on the quantum, we first discussed a new material transport method, and designed trolleys that would carry a POLCA card on a hanger (Figure 14.4). Next, based on the amount of parts that could be accommodated on a trolley, we calculated that a quantum of two hours of machine time would be appropriate in most cases to accommodate parts on one trolley. Finally, based on this quantum we used the procedure to calculate the

Figure 14.4 Design of the transport trolleys with hooks for POLCA cards. The picture on the left shows the concept, and on the right, examples of trolleys in use with various POLCA cards.

initial number of POLCA cards (see Chapter 5) and arrived at a total of 79 cards in the whole system.

We also checked this calculation with the current situation. We observed the WIP in the factory and converted it to the equivalent number of POLCA cards, using the quantum. Interestingly, the current WIP amounted to around 132 cards, showing that the POLCA system had the potential to immediately reduce the WIP by around 40%! Even more interesting, the workers on the shop floor were still complaining that they did not have enough pallets, but at the same time they did not have the right jobs. So, we knew from these initial calculations that we were on the right track with moving to POLCA.

Authorization Method

For the Authorization portion of POLCA, we decided that this would be done by the Cutting supervisor, prior to the first cutting process. This supervisor had access to all accepted orders with calculated process times and based on the shipping date and his experience he would assign the Authorization Date for Cutting. From Cutting onward the system would operate on the Release-and-Flow POLCA rules.

Simplifying the Design: Moving to POLCA "Lite"

After we completed the initial design of the entire system, we explained it to the key managers, production supervisors, and operators. The feedback from the group consisted of two main points. First, people thought the system was too complicated and felt that 24 different loops would be too many to manage. And second, they were concerned about day-to-day variability of the workload across different cells. We had calculated the number of POLCA cards based on average demand for a quarter. However, their experience was that the high variability in mix of jobs could put significantly more short-term load on some machines, and the lack of POLCA cards could starve these machines and/or downstream operations.

Following some brainstorming sessions, the idea that emerged was to combine upstream machines of a given type into a "Sector" that still supplied individual downstream machines, and to have the cards returned from the downstream machines to the Sector Board. This would give a larger pool of cards so that if a given downstream machine had a higher workload, it could draw from this bigger pool of possible jobs.

Drawing from popular terminology, we decided to call this modification "POLCA Lite!" Figure 14.5 uses an example to illustrate the difference between classic POLCA and POLCA Lite. In the classic POLCA, each upstream cell would have an individual loop to all its possible downstream cells. In POLCA Lite, as one example, the cells NH1 through NH4 were combined into Sector NH, and the sector as a whole had a POLCA loop to each of the downstream cells. For example, consider the loop shown to Cell SH2. In a given week, if there was a lot of work for SH2, then all the upstream NH Cells together could use the sector cards to send jobs to SH2.

In a similar fashion, we combined the Cutting area into one sector (R) and created POLCA loops from Sector R to the NH operations. During this change it was also decided that the NH sector should be separated into six downstream cells. Figure 14.6 shows the POLCA Board used for the Sector R and NH connections with POLCA cards from Sector R to the six NH cells (BC, BL, DEL, CNC, KR, and FAZ).

These design modifications reduced the number of loops from 24 to 10, and the number of POLCA cards from 78 to 54. The simplified POLCA Lite system was approved by the Szklo team and we moved on to the implementation.

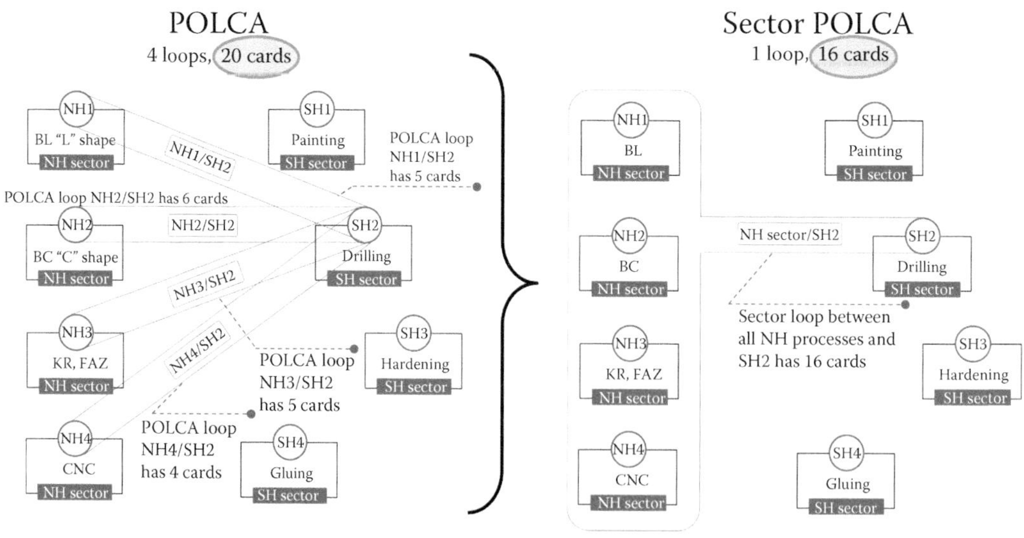

Figure 14.5 Illustration of the concept used to simplify the design and create the POLCA Lite system.

Figure 14.6 POLCA Board for Sector R to the six cells in Sector NH. For example, the BL column shows that there are two cards available for jobs going from Sector R to Cell BL.

Results of the POLCA Implementation

Within just a few months of implementation, the POLCA system was already beginning to show significant results. The following results were observed in less than one year, and have continued to be realized:

- The system has stabilized, with an average of 58 POLCA cards in a planning period, which translates to a reduction of 56% from the WIP levels before POLCA—in short, the WIP has been slashed by more than half!
- The reduction in WIP also means that the company no longer has to store parts outside the factory, thus also reducing problems with damage and quality issues.
- The large reduction in the number of pallets allowed the company to justify investing in specialized trolleys (as shown in Figure 14.4). The customized design of the trolleys combined with the limit imposed by the

quantum has meant a reduction in damage due to lifting and setting down of parts, and also due to the previous moving back and forth of pallets.

- The reduction of WIP by more than 50% has also translated into a reduction of production lead time by more than 50%. This has also meant that orders are released later to the shop floor, and there is more time for customers to make changes before the job is actually started.
- The shorter lead time also means that they don't need to plan as far ahead and so there are fewer hot jobs and fewer schedule changes.
- There is less of a concern about machines not having any work to do. POLCA keeps the jobs flowing without creating excessive WIP. The logic in the POLCA system makes sure that jobs get to machines in time when it is necessary, and if a machine has no work at a given moment, people now trust that it is okay with the current conditions, and they don't run around trying to find work. The visualization in the POLCA system also helps workers and production leaders understand and trust the system in such situations.
- As a result of the previous two points, production leaders now spend very little time on chasing upstream jobs or re-scheduling their areas, and they have more time for coaching and improvement of the processes.
- With all the above factors, the company's on-time delivery performance has improved by 30%.

As has been the case with other POLCA implementations, Szklo also experienced benefits with the buy-in from the people on the shop floor. In the words of the production manager, compared with the past situation where supervisors and managers were involved with expediting jobs, "The POLCA system has now involved our front-line operators to take responsibility for the flow of materials."

About the Author

Karol Bąk graduated from the Technical University of Gdansk with a Master of Science degree in Management and Production Engineering, and began his career in industry as a manufacturing and supply chain manager. Since 2012, he has worked as a consultant with 4Results, Warsaw, and been involved in over ten QRM implementations. He is one of the pioneers of QRM in Poland, and the author of several technical papers and training materials related to QRM and POLCA. He also works as a lecturer on QRM-related subjects at the Technical University of Gdansk.

Chapter 15

From Germany: Yes, Even Small Companies Can Benefit from Implementing POLCA

Guest author: Markus Menner

In 2014, we were approached by *Preter CNC Dreh- und Frästechnik GmbH und Co.KG,* a small company in the Black Forest region that produces drilled and milled parts on CNC machines. As the pioneer of Quick Response Manufacturing (QRM) in Germany, we at *axxelia* help companies reduce their lead times by implementing QRM principles. POLCA is one of the tools that we use regularly, based on our software *timeaxx*, which incorporates the POLCA logic. Preter had heard about our successful implementation of software-based POLCA in some other companies and their initial idea was to utilize the same software right away in their factory. As we spoke on the phone, I explained that there is a strategy (QRM) behind our software and that simply cutting and pasting the same implementation from other companies was not likely to result in the desired improvements. Preter's main issue at that time was related to their poor on-time delivery performance. They had already tried various approaches, including the MRP functionality included in their ERP system, but, to their disappointment, instead of improving their delivery performance, this actually degraded it. The management felt that they had to work on their planning processes in order to be able to deliver on time. "Let's see," I said, and then I made an appointment with the management on site.

Evaluating the Company's Situation

Preter was founded in 1986 as a one-person company, and in the beginning it had only conventional turning and milling machines at its disposal. Over the years the company has grown to twelve people and it has become a specialist for single-piece and small-batch manufacturing. Today, the company manufactures custom metal parts on around 10 CNC machines. Typical operations required by the products include drilling, milling, outside processes at subcontractors (for example, coating) and finishing processes such as assembly operations. A major issue faced by the company is that qualified workers are hard to find due to the location of the factory. Consequently, improving processes and dealing with existing capacity in an optimal way is key to Preter's ongoing success.

Upon reviewing the company's situation, I felt that the small-batch and custom environment meant that QRM would be well-suited as a strategy for Preter. At our first meeting, I presented the QRM strategy and its benefits for this type of business. They had never heard about QRM and some of the concepts looked odd to them. However, the concept of POLCA seemed clear and attractive to them. With POLCA, they thought that they would be able to improve their performance significantly. They understood the principle of not pushing too much material onto the shop floor and to downstream resources, and that this would help reduce work-in-process (WIP) and shorten lead times. Again, they pressured me that their goal was to get software that would solve their problems, which were primarily a lack of transparency, inadequate planning, and no follow-up control on the shop floor. However, I convinced management to invest in a few days of analysis to develop a plan for how to move forward prior to jumping into the implementation of a software package.

Our initial analysis revealed numbers that are typical of this type of business. The average lead time, as measured by the MCT metric (see Appendix A) ranged from 20 to 70 days, with about 40 days as an average. Management had a strong focus on maximizing the utilization of machines (aiming for 90–95%), avoiding setup times through batching, and making products to stock based on forecasts. As a result, the touch time on orders (more formally, the gray space, as defined in Appendix A) never exceeded 2% of the MCT—this meant that during 98% of the time, products were just sitting on the shop floor! Finally, the metric of most importance to the customer, namely the on-time delivery, was below 60%.

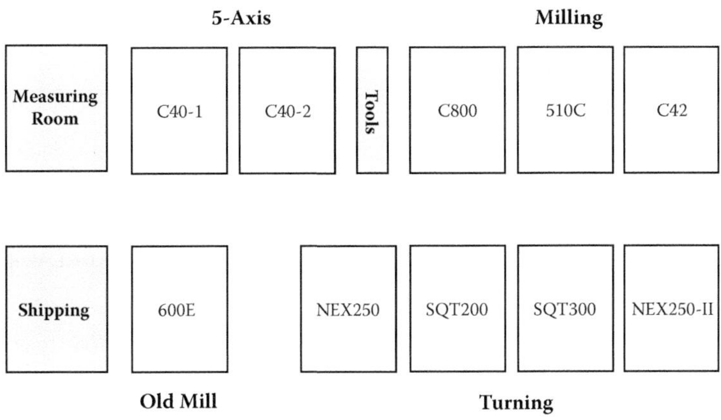

Figure 15.1 Shop floor layout at the Preter factory.

We also reviewed the shop floor layout (Figure 15.1) and realized that there were no organized flow patterns. Because of the custom nature of the business, each work order would just go from one machine to any other machine, based on the operations needed and which machine had the right capability as well as the availability at that moment. Hence, job routings crisscrossed all over the shop floor, creating the classic "spaghetti flow" that is discussed in the literature on job shops.

Opportunities and Challenges for Implementing POLCA

Based on our review of the lead times, on-time delivery, and shop floor layout, we felt there was definitely opportunity for applying both QRM and POLCA. However, we felt that some challenges would have to be overcome for this type of company. The first challenge was the small size of the company, which meant there were few machines and seldom duplicate resources. This limited the possibilities for restructuring into cells. Nevertheless, we tried to determine some families of orders (known as FTMS in the QRM terminology, see Appendix B) around which to create cells. We found that in order to keep their customers happy, the company accepted many quick turnaround jobs, which they just added onto their existing heavy schedule! As expected, these jobs jeopardized the delivery of their regular orders. So, we thought it made sense to dedicate some capacity (specific machines) to these quick-turn jobs. Thus, we created just two FTMSs, one for the regular orders and one for the quick-turn jobs. We used this approach to create cells for the processing of these two types of jobs. Due to the small number of machines, sometimes a "cell" would only be one machine, but as explained in Chapter 2, POLCA can be used between standalone machines as well.

The second challenge we faced was that some orders had few operations, i.e., very short routings. We knew that the benefits of POLCA increase with the complexity resulting from the number of cells and the network of routings, and the management of the company wondered if POLCA would have a significant impact with all the simple routings, and this also made them a bit hesitant about going forward. Nevertheless, I was pretty sure that the move to cells—including planning on a cell level (which includes the combination of several operations), dedicating capacity to the different types of jobs, having a clear policy on spare capacity, and using POLCA to manage and control the flow—would still provide positive results. I was able to convince the company to move forward with the implementation.

Cell Formation and POLCA Launch

Our approach was to start with basic QRM training for everybody in the company, and then to brainstorm together with everyone to create the cell structure that would support the two FTMSs. Finally, we would implement POLCA using our *timeaxx* software.

Through the brainstorming we came up with three cells: a 5-Axis Cell consisting of machines C40-1 and C40-2; an NEX Cell with machines NEX250 and NEX250-II; and a Mill-Turn Cell consisting of machines C42 and SQT300. Ideally, we would have liked to place these sets of machines physically together, but management at Preter was reluctant to move machines since they had just moved to their new building and installed every machine with expensive foundations. So, we agreed to go for a combination of physical and virtual cells. Figure 15.2 shows the three cells: you can see the 5-Axis Cell is a physical cell, while the other two are virtual cells, which means that the machines are not collocated, but they work organizationally as one planning unit. We also dedicated an older machine (600E) to be used for spare capacity. The remaining machines were left as standalone machines (or "one-machine cells").

Finally, we installed the *timeaxx* software, which began operation in March 2015. The POLCA loops, cards, and rules are implemented through the software. Because the company makes so many different parts and there

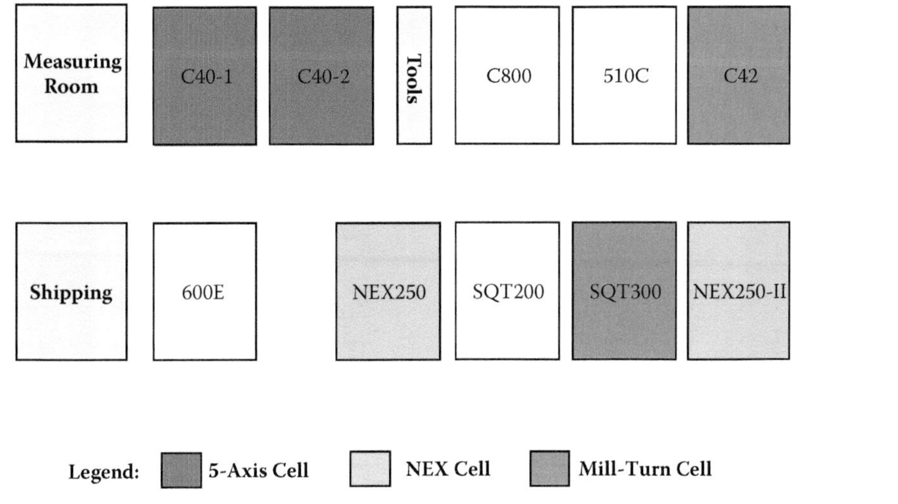

Figure 15.2 Cell structure and standalone machines.

are constantly new parts, there is a POLCA loop from almost every cell to every other cell; these are not shown on the figure as it would become very crowded. We use the Release-and-Flow version (RF-POLCA) at Preter, and the POLCA decisions are communicated to the cells by means of simple "traffic lights" on the computer screens. For example, for the initial release of jobs, *timeaxx* goes through the standard Decision Time Flowchart (Chapter 2) for each job and sets the traffic light to green if that job can be released to the first cell. The Authorization Date for this first step is calculated by *timeaxx* using sophisticated finite-capacity scheduling starting from the promised delivery date while also considering the real-time status of each job in the company. At subsequent cells, the RF-POLCA rules (Chapter 5) are used to set the traffic light to red (don't launch) or green (okay to launch) for each job.

Figure 15.3 shows some of these features of *timeaxx*. The traffic lights are indicated in the left-most column. Jobs that can be started show this column in green, while those that cannot be started have this column in red. Jobs already launched in the cell are in yellow. (There is one additional color used—orange indicates a job that is in process but temporarily suspended, for example, if an operator has gone on break.) Also, in *timeaxx*, since we are using the computer to manage the POLCA cards, we decided on a quantum of one minute, which allows us to manage capacity at a highly refined level. However, this means that in some loops we might have thousands of POLCA cards, so we decided that instead of indicating the absolute number of cards available, we would indicate the percentage available

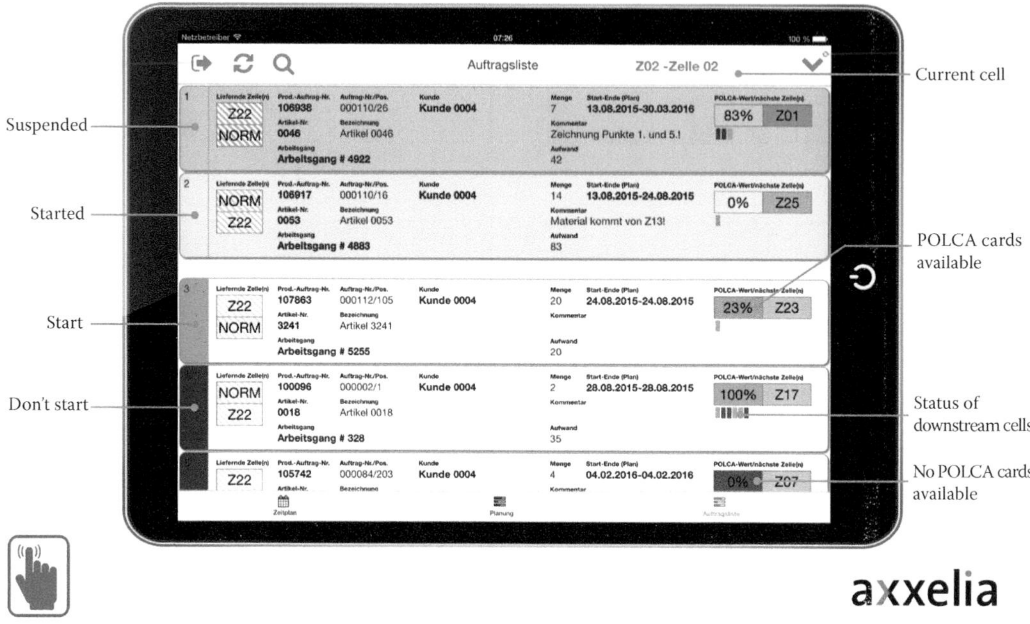

Figure 15.3 Example of *timeaxx* screen for jobs at a cell.

(as a percentage of the total cards in the loop). This gives the workers and management a quick intuitive overview of the downstream capacity available in a given loop. You can see these percentages in Figure 15.3. Note that Item 5 has no cards available, so the traffic light is red. On the other hand, Item 4 does have cards available, but the traffic light is still red because the Authorization Date has not yet been reached. Finally, the small vertical bars under the percentages indicate the status of downstream cells. The idea is that if multiple jobs have the same Authorization Date and available POLCA cards, then these bars give the operator some visibility to decide which job would be better to start based on the status of downstream cells.

Results from the Implementation

From other projects in much larger organizations we have seen that the more complex the structure of the shop floor, the larger the potential effect of implementing POLCA. Therefore, the project in a small organization such as Preter was challenging because of the limited room for reorganization. Nevertheless, this project has been very successful. Within six months of starting the use of *timeaxx* and POLCA, the company saw a 30% reduction in the MCT metric for the shop floor. This outcome continues to be achieved at the time of writing this case study. The POLCA implementation has also led to a significant reduction in customer complaints and hence higher customer satisfaction, resulting in increased customer loyalty. Therefore, we have delivered practical proof that POLCA can bring positive effects even in smaller organizations.

Besides the hard numbers, the company has benefitted from several people-oriented effects. The POLCA system along with the *timeaxx* features has dramatically increased transparency for both the planners and the shop floor workers. What you see can be controlled much better than using mere numbers.

Planners at Preter now feel confident about the actual situation within the company. Compared to the uncertainty in past decisions, planners now see whether a job should be started or not with the help of the simple traffic lights as described above. This is reinforced by the fact that the planners have found the starting dates from *timeaxx* to be very reliable for achieving the promised delivery dates, thus increasing their trust in the traffic light signals. Furthermore, planners are able to keep track of the operations being executed and the status of work orders, so they can quickly respond to queries from customers.

Even the most experienced shop floor employees have accepted the new way of working. On their tablet devices running *timeaxx*, they can see in real-time which order needs to be processed using the same traffic-light methodology mentioned above.

The buy-in from the people in the company can be seen in the words of one of the planners: "The daily struggle of scheduling, rescheduling and frequently expediting jobs has finally come to an end. Even if the schedule is pretty tight, the system now provides the signals and the transparency to help our people understand what needs to be done and to adapt their operations and activities in order to meet expectations."

A final indicator for the success of our implementation is one that should not be forgotten: top management at Preter is very satisfied with what has been achieved.

About the Author

Markus Menner holds a degree in business administration in addition to his education as a specialist in computer science. He began his career as a software architect, development manager, and consultant in the area of ERP systems. He has worked as an R&D Manager at the Laboratory for OLYMPUS, as well as for Beckman Coulter, both global players in the medical device industry. In 2013, he founded axxelia with the focus on Quick Response Manufacturing (QRM) and its rollout in German-speaking markets. At the same time, he designed and developed the QRM and POLCA-based software system *timeaxx*, which is now in use by many customers. These customers include large, medium-sized, and small companies that use a variety of ERP-Systems including SAP, Microsoft Dynamics, SAGE, and Infor, together with *timeaxx*.

Introduction to the Appendices

The final part of this book consists of nine appendices, which fall into two categories: (1) those that describe concepts that are complementary to POLCA, in the sense that they have supported the implementation of POLCA at several companies; and (2) appendices with more details on particular topics related to POLCA. These appendices will be of interest to academics and researchers looking for more details on some aspects, as well as to practitioners or consultants that have a deeper interest in some of these topics. Four of the appendices have been contributed by guest authors who are experts in their fields, as described below.

We start in Appendix A with a concept called MCT because in many of the case studies in Part III companies have used MCT as one of their metrics for judging the progress of their POLCA implementation. MCT stands for *Manufacturing Critical-path Time*, and it is a formal approach to measuring lead time. Since lead time is one of the important metrics for judging the success of POLCA, this first Appendix gives a short tutorial on MCT for interested readers. In a similar vein, several of the case studies involve companies that put in POLCA as part of their overall implementation of *Quick Response Manufacturing (QRM)* strategy. Again, for readers interested in some background on this strategy, Appendix B provides an overview of QRM. For managers or other readers who need more convincing as to why a new system such as POLCA is necessary, Appendix C goes over some existing systems. In particular, it discusses traditional Material Requirements Planning (MRP) as well as a popular alternative today, which is the card-based Kanban system, which originated with Toyota and is a key component of Lean Manufacturing. This appendix explains the issues that arise with using such systems in the high-mix, low-volume and custom (HMLVC) environment. Several other card-based systems have also been suggested as alternatives to Kanban; many of these are rather academic, but two that are more practically oriented are CONWIP and COBACABANA, and this appendix also provides a perspective on them relative to POLCA. Appendix D includes a novel contribution from guest authors about the concept of *Capacity Clusters*. Companies have seen the benefits of implementing cells and simplifying their MRP systems to work at a higher level, but this raises the question of how to perform effective capacity planning without looking at individual operations in detail. A Capacity Cluster provides such a solution, particularly effective for the HMLVC environment, and we are pleased to state that this idea appears for the first time in archival form in this book.

As you read through the details of POLCA, you will see throughout this book that a core strength of the system is that it gets a strong buy-in from people on the front line. In Appendix E, another guest author, a recognized expert on sociotechnical systems, explains using formal sociotechnical theory why POLCA is appealing to people on the shop floor. Continuing with the theme of involving people, POLCA simulation games have been instrumental in showing people how POLCA works and also training employees at companies that have implemented POLCA. Appendix F provides a description of a typical POLCA game used by one training organization along with their experiences in using it with various audiences; this appendix is also written by a guest author, one of the POLCA experts at the training organization.

The next two appendices get into more technical and academic details. Appendix G provides a mathematical explanation of the formula for calculating the number of cards in a POLCA loop. Next, recognizing that there is extensive literature on production planning and control, and that POLCA is just one more contribution in a long history of this field, we felt that this book needed a section that more rigorously placed POLCA in the context of this large body of literature and also provided references to the academic literature on POLCA. We are pleased to have Appendix H contributed by a guest author who is an international expert in production planning and control systems and has also made significant contributions to the research literature on POLCA. Finally, Appendix I provides a guide to resources around the world on both POLCA training and POLCA games.

Appendix A: Introduction to MCT: A Unified Metric for Lead Time

In today's world, lead time is critical: offering products or services with short lead times provides competitive advantage in most industries. Hence, many improvement methods aim to reduce lead time as part of their overall goals. More important is the fact that, properly measured, lead time can be a strong indicator of total enterprise-wide waste. Thus, lead time reduction can be a powerful approach to reducing organizational waste. In addition, reducing lead time results in improvement of key performance metrics including cost, quality, and on-time delivery.

MCT (the acronym is explained in the next section) is a time-based metric that defines lead time in a precise way so that it properly quantifies an organization's total system-wide waste. At the same time, when suitably used, MCT does not need to be data-intensive and can be relatively easy to apply. Thus, MCT provides a simple yet powerful metric for measuring improvement. This has been proven in practice, with hundreds of industry projects spanning over a decade (see "For Further Reading").

Applying MCT in Both Manufacturing and Non-Manufacturing Contexts

Historically, the acronym MCT comes from the phrase "**M**anufacturing **C**ritical-path **T**ime." Although in this book we are concerned with MCT only in the context of manufacturing, in actual fact the scope of MCT extends beyond manufacturing. MCT has been used in several contexts, including in office operations such as quoting and engineering; in new product introduction; in supply chains; in healthcare operations; and in insurance companies.

Thus, we have found it best to just use the acronym MCT (pronounced "em-see-tee") as a metric. For example, in looking at the process for making quotations, we can use the term "MCT-quoting" to label the metric for the time to make the quotations. Similarly, we could use "MCT-hiring" to denote the metric for the time it takes to hire new employees, and so on.

Need for MCT Definition

Why is a precise definition needed for a seemingly obvious concept such as lead time? There are *three* main reasons:

1. *"Lead time" can mean different things.* Consider a supplier of bearings that stocks its standard bearings in distribution centers (DCs) around the U.S. The lead time for this supplier could be either (i) the lead time seen by U.S. customers, typically three days for the bearings to be picked from the DC and arrive at the customer's site; or (ii) the lead time to make a fresh batch of bearings, typically 85 days from starting the first process to arrival at the DC. Now suppose the supplier receives an order for *custom* bearings (not stocked at a DC). There are two more possible lead times: (iii) the lead time quoted by the sales department; and (iv) the lead time actually achieved by the supplier in delivering these custom bearings. We could continue with examples of other "lead times." There is clearly need for a standard, because you can't have a goal of reducing something until you can agree on how to measure it.
2. *Traditional definitions of lead time focus only on the result.* A typical definition of lead time is: "The time from when an order is transmitted by a customer until the order is received by that customer"—but this focuses only on a *result.* It does not give any indication of *how* order fulfillment is achieved and specifically does not capture waste in the process of how that lead time is achieved.
3. *MCT quantifies enterprise-wide waste.* Other improvement methods identify individual instances of waste, such as: producing defective parts, or a worker needing to walk around the factory to find a missing tool. However, such wastes measure the *micro* impact of operational problems—they do not give insight into *macro* system-wide waste.

As will be shown in this appendix, MCT does indeed provide insight into enterprise-wide waste. Later sections will also show that MCT extends naturally to an organization's supply chain, and thus also provides insight into waste in the extended enterprise.

In summary, MCT provides management with a precisely defined metric that gives a unified measure of system-wide waste in a single number. Thus, examining the before-and-after MCT values for an improvement project makes it clear whether improvements have been achieved.

MCT Definition and Application

This is the central concept in this appendix, so it is highlighted here.

> **MCT** The typical amount of calendar time from when a customer submits an order, through the critical path, until the first end-item of that order is delivered to the customer.

Key terms in this definition are now explained briefly.

typical amount The purpose of MCT is to highlight the biggest opportunities for improvement. For effective use of MCT, don't get bogged down with trying to determine details; limit the scope of the MCT analysis and keep data gathering simple.

calendar time Customers view delivery in terms of a specific calendar date, and don't care which days your organization is or is not working. MCT measures in *real time*, not *working time*, and keeps the focus on the calendar date.

customer submits an order This should be when the clock starts as far as the customer is concerned.

through the critical path This requires the following rules to be observed:

1. All necessary activities must be completed "from scratch"—for example, if components need to be fabricated, then pre-built stocks of these items cannot be used to reduce the MCT value; you must include the time to make the components.

2. Include all the normal queuing, waiting, and move delays that jobs incur—do not use values for rush (hot) jobs.
3. Time spent by material at any stage, including all inventory holding points, must be *added* into the MCT value.
4. If there are multiple paths involved, e.g., where components must be completed, or sub-tasks must be performed before a main task, then MCT is the value for the *longest path* from start to finish.

the first end-item Even though the customer might have ordered a batch of parts, this sharpens the focus on delivery of the first end-item. Here, end-item should be interpreted as the first usable set of items for the customer. Example: the customer needs to mount a support strut on a machine, and the supplier provides the strut and two mounting brackets for this operation. For the customer to complete one assembly operation they must receive one strut and two mounting brackets. Hence, this set is the "first usable set of items."

is delivered MCT ends when the order is delivered to the customer's point of receipt. This can be interpreted as needed for different business contexts, as shown below. In the specific case where you are analyzing the supply of components to a customer, MCT must include all logistics times (it is important to quantify the impact that logistics time has on a firm's ability to respond to customers).

Detailed explanations of all the above concepts along with calculation rules and tips for calculation can be found in the *MCT Quick Reference Guide* (see "For Further Reading").

Applying MCT in Various Business Contexts

MCT is not just a metric for manufacturing and part-production. The words "customer," "order," and "delivered" can be interpreted freely for non-manufacturing situations. For example:

- Quotation process: "order" can be a request for quotation.
- Engineering change process: "customer submits an order" can be the sales department requesting an engineering improvement, and "is delivered" can be when the improvement has been incorporated into the first shipped product.

- Employee hiring process: "customer" can be a manager who needs an employee; "order" can be the manager's request to Human Resources for hiring the employee; and "is delivered" can be when the employee arrives for his/her first day at work.

Preview of MCT Map, Gray Space, and White Space

An MCT analysis is typically conducted for a portion of an enterprise—a subset that has been targeted for improvement. The initial goal is to produce an *MCT Map* for that subset and to calculate the corresponding value of MCT. It will be shown that the MCT Map provides a high-level picture of opportunities for improvement, and the MCT (value) provides a benchmark to gauge subsequent improvements. This section defines some terminology related to MCT Maps.

MCT Map This is a graphic representation of the flow of an order through the specified subset of the organization. A simple example illustrates an MCT Map and the corresponding MCT.

Figure A.1 shows an MCT Map for an order from receipt at a company until the order is loaded onto a truck for shipping. The flow in Figure A.1 is from left to right and the representation is intuitive, as seen in the figure. In this example, sales activities and processes prior to receipt of order are not shown; also not shown are any processes used in shipping and logistics after the order leaves the company. This illustrates that the scope of an MCT Map is typically limited to a subset of an enterprise, as required for the goals of a specific project. In Figure A.1, the MCT for this subset is 14 days (= 2 + 9 + 3).

Gray Space This illustrates the total time when *something is actually happening to an order*: for example, someone is working on it, or it is being machined, or it is in an oven. This total time is shown by the

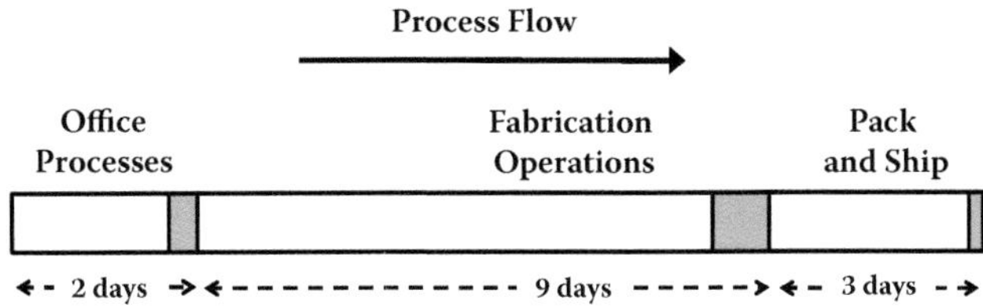

Figure A.1 Example of an MCT Map.

rectangles with the shaded gray. These are customarily placed at the *end* of the operations for which this time occurs. For instance, the order spends nine days in Fabrication Operations, and the Gray Space can be seen at the end of this nine-day segment. Note: while Gray Space might resemble the term "touch time" used in other approaches, it is *not* the same! There are specific rules for calculating the amount of the Gray Space: more details can be found in the *MCT Quick Reference Guide*.

White Space This illustrates the remaining time spent by the order in a particular area; this is the time when *nothing is happening to the order*! This is customarily placed before the Gray Space for the same area. Looking at the Fabrication Operations again, you can see the White Space preceding the Gray Space.

Insights from MCT Maps

While an MCT Map seems simple, it is a powerful tool that can provide many insights.

- The MCT Map is drawn roughly to scale, so that the magnitude of the various elements is apparent. This provides a visual overview—you can quickly see where the greatest opportunities lie.
- For initial insights, times within various segments can be aggregated. In Figure A.1, say the Fabrication operations include three processes: turning, milling, and drilling. In reality, each process has its own White Space and Gray Space. These details may not be readily available, but the total time for an order to go through Fabrication may be known roughly. So, in Figure A.1 we add together the working times for the three operations to get the total Gray Space in Fabrication, and the remainder of the time in Fabrication is aggregated into one White Space. This approach is acceptable for initial analysis, following which it can be decided if more detailed data are needed in any area.
- Figure A.1 illustrates the typical situation, in that Gray Space (working time) is usually a very small proportion of total MCT.

Experience with hundreds of projects in manufacturing companies has shown that Gray Space is typically less than 5% of MCT. In some instances, particularly with companies making large batches of products, the Gray Space can even be less than 1%. In other words, only during a very small amount of a job's total MCT is someone actually working on that job! Projects in other industries, such as insurance and healthcare, have shown that the Gray Space is often less than 20%—and many times less than 10%—in other types of businesses as well.

Thus, right off the bat, an MCT Map usually provides an important insight to the management at a company. To see this, consider that traditional cost-reduction or efficiency-improvement approaches focus on reducing the working time for processing jobs. Since the Gray Space is a tiny fraction of the MCT, *these traditional approaches typically have limited impact on lead time.* The MCT Map therefore shows that management needs to think outside the box of traditional cost/efficiency approaches if it wants to significantly reduce lead time. (One such approach that targets lead time reduction using methods that are different from the traditional cost/efficiency approaches is Quick Response Manufacturing or QRM; see Appendix B.)

Not the Same as a Gantt Chart or PERT/CPM Analysis

Traditionally, stocks of raw material or pre-built components are used to *reduce* lead time to the customer, so a point that confuses people is why time spent in inventory *increases* the MCT value. Another confusion arises because one of the MCT rules states that components need to be built "from scratch." So, if we are building those components, why do we need to add the time sitting in inventory? A third misunderstanding is that people familiar with project management methods see the words "critical path" in MCT and assume that it is a variation of the Critical Path Method (CPM) or the related method of PERT (Program Evaluation and Review Technique). A final misconception that occurs is when seasoned Industrial Engineers take a cursory look at an MCT Map and say, "This is the same as a Gantt Chart, so what's new here?"

The answer to all these lies in the fact that *MCT is not a metric for physical lead time to a customer.* Since CPM, PERT, and the Gantt Chart are all tools used for scheduling or project management, they have to work with actual, physical lead times. MCT is indeed a time-based metric, but it is a metric that quantifies the total system-wide waste and displays all the opportunities for improvement. Everyone knows that inventory consumes capital and occupies space, but as shown in the next section, "The Business Case for MCT" the waste due to MCT is much more and involves many overhead activities.

An example shows how both the above points (stock and "from scratch") *add together* in a worst-case situation. Suppose your customer's products are failing in the field and the root cause is found to be a component supplied by you. Analysis by your engineers shows that this component cannot be reworked; it has to be made with a new design. Now, all existing stocks will need to be scrapped. Plus, you will have to go through all the fabrication processes "from scratch." Hence, by adding together the values of the stock and the "from scratch" times, the MCT metric quantifies the full magnitude of this entire effort.

Note that since there is no work being done on parts while they are sitting in any type of inventory, the time spent in inventory is always classified as White Space.

The Business Case for MCT

While everyone has heard that "time is money," time (as measured by the MCT metric) is actually *a lot more money* than most managers realize! Long MCTs add layer upon layer of overhead and indirect costs, much more than managers anticipate. Industry projects have shown that reducing MCTs results not only in quick response to customers, but also in lower costs and improvements in other metrics.

This happens because MCT is a strong indicator of enterprise-wide waste. Consider an extreme situation: if all the MCTs in your organization were reduced by 90%, what are all the activities, resources, and systems that are in place today that could be reduced, simplified or eliminated altogether? Also, what are the new opportunities that might be available to your business? Stop here for a moment and make a list of answers to these two questions for your organization; then read further.

Activities/resources that could be reduced or eliminated include planning, forecasting, warehousing, inventories, expediting of late jobs, rescheduling, and more. If these items could be reduced or even eliminated if your MCTs were shorter, they represent "waste" in your enterprise because of long MCTs. New opportunities from short MCTs could include increasing sales through rapid delivery of existing products, and gaining market share through rapid introduction of new products. Again, these opportunities are being missed (or "wasted") today, because of long MCTs.

Does reducing MCT really help to trim down these "wastes"? Experience with hundreds of projects has shown that the waste reduction is substantial, and the resulting improvement in performance metrics is very significant. These results have been documented in conference presentations, industry publications, and books. Following are some examples (see "For Further Reading" for additional references).

Impact on On-Time Performance

Long MCTs have a dysfunctional impact on the accuracy of planning, forecasting, and scheduling, resulting in poor delivery performance. MCT reduction can dramatically improve on-time delivery performance.

A supplier of hydraulic valves reduced its MCT by 93%, and its on-time delivery improved from 40% to 98%. A manufacturer of wiring harnesses reduced MCT by 94% and its on-time delivery rose from 43% to 99%. (These examples also illustrate the huge opportunity for MCT reduction that exists in organizations—over 90%!—because most of the MCT consists of White Space.)

Impact on Quality

A long MCT hides numerous quality problems throughout your enterprise and supply chain. Conversely, as MCT is reduced, quality issues are discovered quickly, root causes can be found, and improvements put in place.

The above supplier of hydraulic valves, after reducing MCT by 93%, witnessed reductions in quality rejects from 5% to 0.15%. A company making seat assemblies reduced MCT by 80% and saw its rework rate plummet from 5.0 to 0.05%. These are not incremental improvements; they represent 30-fold to 100-fold reductions in defect rate!

Impact on Cost

As enterprise-wide waste is eliminated, this impacts overall operating costs. Less time is spent on planning, scheduling, expediting, and so on. Fewer resources are involved in warehousing and material handling. There is less investment in inventory, space, and possible obsolescence of materials. As these results cumulate over time, companies see significant reductions in cost. (Here "cost" refers to total product cost, including overhead.) The wiring harness manufacturer documented a 20% reduction in product cost, and the seat assemblies company experienced a 16% reduction.

National Oilwell Varco (NOV) makes oil drilling equipment, with annual sales of over $20 billion. At its factory in Orange, California, NOV reduced the MCT of a product line from 75 to 4 days. When the project was evaluated, management found that the cost of that product line had also been slashed by 30%!

Note that traditional accounting methods do not adequately take into consideration the cost impact of MCT, which can therefore impede the financial justification and hence implementation of strategies to reduce MCT. However, a recent book, *The Monetary Value of Time*, provides a novel framework for assessing the value of MCT in terms of organizational strategy and competitive advantage (see "For Further Reading"). This framework enables companies to develop MCT-based metrics and accounting methods to ensure that their cost accounting and financial justification approaches support MCT reduction efforts.

Impact on Profitability

Nicolet Plastics (Mountain, Wis.) decided to focus on MCT reduction as a key strategy. The results, in just three years, were nothing short of amazing. According to Joyce Warnacut, Chief Financial Officer, "The most impressive impact … has been on Nicolet Plastics' earnings before interest and taxes (EBIT). The EBIT for 2012 alone was roughly equal to the EBIT for the 10-year period from 2000 through 2009 combined!"

Impact on Productivity and Market Share

RenewAire in Madison, Wisconsin, makes Energy Recovery Ventilation Systems. Chuck Gates, President, decided to focus on MCT reduction, and with his team he reduced MCT by over 80%. Even though RenewAire, a small company, was competing with international giants, it was able to increase its market share by 42% over five years. At the same time, the company significantly improved its productivity; while its revenue grew by 140%, it required only a 73% increase in total employees for this growth in sales.

Phoenix Products Company, Milwaukee, Wisconsin, manufactures industrial lighting for applications that include lighting of mines, shipyards, and monuments. Phoenix was struggling with long lead times, late deliveries, and rising costs, all of which provided opportunities for competitors from low-cost countries. In 2004, Scott Fredrick, CEO, decided to focus on MCT reduction. By 2013, MCTs across all product lines had been reduced by 50% along with impressive improvements in other metrics: compared with 2004, by 2013 Phoenix had achieved a 70% increase in revenue per labor hour, and a 30% reduction in overhead. During the same period Phoenix gained substantial market share: its sales grew at an average annual rate of 12.4%, versus 2% for the industry as a whole.

Impact on Space and Office Productivity

Elimination of wasted resources and activities can result not only in reduced costs but also increased output with the same resources. Alexandria Industries in Alexandria, Minnesota, provides custom aluminum extrusions. In 2002, the President, Tom Schabel, decided to focus on MCT reduction. By 2012, Alexandria Industries had reduced its MCT for extrusions by 83%, from six weeks to five days, and realized a 58% increase in revenue per square foot.

As mentioned in several places in this appendix, MCT is not just a metric for parts production and shop floor operations; it can be applied to other business processes. Alexandria Industries reduced MCT in the office by 50% to 75% for processes such as estimating, quoting, and order processing, and over the same time period it experienced a 62% increase in business handled per office employee in those areas.

MCT Maps Help Visualize Data for Complex Operations

An MCT map is a powerful way to communicate insights from an MCT analysis. The MCT map can also help with getting buy-in from management and employees on improvement opportunities.

We now show how to make an MCT Map for a more complex operation, along with the insights that this map provides. Madison Ventilation Products (MVP), a hypothetical company, is engaging in improvement projects. For one of the projects, the improvement team decides to focus on the product line "MVP Industrial Fans," including the main fabrication/assembly operations and the first-level supply chain for three purchased items. Following are the results of data gathering. Refer to Figure A.2, starting at the top-right corner. Incoming orders are processed through Office Operations, which take an average of nine days, of which one day is working time. We will use the abbreviation [9(1) days] from here on, to denote [total (working) days]. (All data have been converted to calendar days here). The supplied components are Forgings [36(3) days], Sheet Metal [9(1) days], and Motors [29(3) days]. Incoming material for these items is stored in a warehouse (WH) for 13, 4, and 11 days, respectively. In-house operations involve fabricating Fan Blades [14(1) days] and Enclosures [9(1) days]. These are placed in a staging area (Stg.) where they wait [3 and 2 days], before going through Assembly [3(0.5) days]. Completed Fans are

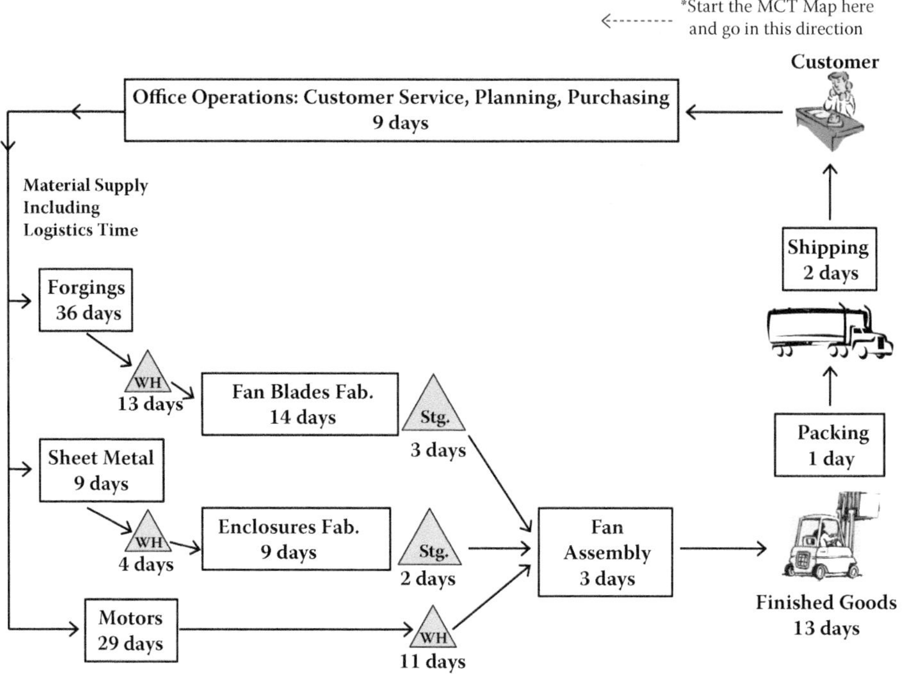

Figure A.2 Process map for MVP industrial fans.

stored in a Finished Goods area [13 days], then Packed [1(0.25) day] and Shipped [2(2) days] to the Customer.

Figure A.3 shows the MCT Map for MVP Industrial Fans. To understand the MCT Map, start with Figure A.2 and trace the three paths going counter-clockwise from the Customer and back to the Customer. Next, trace each of these paths starting from the left side of the MCT Map in Figure A.3 and follow each of the bars for the components. Note that Office Operations are common to each path and are repeated in each bar. The vertical line indicates that these components are supplied to Assembly (it helps visualize the bill-of-materials). On each path, working times are shown as Gray Space and the remaining times as White Space. The MCT Metric for MVP Industrial Fans is derived from the longest path and equals 94 days. To see this, trace these numbers in Figure A.2 for the path with Forgings: 9 + 36 + 13 + 14 + 3 + 3 + 13 + 1 + 2 = 94. As explained earlier,

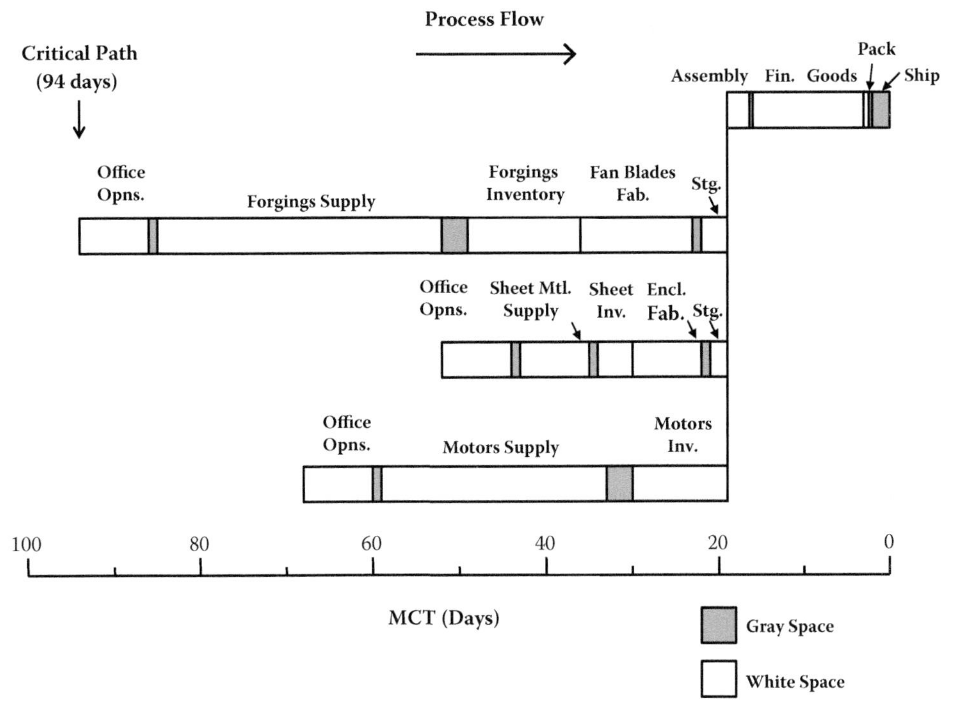

Figure A.3 MCT map for MVP industrial fans.

the MCT Map does not need to cover details for the whole company, it can be used to look at a subset of the operations; here, this map is for MVP Industrial Fans only.

Comparison with Value Stream Mapping (VSM)

Value Stream Mapping (VSM) is a popular tool used for improvement projects. However, in the initial project stages, MCT Mapping has several advantages over VSM:

- An MCT Map is simpler, and provides an easy-to-see, high-level view of the operation: compare the example of a Value Stream Map in Figure A.4 with the MCT Map in Figure A.3.
- An MCT Map represents time proportionally. In the VSM in Figure A.4 it is not easy to pick out the largest segments of time, while in the MCT Map these segments jump out at you. So, the MCT Map helps to target the largest opportunities (White Space).

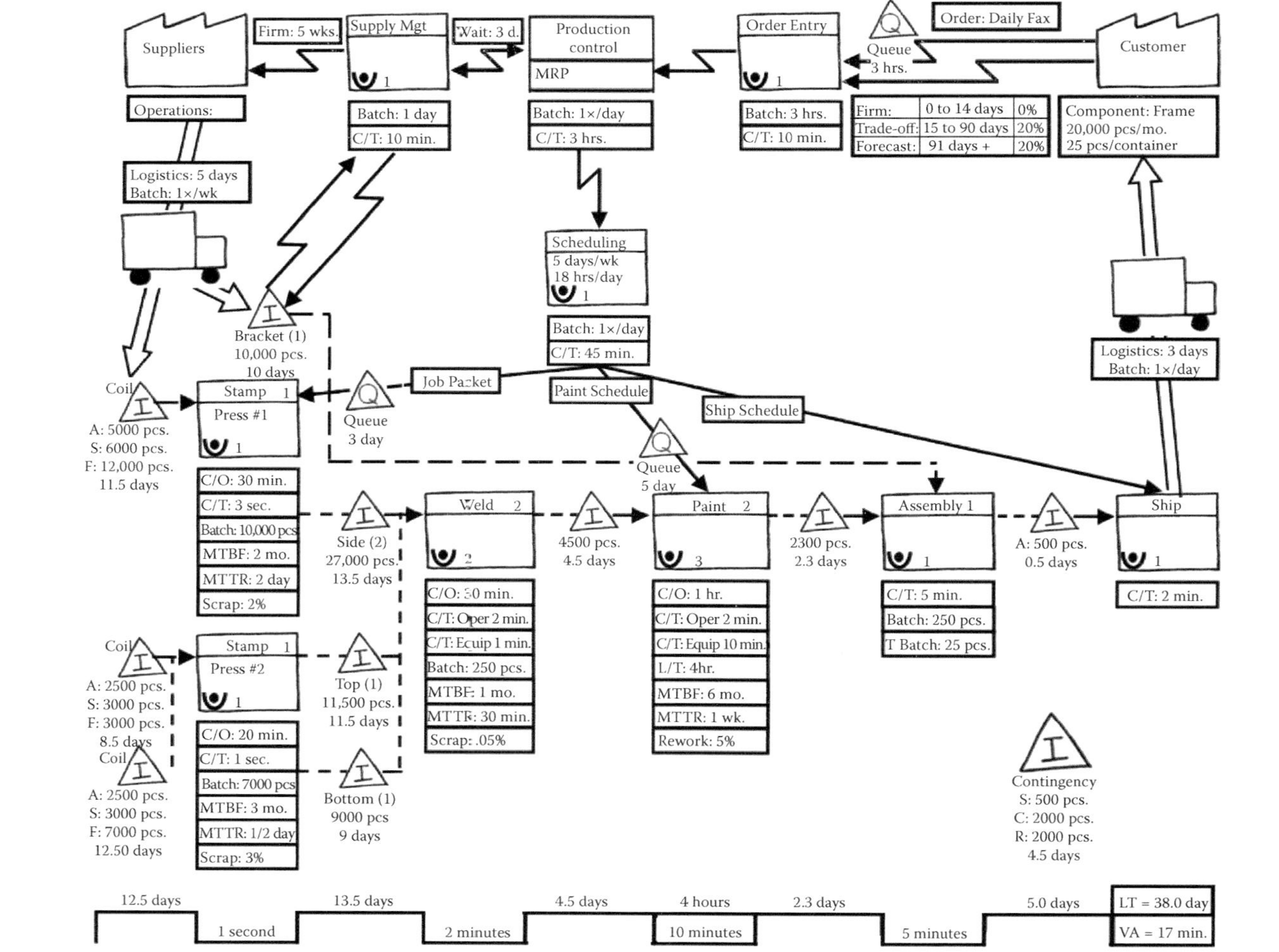

Figure A.4 Example of Value Stream Map (VSM).

- An MCT Map visually highlights the large amounts of White Space versus the small amounts of Gray Space, emphasizing the need to focus on the White Space.
- An MCT Map identifies the critical path, ensuring that improvements are targeted at processes that will make a significant difference. Note: the scale (X-axis) goes from right to left for a reason: you can easily read off the lengths of the other paths to know when another path might become critical.

Note, however, that the two approaches are not contradictory; in fact, they can be complementary. You should begin with MCT Mapping because it shows you the forest before you look at the trees. But once you determine that you need to focus on an area, you could develop a detailed VSM for that area. Or, if you already have a VSM, you can convert the data to quickly make an MCT Map (see "For Further Reading"); this will help you get insights into where to focus your project.

Using MCT in Supply Management and in Other Industries

Modern approaches to supply management emphasize not only supplier performance but also supplier improvement, so it is important to have metrics that support both these aspects. Surprisingly, the traditional metrics are not as effective as one might think! When added as a metric for suppliers, MCT can complement the long-standing trinity of supply management metrics of Quality, Cost, and Delivery (QCD), and assist with strategic supply management initiatives (see "For Further Reading"). Suppliers with long MCTs should be encouraged to engage in activities to reduce their MCT. These MCT reductions will result in higher quality, lower cost, and improved on-time delivery; all of these will benefit the customer as well.

MCT should also be a factor in sourcing decisions. A supplier with short MCT should be given some preference over one with a long MCT, even if the latter has lower cost. This is because a supplier with long MCT results in a number of hidden costs, both in its own operation as well as in the customer's operation. Since MCT includes logistics time, this also implies giving preference to local suppliers over those that are half-way around the world.

Although MCT originated in manufacturing, it can be applied in other industries. MCT Maps have been used for insights into Healthcare operations, for example, Patient flow through an Emergency Room (ER), and in the Insurance Industry, for example, to illustrate the flow for Claims Processing.

Motivating Teams to Drive MCT Reduction

Earlier in this appendix, we described how reducing MCT has resulted in manifold benefits at companies. Hence, as part of their continuous improvement programs, organizations should encourage their teams to reduce MCT. An MCT-based metric that helps motivate MCT reduction is the QRM Number, defined as follows:

$$QRM\ Number = 100 \times (Base\ Period\ MCT / Current\ Period\ MCT)$$

The two quantities on the right side of the equation denote the average MCT measured over an initial (base) period and the average MCT achieved during the current performance period. (The acronym QRM comes from Quick Response Manufacturing; see Appendix B.)

Using the QRM Number has several advantages over using the raw MCT values for teams, as illustrated by an example. Suppose an improvement team starts with an average MCT of 12 days. During the next three evaluation periods it reduces this to 10 days, then 8 days, and finally to 7 days. Table A.1 shows the values of both MCT and the QRM Number, along with the calculation details.

The advantages of using the QRM Number over the "raw" MCT values for the purpose of motivating teams are:

1. *Reducing MCT results in an increasing QRM Number*. Figure A.5 graphs a team's progress over time as measured by both raw MCT and the QRM Number. Improvements by the team result in the MCT curve going down, but the QRM Number curve going up. Literature on motivation shows that people react more favorably to a graph going up than to one going down. Thus, the QRM Number graph better motivates the desired behavior (MCT reduction).
2. *Equal reductions of lead time in the future result in larger increases in the QRM Number*. The team reduced MCT by two days in the second period, and again by two days in the third period (equal improvements in MCT). However, the team was rewarded by a 20 "point" increase in

Table A.1 QRM Number Calculation

Period	*MCT*	*Calculation*	*QRM Number*
1	12 days	(12/12) × 100	100
2	10 days	(12/10) × 100	120
3	8 days	(12/8) × 100	150
4	7 days	(12/7) × 100	171

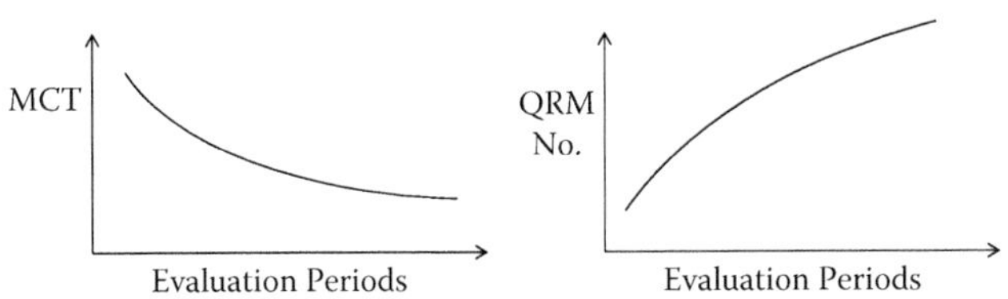

Figure A.5 Graphs of MCT and QRM Number for a team.

QRM Number in the second period but 30 points in the third period. This makes sense because (i) proportionally, there is a larger reduction in the third period, but also (ii) at first it is easy for teams to find a lot of "low-hanging fruit," but then it gets harder to find ideas for improvement. This remark leads to the next advantage.

3. *As MCT reduction gets increasingly difficult, the QRM Number continues to motivate teams.* Figure A.5 shows that while the MCT graph starts to flatten out, the QRM Number graph continues to rise significantly, thus motivating the team to keep improving. Table A.1 shows that although the team got 20 points for the second-period improvement of two days, it actually gets more points (21) for the fourth-period improvement of just one day!
4. *The QRM Number provides a single measure that can be used throughout the organization regardless of the type of work.* For example, if an organization has improvement teams working in various office operations as well as several shop floor areas, the progress of all the teams can be easily compared by tracking their QRM Numbers.

The QRM Number can also be used in Supply Management initiatives as described in the previous section. Using the QRM Number for suppliers helps to support supplier improvement projects using a time-based metric.

Summary: MCT Map and Metric Provide Clear Goal

Improvement projects can lose focus because they have multiple goals such as targets for cost, quality, efficiency, on-time delivery, and so on. These goals might conflict, and thus drive teams in different directions. In contrast, the MCT Map and MCT Metric supply a clear goal for a project: the MCT Metric provides a single number, and the goal is simply to reduce this number to a target derived from opportunities seen in the MCT Map. The justification for using this simple and unified goal was provided in the earlier section on "The Business Case for MCT"—namely, when MCT is reduced, key performance metrics improve.

For Further Reading

As background to the additional readings below, I would like to acknowledge the contributions of Paul Ericksen, who provided the original insight that resulted in the MCT metric. Paul worked with this author and the author's students at the University of Wisconsin to lay the foundations for MCT and prove its use. Numerous students, professionals, and companies were involved with applying MCT and providing practical experiences that helped sharpen its definition and use. The following two-part series provides a perspective on the origins of MCT, how it proved its worth at a Fortune 500 company, and results from hundreds of projects.

- "Lean's Trinity," by P.D. Ericksen, *Industrial Engineer* (Part I, October 2013, pages 39–43; Part II, November 2013, pages 29–32).

In the context of global supply chains and overseas sourcing, the next article explains the need for shifting to time-based supply management and the importance of using MCT as the key metric in this effort.

- "Filling the Gap: Rethinking Supply Management in the Age of Global Sourcing and Lean," by P.D. Ericksen, R. Suri, B. El-Jawhari, and A.J. Armstrong, *APICS—The Performance Advantage*, February 2005.

This booklet is a brief but comprehensive reference for practitioners on how to calculate and use MCT:

- *MCT Quick Reference Guide*, by R. Suri, C&M Printing, 2014.

The next report justifies the use of MCT as a robust, unifying, and enterprise-wide metric to support order fulfillment and drive improvement projects. It contains rules for correctly converting Value Stream Maps to MCT Maps.

- "Manufacturing Critical-Path Time [MCT]: The Enterprisewide Metric to Support Order Fulfillment and Drive Continuous Improvement," by N.J. Stoflet and R. Suri. Technical Report, Center for Quick Response Manufacturing, University of Wisconsin–Madison, 2007.

The following book provides a novel framework for assessing the value of MCT in terms of organizational strategy and competitive advantage. This framework enables companies to develop MCT-based metrics and accounting methods to ensure that their cost accounting and financial justification approaches support MCT reduction efforts.

- *The Monetary Value of Time: Why Traditional Accounting Systems Make Customers Wait*, by J.I. Warnacut, Productivity Press, 2016.

Appendix B: Overview of Quick Response Manufacturing (QRM)

As explained in the introductory chapters of this book, in today's world, companies are increasingly seeing demand for smaller batches, higher variety of products, and even customized products that are tailored to each order. It is clear that with the application of computer-aided design and manufacturing, this trend—commonly referred to as "mass customization"—is only going to get more pronounced. In this book, we have used the acronym HMLVC (high-mix, low-volume and custom) for this production environment. In addition to this trend, customers—both Original Equipment Manufacturers (OEMs) and end consumers—are expecting shorter and shorter delivery times for these lower-volume or customized products.

These considerations led to the development of Quick Response Manufacturing (QRM), a companywide strategy for lead time reduction throughout the enterprise. Using QRM, companies with HMLVC production have been able to reduce their lead times by 80–90%. As a result, these companies have not only seen large increases in market share, but also experienced 15–20% cost reduction and huge improvements in quality and on-time delivery. Companies have also found that the lead time and cost reductions through QRM have enabled them to compete effectively against low-cost countries.

What if you have already invested in other strategies such as Six Sigma or Kaizen? Adopting QRM does not require you to back away from them; QRM builds on these strategies and unifies them under one overarching goal—reducing lead time. If you are already implementing Lean, again QRM will enhance your Lean program and take it to the next level. The origins of Lean are in high-volume, repetitive production, and the core tools in Lean such as Takt times, standard work, and level scheduling are designed to eliminate variability in operations. Similarly, the Six Sigma tools are aimed at reducing variability. However, eliminating variability may not be the right strategy for HMLVC companies. To make this clear, QRM defines two types of variability:

- **Dysfunctional variability** caused by errors and poor systems. Examples are: rework; constantly changing priorities; and "lumpy" demand due to poor interfaces between sales and customers.
- **Strategic variability** introduced by a company to maintain its competitive edge. Examples are: serving markets with highly unpredictable demand; offering customers a large variety of options; and offering custom-engineered products.

The core Lean and Six Sigma techniques aim to eliminate *all* variability in the manufacturing system. The QRM approach is aligned with Lean and Six Sigma in getting rid of dysfunctional variability. However, you may not want to eliminate strategic variability if it is the basis of your competitive advantage—in fact, HMLVC companies clearly incorporate a high degree of strategic variability. So, in QRM you do not *eliminate* strategic variability, instead you *exploit* it! This is done by designing the QRM organization to effectively cope with this variability and still achieve quick response. QRM includes a detailed methodology and an extensive set of tools to achieve these goals. Hence QRM takes Lean and Six Sigma strategies to the next level (Figure B.1).

Challenges to Reducing Lead Time

In principle, managers understand the importance of quick response to customers. However, in practice, initial experiences in lead time-reduction projects with industry partners showed that there are many misconceptions about how to reduce lead times, which prevent successful results. This observation is supported by statistics from a quiz given to a large number of managers and documented in the book *Quick Response Manufacturing* (see "For Further Reading"). So how can a company exploit strategic variability and also succeed in reducing its lead times?

To address this question the Center for QRM was founded at the University of Wisconsin–Madison as a partnership with industry to develop

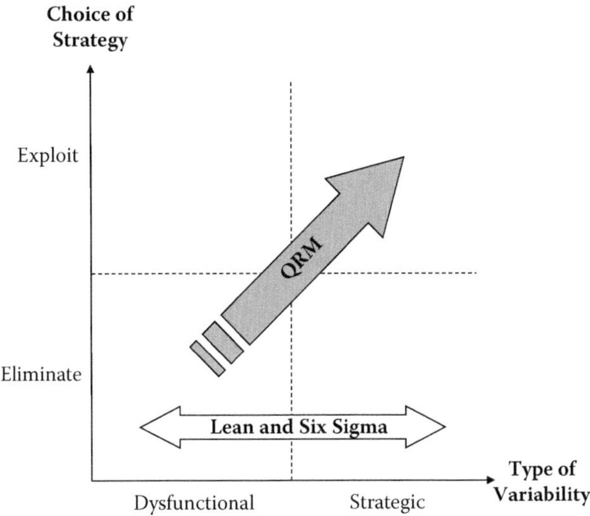

Figure B.1 QRM strategy enhances Lean and Six Sigma programs.

and implement principles for lead time reduction. Over 300 companies have worked with the Center for over two decades to help develop QRM strategy. The first detailed documentation of QRM can be found in the book *Quick Response Manufacturing* mentioned above. Since then, many companies around the world have implemented QRM and helped to further refine the QRM techniques and tools.

This appendix summarizes the core concepts of QRM. If you are interested in more details on QRM, we suggest starting your additional reading with a more recent book, *It's About Time* (see "For Further Reading"). As described in *It's About Time*, QRM is based on four core concepts:

1. Realizing the Power of Time.
2. Rethinking Organizational Structure.
3. Understanding and Exploiting System Dynamics.
4. Implementing a Unified Strategy Enterprise-Wide.

We now explain the four core concepts of QRM.

QRM Core Concept 1: Realizing the Power of Time

Everyone knows that time is money, but time is actually a lot more money than most managers realize! Chuck Gates, President of RenewAire, came to this realization after attending a QRM workshop. Then, using QRM principles, he reduced his product lead times by over 80%. As a result, RenewAire, a Madison, Wisconsin manufacturer of customized Energy Recovery Ventilation Systems, gobbled up market share; this tiny company competing with industry giants multiplied its revenue by 2.4 in five years. At the same time, the company significantly improved its productivity, requiring only a 73% growth in total employees for this 140% increase in sales.

These numbers highlight the point that as companies reduce their lead times, they also see significant reductions in unit costs of their products, often 25% or more. This counters a concern for companies in the United States and other developed countries: employees live in fear of their operation being outsourced to low-cost countries such as China. But the fact is, for a typical product made in a developed nation, direct labor accounts for less than 10% of its cost. Moreover, in terms of the selling price of a product, the number is lower: less than 7% of the price to the customer is attributable to direct labor. Thus, if you use QRM methods to reduce cost by 25%, you wipe out the labor-cost advantage of low-cost countries. When you consider that overseas competitors need considerable lead time for shipping, your short response time makes it impossible for them to compete on the same terms. You can compete against anyone, making products anywhere.

Why is lead time more significant than managers realize? Ponder this question: What is the waste in your enterprise due to long lead times? Do this by imagining a "blue sky" situation: suppose your company's lead times were 90% shorter than they are today: what are all the activities and tasks done today that could be reduced or eliminated? What investments in materials or resources could be reduced or eliminated? (If these items could be reduced, they are truly "waste" in your enterprise—they are there only because of your long lead times.) Also, what new opportunities would be available to your company? (These are also part of the "waste" because your long lead times are resulting in wasted opportunities for your company.)

To help drive home this point, before you read on think about these questions and make a list of "waste due to long lead times" for your enterprise. Then review Figure B.2, which shows items listed by managers and employees that have attended QRM workshops.

Note: To enhance your learning, think about the questions above before looking at the details in Figure B.2.

As you review Figure B.2, you will surely see some items that you listed but also some that did not occur to you. Managers find this exercise to be

Examples of activities and costs that are incurred today but would shrink or be eliminated if lead time were reduced substantially:

- Expediting of hot jobs or late orders: requires systems, unplanned air freight, shop floor and office personnel to manage and execute the changes, even top management time to negotiate priorities between multiple hot jobs
- Production meetings required to update priorities and change targets
- Overtime costs for trying to speed up late jobs
- Time spent by Sales, Planning, Scheduling, Purchasing and other departments to develop forecasts and frequently update them
- WIP and Finished Goods holding costs and space usage
- Resources used to store and retrieve parts repeatedly during the long lead time, plus potential damage to parts due to the repeated handling
- Obsolescence of parts made to forecast and stocked but not used
- Quality problems not detected till much later, resulting in large amounts of rework or scrap
- Time to deal with delivery date and quantity changes, and with feature and scope creep (job specifications keep changing during the long lead time, constantly causing rework)
- Order cancellations or loss of sales to competition
- Sales time devoted to expediting and to explaining delays to customers
- Investment in complex computer systems and organizational systems required to manage this dynamic environment

Examples of opportunities that are lost because of long lead times:

- Opportunities to gain market share by offering shorter lead times for current products
- Opportunities to beat the competition to market and gain market share through rapid introduction of new products with improved functionality

Figure B.2 Enterprise-wide waste due to long lead times.

an eye-opener and realize there is far more waste in their enterprise due to long lead times than they initially thought.

Looking at Figure B.2, very few of these costs relate to direct labor—most of them are in the category of overhead and other indirect costs. In a typical U.S. factory, overhead accounts for 40% of the cost of goods sold (COGS), and raw materials and purchased parts account for around 50% of COGS. The remaining 10% is direct labor, as mentioned already. In addition, indirect costs such as selling, general and administrative (SG&A) expenses, and Research and Development (R&D) expenses are accounted for separately from COGS and can add another 30% on top of the total of the COGS expenses.

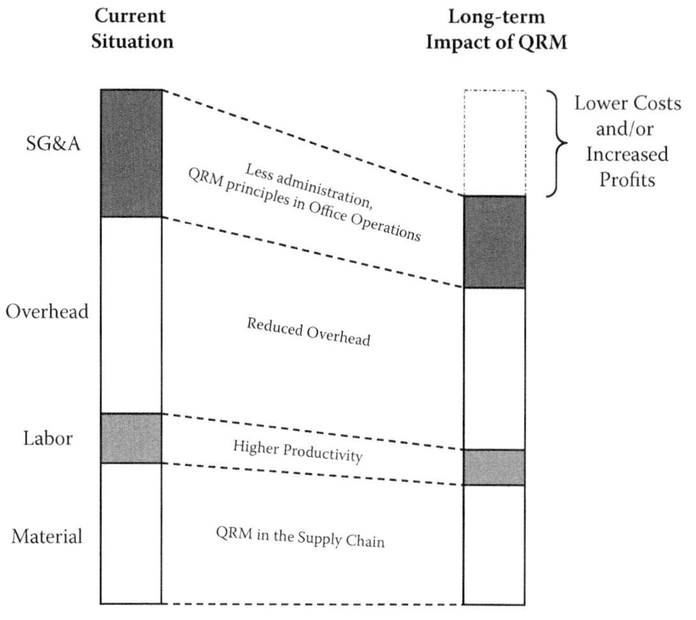

Figure B.3 Long-term impact of QRM on total costs and expenses.

For companies making low-volume and custom products, QRM has impacted all these costs significantly. Reduction of the waste in Figure B.2 has lowered both overhead and SG&A expenses. Using QRM in the supply chain has reduced material costs. The QRM organization (described in the next section) has improved both office and shop floor productivity. The net result of these has been the 25% or greater cost reduction described earlier (see Figure B.3). And the beauty is that this cost reduction does not come at the expense of other performance measures, because at the same time companies achieve lead time reductions of 80 to 90% and huge improvements in both on-time delivery and quality (see Table B.1).

Table B.1 Impact of Lead Time Reduction on Quality and On-Time Performance

Company (Product Type)	*% Reduction in Lead Time*	*% Rework/Rejects (Before → After)*	*% On-Time Performance (Before → After)*
Seat Assemblies	80	5.0 → 0.05	40 → 95
Hydraulic Valves	93	5.0 → 0.15	40 → 98
Wiring Harnesses	94	0.3 → 0.05	43 → 99

Accounting Systems Miss the Connection

Why are managers not aware of this huge impact of lead time? A key reason is that accounting systems miss the connection: they simply do not identify a link between lead time and various activities. Instead, the costs of all indirect activities go into a general overhead pool where they are comingled with other costs and disconnected from their root causes. Then this overhead pool is applied across all products. Also, SG&A and R&D expenses are separately reported and not connected with root causes. Thus, there is no easy way for the accounting system to predict the benefits of lead time reduction. And since cost systems form the basis for much decision-making, managers too miss the connection. *In actual fact, squeezing out time throughout your enterprise leads to numerous improvements in cost and other measures seen in Figure B.3 and Table B.1.*

This lack of causal understanding is compounded by the fact that companies don't do a good job of measuring lead time, especially when it comes to internal activities. QRM theory also provides a precise metric for lead time, called Manufacturing Critical-path Time (MCT). We will not spend time on this here; however, Appendix A provides an introduction to MCT. A promising development that ties together MCT and accounting systems is that a recent book, *The Monetary Value of Time,* provides a novel framework for assessing the value of MCT in terms of organizational strategy and competitive advantage (see "For Further Reading"). This framework enables companies to develop MCT-based metrics and accounting methods to ensure that their cost accounting and financial justification approaches support MCT reduction efforts.

In summary, the first core concept in QRM shows managers the enormous impact of time on their operation, and why reducing lead time can be so beneficial. To support this view, the driver in QRM is elimination of lead time (as defined precisely through the MCT metric). This contrasts with Lean where the driver is elimination of waste. The Lean view (typically involving seven types of waste) results in more of a local view of waste, while the QRM approach focusing on time encourages a global view of waste throughout the extended enterprise.

QRM Core Concept 2: Rethinking Organizational Structure

Reducing lead time requires rethinking your organizational structure. Why? An obvious reason is that you won't get 80–90% reductions by fine-tuning what you are doing today. But a deeper explanation stems from how most enterprises are organized. Figure B.4 shows the progress of an order through a Midwest manufacturing company (the data are sample averages of actual orders). A typical order spends five days in the Order Entry department, then it takes 12 days for components to be fabricated, nine days for assembly to be completed, and eight days until the order is packed and shipped—for a total lead time of 34 days within the company. Figure B.4 also shows the "touch time"—the gray space in the rectangles—when someone is actually working on the job. This accounts for under 20 hours, so based on an eight-hour day, the touch time is less than 2.5 of the 34 days. The rest of the time is the "white space" in the rectangles, where nothing is happening to the job. This ratio is not unusual at all—from hundreds of projects at manufacturing companies we have observed that touch time typically accounts for less than 5% of lead time, and in some cases even less than 1%!

Traditional efficiency notions focus on reducing the touch time (gray space). This focus is promoted by costing systems, which assume that product cost is driven by direct labor and/or machine times. Taking the company in Figure B.4, management might target what appears to be the largest cost driver for this job, namely the 12 hours of labor in fabrication. An improvement reduces this to nine hours—a 25% reduction in labor for fabrication, an apparently big success by traditional measures! But what effect does this have on the lead time of the job? The three-hour reduction is barely a dent in the 34 days and would not even be perceptible to customers.

To reduce lead time, companies need to shift from cost-based to time-based thinking. Cost-based thinking stems from mass-production methods where jobs are divided into many small tasks and work on each task is done

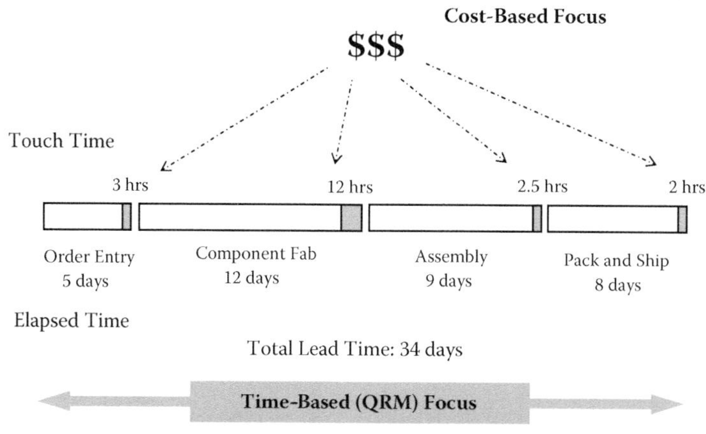

Figure B.4 QRM approach is different from traditional cost-based focus.

by people who specialize in that task. This creates many functional departments with lots of handoffs to process each job. Also, the pressure for cost reduction means that managers minimize the number of resources in their department, so both people and machines end up being highly utilized. From our personal experiences (e.g., with highway driving, or standing in line at supermarkets) we know that high utilization creates long queue times. So, the high utilization of machines and people in each department means that there are large backlogs of work in the departments. When combined with all the handoffs from department to department, the result is long lead times. Now all the factors in Figure B.2 (waste due to long lead times) mount up, resulting in poor quality and high costs.

Four Keys to Organization Structure for Quick Response

In contrast to the cost-based approach, which focuses only on the touch time (gray space), the QRM approach targets reduction of the total lead time (gray space plus white space, from start to finish). In order to reduce this lead time in the face of unpredictable demand and an environment of low-volume or custom products, you need to make four changes to your organizational structure:

- *From functional to cellular.* You must transform the organization of functional departments into one comprised of "QRM Cells." Although the cell concept has been in use for some time, QRM Cells are more flexible, more holistic in their implementation, and are also applied outside the shop floor. QRM Cells are designed around a collection of processes or jobs that share similar characteristics and where there is an opportunity for benefit through lead time reduction. This collection is called a Focused Target Market Segment (FTMS), defined more precisely in the book, *It's About Time.* A QRM Cell is a set of dedicated, collocated, multifunctional resources selected so that this set can complete a sequence of operations for all jobs belonging to a specified FTMS. The set of resources includes a team of cross-trained people that has complete ownership of the cell's operation. The primary goal of a QRM Cell team is reduction of the cell's lead time (measured via the MCT metric).
- *From top-down control to team ownership.* Instead of managers or supervisors controlling departments, QRM Cell teams manage themselves and have ownership of the entire delivery process within their cell.

- *From specialized, narrowly focused workers to a cross-trained workforce.* In contrast with the approach of having each person do one task efficiently, people are trained to perform multiple tasks. While companies talk about cross-training, managers underestimate its benefits and thus do not invest enough in it. On the other hand, companies implementing QRM have seen significant increases in quality and productivity as a result of combining cell structure with cross-training and team ownership.
- *From efficiency/utilization goals to lead time reduction.* To support this new structure, you must replace the traditional cost-based goals of efficiency and utilization with QRM's goal which is a relentless focus on lead time reduction. (More precisely, the goal is based on reduction of MCT.)

Unlike many cells implemented in industry today, QRM Cells do not require linear flow; they accommodate a variety of job types with different routings and the team owns and manages the flows within the cell. Also note that nowhere in the definition is there any mention of the Lean concept of Takt times in the design of the cell—we will elaborate on this point in the third core concept below.

The effectiveness of QRM Cells is illustrated by numerous industry case studies in the books *Quick Response Manufacturing* and *It's About Time.* Additional case studies and publications can also be obtained from the Center for Quick Response Manufacturing (see "For Further Reading").

The organizational structure using QRM Cells is critical to QRM implementation; however, it alone will not ensure success. A manufacturer of specialized transmissions converted its entire operations to cells, yet its quoted lead time was still around six months, and even with this long quoted lead time it had an on-time delivery record of just 40%. Thus, simply installing cells will not guarantee short lead times. The cells need to be complemented with other QRM policies described in the next two core concepts.

QRM Core Concept 3: Understanding and Exploiting System Dynamics Principles

This core concept helps managers understand how system dynamics impacts lead time. The need for this understanding is illustrated by a common management misconception: "To get jobs out fast and operate efficiently we must keep our machines and people busy all the time."

This misbelief stems from cost-based thinking: to minimize cost you should ensure that each resource is used as much as possible so that you can make do with the least number of resources. So, what is the fallacy in this reasoning? As your resources get busier, you create increasing waiting times for jobs—the opposite of the quick response that you are trying to achieve.

The QRM principle that replaces the traditional belief is quite different: "Strategically plan for spare capacity—the planned loading of your resources should be under 85%, or even under 75% in very high-variability environments."

Most managers' first reaction to this is: "We can't afford to do that! Our costs will be much higher than our competition that uses fewer resources." QRM tackles this by using system dynamics theory, which tells us that lead times increase greatly as resource utilizations approach 100%. Worse, now small miscalculations in capacity, or any other disturbances such as hot jobs or machine breakdowns, cause an enormous increase in lead times, as seen from the first graph in Figure B.5. The figure shows the QRM way of explaining this theory to managers in nontechnical terms by calling it "The Magnifying Effect of Utilization."

In similar nontechnical terms, QRM teaches managers about "The Miraculous Effect of Spare Capacity." The second graph in Figure B.5 shows that when you are operating at high utilization (i.e., very little spare capacity), a small investment in more spare capacity (depicted by the horizontal arrow) results in a large reduction in lead time (as seen from the vertical arrow). As a concrete example, if a resource has 90% utilization, by adding just 10% of spare capacity you can reduce the queue time at that resource by 55%!

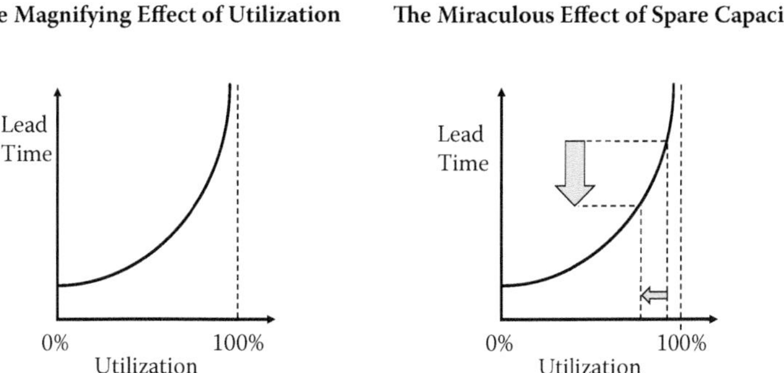

Figure B.5 Effect of utilization and spare capacity on lead time.

But what about the cost of this spare capacity? This is where the first core concept ("Realizing the Power of Time") comes back into play. While it may cost more to operate an area with a little more labor or equipment, the shorter lead times result in lower system-wide "waste" and the reduction in these costs outweigh the cost of the additional resources—review Figure B.2 to be reminded of these system-wide wastes and associated costs. When you add to this the potential increases in sales, you understand that companies have found their investment in spare capacity paid back many times over.

Since in QRM you do not eliminate strategic variability, it is important to design your system to cope with some variability. The higher the variability you are designing for, the more spare capacity you need to incorporate, and QRM uses calculations to help managers with such decisions. QRM also uses insights from system dynamics to make batch-sizing decision; these batch sizes differ from those based on traditional economic order quantity (EOQ) calculations.

Incorporating system dynamics into its core concepts is a key aspect of QRM. Other approaches base system designs on simplistic assumptions, ignoring this issue altogether. Lean uses the concept of Takt time: a fixed interval within which a resource must complete each job. Takt time is calculated solely from production targets. However, QRM shows that both lead time targets and variability need to be included in the calculations for capacity planning.

QRM Core Concept 4: Implementing a Unified Strategy for the Whole Enterprise

Managers are excited to learn that QRM is a strategy that goes beyond just optimizing the shop floor, and it can be used to improve the entire organization. The same time-based mindset and QRM principles extend to all these areas:

- *Office Operations.* Operations such as quoting, engineering, scheduling, and order processing tend to be neglected as a source of improvement in manufacturing companies. Yet they can significantly extend your lead times and increase your overhead costs and SG&A expenses. Using tools geared to office operations, QRM extends the cell concept to the office environment, called a Quick Response Office Cell or Q-ROC (pronounced "queue-rock"). Q-ROCs have enabled companies to reduce office lead times by 80% or more.

- *Material Requirements Planning (MRP) System.* QRM theory shows how the planning logic in a traditional MRP (or ERP) system can result in a spiral of increasing lead times. QRM restructures the system by simplifying it to support the cellular organization. This simpler system is called a High-Level MRP (HL/MRP) system (also see the discussion in Appendix C). When supplemented by the POLCA shop floor control system described in this book, it results in much shorter lead times.
- *Supply Management.* By including MCT as a primary metric in the supply chain, instead of just cost, and by making executives aware of the full cost of lead times on their operations, QRM makes two fundamental changes to supply management: it uses MCT as a primary focus of supplier improvement programs, and it impacts the way sourcing decisions are made. For example, for certain types of parts QRM encourages the use of local suppliers rather than low-cost suppliers half-way around the world. Despite the shift from cost to time, companies have found that MCT reduction helps reduce overall supply chain costs by 10–15%. There are other benefits too: one equipment manufacturer reduced MCTs by an average of 78% across its supply chain and this resulted in a five-fold reduction in supplier quality defects and late deliveries.
- *New Product Introduction (NPI).* With today's fast-paced changes in technology and markets, new products are the lifeblood of a manufacturing business. There are many proven techniques for NPI, such as concurrent engineering and quality function deployment. Even so, QRM further improves the NPI process. The key again is awareness of the impact of NPI lead time on your business, and rethinking conventional decisions in terms of their impact on this lead time. For example, QRM's time-based approach results in new tradeoffs during prototype construction and novel ways of thinking about product options during design. The combined impact of these changes can be substantial. By training its NPI teams in QRM, a manufacturer of medical instruments reduced its NPI time from two-and-a-half years to less than six months.
- *Shop Floor Control.* As part of the Lean toolkit, Kanban systems are popular for shop floor control. Indeed, Kanban is simple and highly effective in higher volume production, but it is not the best system for HMLVC situations, as explained in Appendix C. Instead, QRM uses the POLCA system described in this book. Note that QRM cells qualify as POLCA-enabling cells, so POLCA works well with the QRM structure of cells. In addition, using HL/MRP enables POLCA to be more effective: this is also explained in Appendix C.

Implementing QRM: From Cost-Based to Time-Based Decisions

In this appendix we have repeatedly talked about replacing cost-based decisions with time-based decisions, but how can managers justify such decisions? QRM helps in a number of ways: it provides rules of thumb to predict the cost impact of lead time; it shows how to move from cost-based to time-based justification of projects; and it provides ways to adjust your accounting system. On the last point, QRM does not require that you change to new accounting practices such as Lean Accounting. In fact, the book *It's About Time* provides a few simple adjustments to your existing accounting system that go a long way toward supporting time-based thinking. A recently published book, *The Monetary Value of Time* (see "For Further Reading"), also provides some simple ways to adjust your accounting system to support time-based thinking.

Securing Your Company's Future with QRM

With the growth of global competition, with the changes brought about by outsourcing of jobs to low-cost countries, and with the difficult economic conditions around the world, companies need to reexamine their competitive strategy. Over the past two decades organizations have implemented strategies like Kaizen, Six Sigma, and Lean. Modern technology has allowed companies to vastly increase the variety of products they can manufacture; at the same time, it has given customers the ability to interact with companies through the internet and to expect higher levels of customization. You need a strategy that will explicitly take advantage of the market shifts that are occurring as a result—and QRM is designed to do just that! The good news is that you do not need to turn your back on improvement strategies that you have already implemented and start over: QRM builds on the foundation created by previous methods and takes your competitiveness to the next level.

The track record of companies that have implemented QRM shows that if you can understand and implement QRM before your competition figures out how to do it, huge market opportunities, improved profitability, and a highly stimulating work environment await your enterprise and your employees.

For Further Reading

Articles on Quick Response Manufacturing began appearing in trade journals during the mid-1990s, but the following is the first comprehensive book to be published on QRM:

- *Quick Response Manufacturing: A Companywide Approach to Reducing Lead Times*, by R. Suri, Productivity Press, 1998.

If you are new to QRM, we suggest starting your reading with the following more recent book:

- *It's About Time: The Competitive Advantage of Quick Response Manufacturing*, by R. Suri, Productivity Press, 2010.

Numerous industry case studies and other publications can be obtained from the Center for Quick Response Manufacturing:

- Center for Quick Response Manufacturing, University of Wisconsin–Madison, www.qrmcenter.org.

The following recently published book provides simple and effective ways to adjust your accounting system and metrics to better support time-based thinking:

- *The Monetary Value of Time: Why Traditional Accounting Systems Make Customers Wait*, by J.I. Warnacut, Productivity Press, 2016.

Appendix C: Perspectives on Applying MRP, Kanban, or Some Other Card-Based Systems to HMLVC Production Environments

Most manufacturing companies today use some type of Enterprise Resource Planning (ERP) system. Such systems include a Material Requirements Planning (MRP) module as an important part of their functionality. The goal of the MRP module is to plan both materials and manufacturing resources to ensure smooth functioning of the production system. An important part of this is planning the capacity of workcenters as well as the timing and flow of products through these workcenters. As jobs are released to production, the planning may be followed by additional functionality for shop floor control of these jobs. Since many companies already have these MRP modules in place, it is useful to understand why a system such as POLCA is still needed for shop floor control.

Another popular approach to shop floor control is the card-based Kanban system, which originated with Toyota. Kanban is a key component of Lean Manufacturing (also based on Toyota's Production System), and with the spread and success of Lean in many organizations over the last several years, it is helpful to understand why Kanban does not provide an effective solution in the HMLVC environment. In response to this point and other concerns, several other card-based systems have been suggested as alternatives to Kanban; many of these are rather academic, but two that are more practically oriented are CONWIP and COBACABANA, and this appendix also provides a perspective on them relative to POLCA.

In fact, the above two types of approaches—namely MRP systems and card-based systems—fall into two broad categories popularly known as *Push* and *Pull* systems. While the popular trade literature uses these terms somewhat loosely, the academic literature has tended to follow a precise definition presented by Hopp and Spearman in 2004. Essentially, their definition states that a Push system does not have an explicit limit on the work-in-process (WIP) in the system, while a Pull system does limit the amount of WIP that can be in the system (see Appendix H for more details and references to the academic literature on this point). Below we will explain more about the operation of MRP systems and card-based systems, and this will also help provide some perspective on the origins of the terms Push and Pull.

In summary, as a manager looking into the potential of using POLCA at your company, or as a consultant advising a client about the possible use of POLCA, it is helpful to understand the attributes and advantages of POLCA relative to the main alternatives that are available, particularly for HMLVC production. The goal of this chapter is to provide you with insight on this point. To further clarify this, the aim here is not to provide a thorough tutorial on MRP or Pull systems—there are entire books devoted to these subjects—but rather, to focus on the key operating features that will help to provide you with the insight just mentioned.

An additional clarification about this appendix is that although there are many other approaches to shop floor control, such as Theory of Constraints and Workload Control (see Appendix H), the comparison in this appendix is limited to discussion of MRP (since that is very widely used) and other card-based systems (since those might be perceived as similar to POLCA, so a deeper understanding of the differences would be helpful). However, for interested readers, Appendix H does include a thorough discussion of other approaches along with references to more details on those approaches.

Using MRP in the HMLVC Context

In general, ERP systems serve many useful functions for a manufacturing enterprise, including non-manufacturing functions, such as management of marketing and sales; financial activities such as general ledger, receivables and payables; and human resources functions such as recruiting, training, payroll, and benefits. In the context of this book, the portion of an ERP system that we will discuss is related to its capabilities of Material Requirements Planning (MRP), and the later development called Manufacturing Resource Planning, which was given the acronym MRP II. This portion of the ERP system is responsible for manufacturing-related areas such as materials and production planning, purchasing, and inventory. For simplicity, below we will use the term MRP to denote the general functionality included in both MRP and MRP II modules.

The MRP module includes several important functions, such as: creating and maintaining bills-of-materials (BOMs); storing the routings for all parts that are made in-house; production planning and scheduling; maintaining stock levels for raw material, components parts and finished goods; and placing orders and dealing with suppliers for raw material and components. While most of these functions are useful and effective, there are however weaknesses in the way these systems deal with production planning and scheduling, particularly in the HMLVC context, which results in the need for a control system such as POLCA. To see this, let's briefly review how the production planning and scheduling portion of an MRP system works.

Consider a product that needs ten manufacturing operations. Based on the ship date for the end product, the MRP system schedules what is called a *Start Date* for each operation. This is the system's estimate for when the job should be started at each operation in order for it to make it through the whole shop floor on time. MRP does this by what is known as *backward-scheduling*: the system starts with the ship date, and subtracts the lead time of the last operation to get the Start Date for that operation. This is repeated for each preceding operation until we get the Start Date for the first operation. Figure 4.3 in Chapter 4 shows an example of this process for a product with three operations. Although the figure illustrates the calculation of Authorization Dates, the logic is the same for the Start Dates in MRP. However, there is an important difference between an Authorization Date in POLCA and a Start Date in MRP. As specified in the Decision Time rules for POLCA (see Chapter 2), the Authorization Date alone is not sufficient for a team to start a job; other conditions need to be satisfied too, and, in particular, one of these conditions depends on the status of downstream operations. On the other hand, in MRP, the goal is that a job should be started on its Start Date, regardless of what is happening in the rest of the system. This explains the origin of the term *Push*: each workcenter in the system completes jobs and "pushes" work to the next workcenter without regard to the status of that next workcenter.

Note: The systems discussed in this appendix typically work at a workcenter level and so we use this term in our descriptions; however, in these descriptions you can replace "workcenter" by "manufacturing cell" if you need to relate the description to how POLCA would work in that situation.

Returning to the backward-scheduling logic explained above, we see that to calculate the Start Dates, the MRP system needs to know the lead time for each operation. This lead time is not calculated by the system, but rather, it is a parameter that is entered into the MRP system by the planning department ahead of time.

The last point above is important since, at the time the lead times are being set in the MRP system, the planners don't know the actual workload that will hit a particular department in a given week. On the other hand, it is important that parts arrive on time at subsequent operations, because this affects the daily or weekly capacity planning for those departments as well as the ability to keep parts on time at following operations. Since the planned lead times are specified regardless of the actual workload, these lead times typically have significant padding—also known as safety time—to ensure that the operation gets done in time for the next step. Since most parts have multi-step routings, and each step has a padded lead time, the net result of this is that jobs are released to the shop floor long before they are due.

An example helps to make the magnitude of this problem clear, particularly for HMLVC production. Let's work with the above-mentioned product with a ten-step routing. Suppose each operation is done at a different workcenter, and on average all the workcenters have a two-day lead time. This would mean that if the product simply went through each workcenter unrestrained, it would have an average lead time of 20 working days, or four weeks. However, in an HMLVC situation, the planners know that both the volume and mix of orders can vary significantly from week to week and these variations are hard to predict. In addition, customized orders may have workloads that are highly variable and also not known precisely. Finally, with the high variety of products as well as the customized manufacturing, the planners know that unexpected problems can occur, such as incorrect drawings, quality issues, rework, and so on. In fact, the planners have observed that if there is high demand or if other problems occur, the lead time for any workcenter can grow to a week or even more. Keeping in mind the previously mentioned goal of getting the product to each subsequent operation on time, let's say the planners decide to enter a one-week lead time for each operation in the MRP system. With 10 routing steps this means that the product would be released to production 10 weeks before its ship date. This is a 150% increase over the average lead time!

Let's continue with some simple numbers to indicate the broader impact of this greatly increased lead time. Suppose the company ships 10 jobs per day—in other words 50 jobs per week—and each job typically has a quantity of 20 pieces shipped to the customer. If, hypothetically, the lead time did not deviate much from the average, then each job could be released four weeks from its ship date and would be in the factory for just four weeks. This would mean that at any time there would be about 200 jobs on the shop floor, and a WIP of around 4,000 pieces. However, with the 10-week planned lead times actually in use, instead we will see about 500 jobs on the shop floor and a WIP of 10,000 pieces. Not only is there a lot of WIP to manage on the floor, but also now we need to babysit 500 concurrently open jobs as they work their way through the factory. Obviously, extra work is required to manage the large number of jobs as well as the congestion on the shop floor. But less obviously, there are several longer-term negative consequences that result from this method of operation, as we explain next.

In the HMLVC context, Sales often receives requests for quotes for products needed in a short lead time: specifically, shorter than the MRP planned lead time for those products. After probing the customer, if Sales determines that landing the order is contingent on committing to the short lead time for delivery, and since Sales knows that the MRP lead time is substantially padded, an exception is made and the order is accepted. Of course, this doesn't just happen once, it can happen frequently; in some companies even several times a week!

The counterpart to this is what happens on the shop floor. Supervisors know there is slack in the schedule, so they don't want workers to start an order based on the MRP schedule, because the next workcenter doesn't really need it at this time. Instead, since supervisors know that often urgent orders are accepted by Sales, or jobs have quality problems requiring rework, they use this slack to work on these urgent jobs or last-minute rework. As a result, workers also start to ignore the MRP generated schedules; there is a lot of slack in the schedule and everyone knows this. So, people don't feel any pressure to start jobs on the MRP listed start date. In effect, the MRP start dates become meaningless, and a "hot job mentality" results instead, for the following reason. As the slack time gets used up by unplanned jobs, it gets to the point that the next workstation is actually waiting for the job that should have been started and the supervisor hears people complaining. At this point, this job becomes a hot job and it gets started at the current workcenter. Over time, workers realize that they should only work on something that is a "hot job" because then they know that it is really needed. The frequent hot jobs—now consisting of both urgent accepted orders as well as the normal planned orders that are getting delayed—start disrupting the schedule to the point where even the planned slack time is not sufficient to keep regular jobs on schedule. The company ends up in a downward spiral of worsening on-time delivery and more and more hot jobs in an attempt to meet delivery schedules.

This is not a hypothetical set of scenarios: consider the examples of the two companies described in the first paragraph of Chapter 1. Before implementing POLCA, both Alexandria Industries and Bosch Hinges had quoted lead times of six to eight weeks, but in spite of these long lead times they were constantly struggling with late deliveries resulting in daily expediting efforts for numerous hot jobs on the shop floor.

To be fair to MRP systems however, we should recognize that they are designed to be *planning* systems. The plans are put in place weeks or even months ahead of a particular job's delivery date. Also, the time periods are aggregated into *buckets*, which can be as large as a week, so there is not high fidelity in the details of timings. Further, the planned lead times are based on what is called the *infinite capacity assumption*, which (in essence) states that the lead times for operations are assumed to be fixed regardless of the amount of capacity being used at any workcenter. All of the preceding factors mean that during the execution of the plan, what is happening at a given moment might be quite different from the plan. Chapter 4 emphasized the difference between planning and control (additional details can be found in Appendix H). MRP system designers and other software vendors have also understood the preceding points and proposed that an MRP system needs to be complemented by some kind of shop floor control system. In fact, MRP systems typically provide their own shop floor control module. In addition, to deal with some of the issues described above, and in particular to counter the infinite capacity assumption, software vendors have created systems known by various terms such as Advanced Planning Systems (APS), Manufacturing Execution Systems (MES), and Finite Capacity Scheduling (FCS) systems.

POLCA is one more alternative to these other control systems in use in industry. Chapter 3 discussed in detail the advantages of POLCA over software-based control systems. Also, the section "Bridging the Gap Between Theory and Practice" in Appendix E provides further insights on the limited success of sophisticated software-based scheduling systems in practice.

As one solution to the issues with MRP, some experts suggest scrapping the MRP logic altogether and replacing it with the Kanban system. However, in a following section we will explain why Kanban won't serve the needs of an HMLVC company either. Others suggest that you invest in more sophisticated scheduling systems to complement your MRP system, such as those listed above. However, if you decide to use POLCA, none of these drastic approaches is necessary. Instead, a review of your shop floor structure, complemented by some simple fixes to your MRP system, and followed by the implementation of POLCA will enable you to improve your delivery record and eliminate hot jobs as seen in the case studies in Part III of this book. Thus, if you have an MRP system in place, with POLCA you can build on the system you already have, rather than needing to replace it. We now elaborate on this point.

Simple Adjustments to MRP to Enhance the Success of POLCA

The first step is to review your shop floor organizational structure, as explained in Chapter 4. For smaller companies, say with less than 20 workcenters, that are already using MRP, this may not be necessary, and they can go directly to implementing POLCA. Chapters 14 and 15 have case studies with examples of such companies. However, for larger companies that have dozens of workcenters and complex job routings, the first step is to create manufacturing cells, as also explained in Chapter 4. If you already have such cells, or if you are going to reorganize into cells, it is important for the success of POLCA that these cells qualify as *POLCA-enabling cells*, defined in Chapter 4.

The next step, as you create cells for various components, is to use this opportunity to flatten BOMs and eliminate operations. Even if the BOM consists of multiple levels, if you bring all the related operations into one cell you can list the components at one level for MRP. Supporting documentation can be used to train the cell team to assemble the components in the right sequence, but this flattens the BOM for the MRP logic and simplifies some of the MRP iterations as well as the padding needed for routings.

Now that you have created cells and flattened your BOMs, you can use a simplified version of your MRP system known as High-Level MRP (HL/MRP). As shown in Figure C.1, you use this HL/MRP system to provide high-level planning and coordination of materials from external suppliers and across

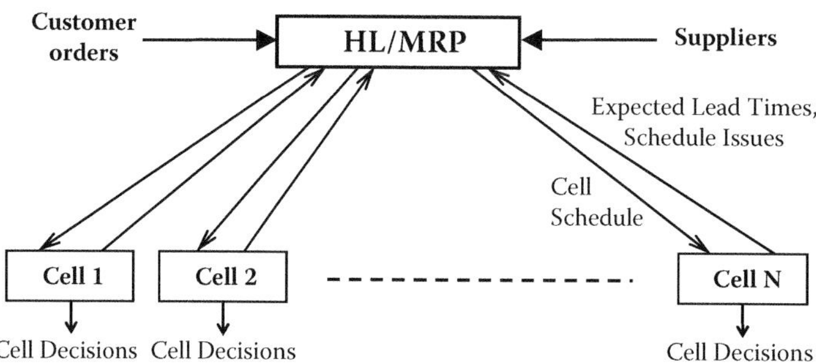

Figure C.1 Illustration of HL-MRP system.

internal cells, in order to meet delivery dates. The system should manage stock levels and reorder material using its standard logic. It should also calculate Authorization Dates using the standard logic by beginning with the delivery date and performing backward-scheduling based on lead times as described in Chapter 4. However, lead times in the system should be set at the cell level, not at the operation level. For instance, if a part goes through six operations within one cell, in the standard MRP system this would be represented by six steps with six lead times. Now, with the HL/MRP system, the entire set of operations performed in the cell would be represented by one step with one lead time, which would be the planned lead time for that cell. (Flattening the BOM will allow additional collapsing of operation steps at multiple levels of the BOM into one MRP step for the cell.) Adjusting your MRP system to the HL/MRP construct also gives you the option to explore using the concept of Capacity Clusters to further simplify your planning (see Appendix D).

Since the cells are POLCA-enabling cells, as described in Chapter 4 the cell teams have complete ownership of the cell's operation. As the cell teams become more experienced at managing their schedules and capacity, then by looking at their upcoming loads they can take a proactive role in communicating with the MRP planners as conditions change. For instance, they can suggest that work be reallocated, schedules adjusted or that their lead times should be modified. This will help to keep the lead times and schedules in the HL/MRP system more realistic and accurate, and prevent some of the MRP-related problems just described. If you have decided to use the Capacity Clusters concept, then this feedback from the cell teams will also help with fine-tuning the parameters for the Capacity Clusters as described at the end of Appendix D.

The final piece of this approach is that as jobs need to go through multiple cells, the POLCA system will coordinate the flow between cells. As already explained, even with the best planning and the simplified HL/MRP system, real-world disturbances such as machine failures or unexpected orders will still occur. As detailed in Chapter 3, POLCA helps to ensure that cells adjust their production decisions to meet end-item delivery schedules.

Even though you are still using the MRP logic, this approach will succeed for three key reasons: (i) There are far fewer routing steps, and therefore fewer padded steps in the MRP routings; (ii) With the shorter lead times there is less variability in the planning and thus less padding is needed for each step; (iii) The use of POLCA along with POLCA-enabled cells and team ownership greatly increases the likelihood that schedules are achieved at each step.

This concludes our overview of key features of MRP systems, and why these systems still require additional control systems in order to function effectively. If you already have an MRP system and are intending to implement POLCA, this section has also provided some pointers to adjust the MRP system so that the combination of MRP and POLCA will provide the best possible results. (Also see the recommendations in the section "Implementation Through Your ERP System" in Chapter 4.)

More explanation of the HL/MRP concept, along with a detailed example of flattening a multilevel BOM, can be found in the book *Quick Response Manufacturing* (see "For Further Reading"). Additional discussion on why the HL/MRP approach will work better than using standard MRP can be found in the book *It's About Time* (see "For Further Reading").

Using Kanban and Other Card-Based Control Systems in the HMLVC Context

As mentioned in the introduction to this appendix, with the spread of Lean Manufacturing, Kanban systems (which are a key component of Lean) have also become a popular approach to controlling the flow of material in a manufacturing enterprise. So, next we discuss the application of Kanban in a HMLVC production environment.

The main feature of a Kanban system is that when a certain quantity of parts (e.g., one bin of parts) is taken into a workcenter to be worked on, a signal is sent to the upstream workcenter to make another batch of the same parts in that quantity. This signal is typically in the form of a physical card, called a Kanban card. The card specifies the part number and the quantity to be made. Thus, these parts will only be made by the upstream workcenter when the downstream workcenter consumes them. This is in contrast to the MRP system operation where the upstream workcenter works on jobs and sends them to downstream workcenters regardless of their current status or needs, resulting in the term Push for this system, as explained in the previous section. In the Kanban system, when a workcenter sends a Kanban signal to an upstream workcenter asking for more parts, the analogy is that it is "pulling" parts from the upstream workcenter based on its current usage of material. This explains the popular use of the term *Pull* for Kanban versus Push for MRP. Some of the strong points of the Kanban system are that it is simple, visual, and the Kanban cards limit the WIP in the system.

Note that a Kanban signal is a *replenishment* signal. You use a certain quantity of a particular part, and then you ask for that quantity to be replenished for precisely that part. Let's understand why this is the case and how the system would work in an HMLVC company. For this we revisit the example of the axle manufacturer in Chapter 3, but we will go over a few more details here.

Consider a company that makes axles for non-automotive applications, such as for construction and mining equipment. This company has a large variety of products and many of them have very low annual demand. The best way to understand the operation of Kanban is to start from the finished goods stock. When a bin with a certain type of axle (let's call it SR384) is shipped out of the warehouse, a Kanban signal is sent to the previous workcenter (i.e., the last operation before the warehouse) to restock that bin. In order to maintain the right stock of all the axles in finished goods, it is important that this signal clearly specifies both the exact type of axle that was shipped (SR384), and the quantity that was in the bin that got shipped out (let's say five axles). This explains why a Kanban signal must specify the part number and quantity to be made.

Now let's follow this signal to that previous workcenter, let's say it is Workcenter Z. When Workcenter Z gets the Kanban signal ("Part Number SR384; Quantity 5"), the operator needs to look in the workcenter's input buffer for a bin of partly finished SR384 axles, which he or she then pulls into Workcenter Z to work on. As the operator does this, he or she then takes the Kanban card attached to this bin and sends it to the previous workcenter, let's say it is Workcenter Y, asking for (or pulling) more of those semi-finished axles from Workcenter Y. Again, this signal must specify the part number and quantity, because Workcenter Z needs Workcenter Y to supply the right semi-finished axles to replenish what was just used. This also emphasizes the point that Kanban is a *replenishment* system: you keep replenishing what you just used. The Kanban signal essentially says, "I used five of pieces of Part SR384, send me another five pieces of (semi-finished) Part SR384."

As each workcenter consumes material to make parts for the Kanban signal it just received, it sends a pull signal to its preceding workcenter, so we can continue to follow this pull signal upstream through all the workcenters until we get to the raw material in the company (or the signal could go directly out to a supplier too). Now note that each of these workcenters along the way needs to have a bin of semi-finished material that will become the finished axle SR384. There may be some point early in the fabrication operations where parts are generic—for example, 1" bar stock—but once you start fabrication operations that are particular to the SR384 axle, then all the semi-finished material from this point on is part-specific. In other words, each workcenter needs to have bins of semi-finished parts that can be used to replenish the pull signal that it receives.

Let's think about this more deeply for an HMLVC company. Suppose the company makes 1,000 different axles. Then at each step of the manufacturing operation, once semi-finished parts become specific to a final product, each workcenter needs to stock material that can be used to replenish a pull signal for any one of these 1,000 axles (Figure C.2). And if the company makes 10,000 different end items (which is not unusual for an HMLVC company) then it needs to stock up to 10,000 items in bins at *each* manufacturing step! Somehow this doesn't seem right: Kanban—a popular system that has as its goal the elimination of waste—is flooding the manufacturing

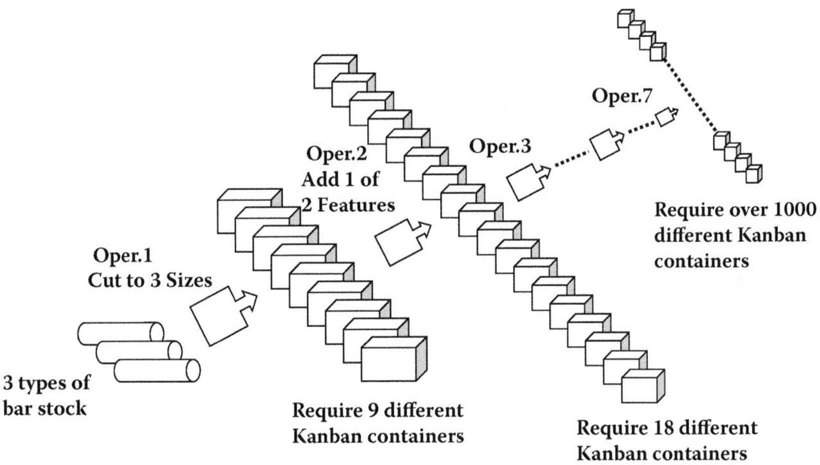

Figure C.2 Inventory needed at each operation to support the functioning of a Kanban system for a company making over 1,000 types of axles.

system with inventory! The explanation is that the Kanban system was designed with Toyota's manufacturing situation in mind, namely higher-volume production with limited variety. In this situation, there aren't that many different bins, plus each bin moves along frequently so the material flow and inventory turnover are reasonable. But let's consider this flow and turnover for a more typical situation faced by our axle manufacturer.

Let's say this company stocks in its finished goods a container with six axles of a certain type: an axle that is used in only a few specialized machines, and so typically the company gets an order from a distribution center only about once a year for one of these containers. Let's see what happens when this order arrives. The container is shipped and a Kanban signal is sent to the previous workcenter. Within a couple of days, the previous workcenter completes the production of the six axles and restocks the warehouse. Now these axles are going to sit in the warehouse for about a year before another order is received. In other words, you have an inventory turnover rate of only once a year for this product! If most of your products have low demand, then your overall inventory turnover is also going to be very low, perhaps two or three times a year at best. In an era when management expects to achieve inventory turns of 20 or more a year, this would be atrocious. But it is actually worse than this, because not only are the finished axles sitting in the warehouse, but along the whole flow path of these particular axles there are partly completed products sitting at various stocking points throughout the shop floor, waiting for a pull signal from an upstream workcenter so that they can be worked on and sent to that workcenter. This is not an unknown phenomenon; in fact, people that design Kanban systems know that these inventories will be needed throughout the shop floor and they even have a name for these intermediate stocks: they are called *supermarkets*. When the production volume is high, there is no need for concern about supermarkets because the items move through the supermarket quickly and the inventory turnover rate is high enough. Thus, Kanban works well in these environments; but in low-volume environments, instead of limiting inventory, it actually adds unnecessary inventory.

Let's pause for a minute here to remind ourselves why POLCA works better in this situation. The key is that, unlike Kanban, in POLCA jobs are not launched just because a card is received. There has to be an Authorization as well. More precisely, the Authorization is checked first, and the check for the right POLCA card occurs only after a job is Authorized. Chapter 3 explains in more detail how the POLCA system combined with the planning for this very low-volume axle would ensure that there is no significant stock throughout the whole system for such a low-volume product.

Also note that we just went over all the semi-finished inventory created by the Kanban system just for one axle. As already mentioned, for an HMLVC company, there might be thousands of such parts, with more inventory not just in finished goods, but at every step in the whole upstream supply chain. Thus, it is important to understand that while Kanban is effective and has been very successful in higher-volume production environments, it is not the right approach for products with low demand.

Further, Kanban is also not the approach for companies with customized products. Let's return to the example of the axle manufacturer, and suppose it receives an order for a custom-engineered axle for a large mining machine. How would the Kanban system work in this situation? It won't work at all. You can understand this via the previous observation that Kanban is a replenishment system: you begin by shipping finished goods and then sending a signal to have them replenished—but you can't have something in finished goods if it has never been made before; in fact, it hasn't even been engineered yet. Indeed, at each step of the whole Kanban shop floor operation, you pick partially completed material from your stocking point and then ask for that material to be replenished. So, this flow cannot operate for a custom-engineered product from the point at which customized operations begin and onward. (Chapter 3 also explained why in POLCA, there is really no difference between the way in which regular products and custom products are treated, so this is not an issue for the POLCA system.)

In summary, if you have a large variety of low-volume products, and/or custom products, Kanban is not an effective approach to production control in your factory. In fact, there are other issues that can arise with a Kanban system in the HMLVC setting which we will not cover in this short overview; additional details can be found in the book *Quick Response Manufacturing* (see "For Further Reading").

The above discussion doesn't mean that you cannot use Kanban at all in an HMLVC company. *In fact, there could be portions of your operation where the use of Kanban might make sense.* These include

- Raw materials or components supplied by vendors, particularly items with higher usage such as bar stock, extrusions, sheet metal, fasteners, clamps, wires, switches, and so on. This can include vendor-managed inventory (VMI) systems.

- Higher-volume components made in-house and used in multiple finished products (hence the higher volume). For this you would need to have a portion of your factory dedicated to these types of parts, and then this part of your factory could be run separately on a Kanban system. This approach could also be used for stocking up on semi-finished products that are common to many end-items, prior to the fabrication operations that distinguish them. For instance, you could have bins of bar stock accurately machined to 1" diameter, cut to a specific length and finished with certain features such as splines and chamfers, which can then be used to make a variety of different axles, and thus the demand for this type of stock could be high enough for it to be made in this Kanban-operated portion of your factory. The Kanban-operated part of your factory would then feed material to the POLCA-operated part of your factory where the products get the distinguishing characteristics that make them particular to each end-item.

In response to some of the limitations of Kanban for HMLVC companies, several other variations of card-based systems have been introduced in the literature. While many of these are rather academic, next we discuss two variations that are more practically oriented. Again, this is not intended to be a comprehensive tutorial on these systems, but rather, a brief overview with a focus on providing insight on the use of these two approaches relative to using POLCA. (For a more detailed and academically oriented review of the relevant literature, see Appendix H.)

The CONWIP System

The first variation we discuss is called CONWIP; the name is derived from "constant WIP" (see "For Further Reading"). The basic CONWIP system uses only one loop that covers the whole shop floor. A job entering the shop floor needs to have a CONWIP card, which stays with the job through all its operations, and when the job finishes its last operation and leaves the shop floor, the card is returned to the entry point. At this time, another job can take this card and enter the shop floor. Thus, the total number of jobs on the shop floor is limited by the number of CONWIP cards. If there are 50 cards, there can be no more than 50 jobs on the shop floor: if all 50 cards are in use, then the next job at the entry point will need to wait until a job leaves and its card is returned. Also, if multiple jobs are waiting to enter the system, there needs to be a rule specifying how the next job is selected from among these jobs. One rule (similar to the Authorization Date in POLCA) could be based on back scheduling from the due date of a job; however, other rules are possible too.

It is clear that the operation of CONWIP is very simple, and this is one of its strengths. Now we remark on some other points related to the operation of CONWIP, in order to provide insights on choosing this system relative to others. These remarks are based on comments made in past literature (see Appendix H for a detailed set of references) as well as in a recent book by Thurer, Stevenson, and Protzman (see "For Further Reading"). CONWIP works well for production systems that can be regarded as *flow shops*, where all jobs flow linearly in the same routing. If jobs have multiple routings through the shop floor, then the CONWIP signals may not provide sufficiently accurate information. As a simple illustration of this, if there are three main routings through the shop floor, and two of those routings visit a machine that is starting to get backed up, while the third routing does not need this machine, then for a period of time jobs leaving the third routing will continue to allow more jobs to enter the system. So, it will not be apparent for a while that there is a bottleneck because jobs on the third routing that are already in the system will continue to be processed and exit and send cards back, and further, if any entering jobs belong to the third routing they will continue to flow through and send signals back for more jobs. Eventually, if all cards are stuck with jobs on the first two routings, and the next job to enter is also on one of these two routings, then no more jobs will enter until the bottleneck is resolved. However, this also means that all workcenters that are only on the third routing will now be starved, even though the bottleneck is elsewhere and they could potentially be working on other jobs. As a result of such arguments, it has been suggested that if CONWIP is to be used in a situation with multiple routings, then there should be one CONWIP loop (i.e., one type of CONWIP card) for each routing. This will ensure that the CONWIP signals are more responsive to bottlenecks on that routing.

In a typical HMLVC company, it is likely that there are many possible routings through the shop floor. Specially with customized jobs, it could be that products need to go from one operation to almost any other operation. Plus, note that the CONWIP loop needs to go through the whole routing, so there needs to be one type of CONWIP card for each possible complete routing. This means that the number of CONWIP loops could grow almost exponentially with the number of workcenters on the shop floor, so that the number of different CONWIP cards and loops being managed would be huge. In contrast, POLCA loops only go between pairs of cells (or workcenters), so the number of loops grows at a much lower rate. Chapter 11 details a calculation at a pharmaceutical company where CONWIP would require as many as 2,240 loops while POLCA needed 147 loops at most.

Another issue pointed out in the literature is that there is a delay in the feedback signal from CONWIP. Since a card has to go through the entire routing before being returned to the entry point, for long routings if there is a problem or some congestion, there could potentially be a long delay in this issue being recognized at the starting point. So, researchers have suggested that CONWIP may work better for systems with shorter routings.

A final point to be made is that for the same reason as above, CONWIP provides system-level feedback, whereas POLCA provides local feedback. For instance, if there is a problem or backup at a particular workcenter, jobs going through this workcenter will be delayed, but there will not be specific information from the CONWIP signal about which workcenter is causing the problem. On the other hand, as seen from the descriptions in Chapters 2 and 3, POLCA provides clear and relatively quick feedback about which workcenter is backlogged. This was another reason cited by the pharmaceutical company for using POLCA (see Chapter 11).

In summary, CONWIP might work very well for flow shops with short routings (not too many steps), and it is simple to implement in such cases. However, it may not be the appropriate approach for HMLVC companies with many different routings, or potentially long routings. This completes our overview of CONWIP for the purposes of providing some insights.

The COBACABANA System

Another variation of a card-based control system is called COBACABANA, an acronym derived from "Control of Balance by Card-Based Navigation." This system was introduced by M.J. Land in 2009 (see "For Further Reading") and is described in detail in a more recent book, *Card-Based Control Systems for a Lean Work Design* (see "For Further Reading"). As stated by the authors of this book the system is aimed at job shops, so in our terminology it could be suitable for HMLVC companies. We now provide a brief overview of how this system works based on the descriptions in that book, which we will henceforth refer to as "the COBACABANA Book."

The COBACABANA Book contains numerous illustrations to help explain the detailed operation of the system. We will not reproduce these diagrams here as the purpose of this appendix is not to be a detailed tutorial on other systems; rather, we will highlight the main points of how the system operates with brief explanations that suffice to lead us to the key insights needed for this appendix.

In COBACABANA, arriving orders are not released directly to the shop floor; they first have to wait in a pre-shop pool and then they are released according to *pool sequencing rules*. The book provides several options for these pool sequencing rules based on various goals. Next, the system puts in place card loops from a central planner to *each* workcenter on the shop floor. Before a job is released to the shop floor, it is assigned several pairs of cards: one pair for each operation. This pair consists of a *release card* for the operation, and an *operation card* for the same operation. When the job is released to the shop floor, all the release cards stay with the planner, but all the operation cards travel with the job. As each operation is completed on this job, the corresponding operation card is returned to the central planner.

As indicated by the full description of the acronym, the purpose of the cards is to control and balance the workload on the shop floor. The first step to doing this is that cards can have varying sizes, and the amount of workload at a given operation is indicated by the size of the card. The next step is that the central planner has a planning board with rows for each workcenter. As each job is released to the shop floor, the release cards for that job are placed on this board as follows. For each operation, the release card is placed on the row for the workcenter where this operation will be performed. The cards are stacked from left to right, so the length of each row indicates the total workload headed for that workcenter. Based on the pool sequencing rule, the next order is considered for release, and all its operation cards are tentatively placed on the planning board and added to the workloads for the relevant workcenters. If no workcenter's limit is exceeded then this job can be released. Otherwise this job is held back and the next order is considered based on the sequencing rule.

The final step is the feedback loop: as each operation is completed on a job, the corresponding operation card is sent back to the central planner. This is a signal to the planner to remove the relevant release card (i.e., the one that was paired with this operation card) from the planning board, and push all the cards for this row to the left, thus indicating that some capacity has become available at that workcenter.

This completes our overview of the system's operation. Although the above-mentioned book contains many more details that take into account various situations that may occur, this brief description suffices for our purpose of providing the main insights on COBACABANA.

There are two important features that COBACABANA provides for an HMLVC company. The first is that, unlike CONWIP, where routings need to be short and it is better if there are not too many unique routings, in COBACABANA any type of routing is possible. The second is that COBACABANA focuses directly on controlling and balancing the workload for individual workcenters. This is different from Kanban, CONWIP, and POLCA, which all control the workload indirectly by limiting the number of jobs in the system (or portions of the system).

Next, let's outline some insights related to the choice of implementing a COBACABANA system. The first point to be noted is that the system requires management to decide on the pool sequencing rule to be used. In fact, there is a large universe of rules to choose from, and the research literature on scheduling contains hundreds of publications comparing the effectiveness of different rules for various situations. It may be difficult for management to make the technical call on which rule to use; and further, if the system doesn't work as well as planned it may not be clear if it is the sequencing rule that needs to be fixed, or other parameters in the system that need to be adjusted. In POLCA the release rule is both simple and clear, based on the Authorization Date for the first operation. This brings up the next point, which is that most manufacturing enterprises today use some type of ERP system along with an MRP module as explained earlier in this appendix. With POLCA, as explained in Chapters 3 and 4, you don't need to have an MRP system but if you do, the way to connect POLCA with this system is clear and also explained in those chapters. The COBACABANA Book does not provide details on how to use it with an existing MRP system; interestingly, the authors give several examples of calculating operation start dates using back-scheduling logic similar to MRP, but they do not take the next step of explaining how to make a successful hookup with an MRP system.

The next insight is that COBACABANA requires a central planner to make most of the key decisions. This is not in keeping with the ideas of teams and worker ownership that are central to today's manufacturing. In fact, it is not in line with key sociotechnical success factors for system design (see Appendix E). In contrast, in POLCA, the planning system creates the starting conditions for jobs to be released, but then the POLCA cards and rules allow the system to operate in a decentralized manner with local decision-making, as well as cell-to-cell interactions when needed, without any central planner being involved. (Appendix E confirms that POLCA is also consistent with the key sociotechnical factors just mentioned.)

The next issue to consider is the complexity of both the card information and the release decision. Each order needs a pair of cards for each of its operations, and the cards need to contain information about the order, the operation, and the workcenter where the operation will be performed. In addition, cards need to be of varying sizes based on the workload. So, standard cards cannot be used, as is the case with Kanban, CONWIP, and POLCA. For COBACABANA this requires preparation of all of these cards for each order. After the cards are ready, the central planner needs to check the release decision for the next possible job. If this job has seven steps in its routing, the planner will need to put seven cards on the planning board and see whether the job can be released. If not, then the planner needs to remove these cards and do the same for the next possible job. So, you need to allow for the time that it will take the planner every day for the card preparation and these release decisions for all the jobs on that day.

The next issue to consider is the effort for returning the cards to the central planner, and also updating the planning board. As each operation is completed, the relevant card needs to be returned not just to a preceding workstation but all the way back to the central planner. In a small factory, this may not be a major issue, but even in a medium-sized factory with 50 workcenters, there may be frequent completions of operations as well as some distance back to the central planner, so you would need to consider how all these cards would be returned in a timely fashion. Also, depending on the frequency with which cards are returning, the central planner might need to devote significant time to removing each of the paired cards from the planning board.

Another insight here is that COBACABANA does not have workcenter-to-workcenter card loops. All the loops go from workcenters to the central planner. Also, the central planner releases jobs based on workload estimates *at the time of release*. Let's understand the implications of this. Suppose a job has seven operations, and at the time of release all operations appear to have capacity. But let's say it takes an average of a day of queue time along with processing time at each step, so the job will not reach its fifth operation until five days from now. During this window of time, many things could have changed. A previous job at this workcenter could have had a problem and taken longer. There could be a machine failure causing a backup. A whole bunch of other jobs could have reached this workcenter just before this job. And so on. As a result of any of these, this workcenter could be experiencing a severe backlog on this particular day. However, there will not be any immediate feedback to help point out or alleviate this problem. In POLCA, all the upstream workcenters (or cells) will know from the lack of cards that this particular downstream workcenter is backed up, and there may be some local decision-making by various teams to help alleviate this bottleneck, as explained in Chapter 3. However, in COBACABANA, as jobs are completed at this bottleneck station, operation cards will continue to be returned and more jobs may be released to it. To summarize this point, in COBACABANA jobs are released based on a planned workload, some of which can be for many days into the future, and there is no detailed consideration of the actual timing of when this workload might hit a given workcenter, nor any feedback if such timing creates a sudden backlog.

As an example, we elaborate on how one of the above situations may occur. Suppose that jobs have been released to balance the workload on Workcenter X, based on the total requirements of all operations going to X. However, since the sequence of operations and detailed timing is not considered, suppose very few jobs go to X for a couple of days, so that not much capacity is used at X, and then a number of jobs arrive on the same day at X. Then there will be a substantial backlog at X on that day.

Another issue that needs to be dealt with is that a job with a large workload at one operation may get stuck in the release process. As an example, suppose an operation is completed at Workcenter Y and after the operation card is returned there is now 10% of capacity available at Y. If Job A requires 20% of the capacity at Workcenter Y it can't be released, and the next job in the pool will be considered. Suppose this next job uses up 8% of the capacity at Y (and it also doesn't exceed the limit for any other workcenters) then it will be released and will reserve this additional 8% of capacity of Y on the planning board. This cycle might be repeated for the next completing job at Y. In this way, many jobs could keep jumping ahead of Job A.

In response to several of the above-mentioned potential problems, the COBACABANA Book proposes a number of strategies. The authors suggest calculating what they call *direct* and *indirect* workload, and also a *corrected aggregate load*. However, these require more complex planning calculations along with some judgement calls in designing these calculations. To mitigate what they call premature idleness (as in the Workcenter X example), and also to overcome the problem of Job A, which was stuck, they suggest a technique of *work injection*: this recognizes that (in their words) you "need to temporarily violate the workload limit" at a certain workcenter. This approach requires a judgement call: someone needs to make a decision when to violate this limit; thus, the rules for using the system start to get fuzzy. A strong point of POLCA, frequently cited by industry users, is that shop floor personnel like the fact that the rules are clear to everyone.

In fact, it is not clear how successful these various mitigating strategies would be in practice. The COBACABANA Book does not provide any industrial examples or case studies such as those provided by us in Part III of this book; in the absence of such application examples we do not have information on how this system would be received by shop floor personnel or by the planner or by management, and nor do we have information on the results that the system has been able to achieve. This makes it difficult to decide on the adoption of various alternative strategies proposed by the authors.

This completes our discussion of some key alternatives for shop floor control along with practical insights on the operation of these alternatives, in order to help readers with their choice of system to be used in their context.

For Further Reading

More explanation of the HL/MRP concept, along with a detailed example of flattening a multilevel BOM, can be found in the following book:

- *Quick Response Manufacturing: A Companywide Approach to Reducing Lead Times*, by R. Suri, Productivity Press, 1998.

Discussion on why the HL/MRP approach will work better than using standard MRP can be found in:

- *It's About Time: The Competitive Advantage of Quick Response Manufacturing*, by R. Suri, Productivity Press, 2010.

There is extensive literature on Kanban systems, but the examples in this appendix are based on the example of a car bumper being replenished at a dealership, and the Bumper Works factory that makes the bumpers, as described in the book:

- *Lean Thinking*, by J.P. Womack and D.T. Jones, Simon and Schuster, 1996.

For descriptions of CONWIP, see:

- "CONWIP: A Pull Alternative to Kanban," by M.L. Spearman, D.L. Woodruff, and W.J. Hopp, *International Journal of Production Research*, Vol. 28, No. 5, 1990, pages 879–894.
- *Factory Physics: Foundations of Manufacturing Management*, by W.J. Hopp and M.L. Spearman, Richard D. Irwin, 1996.

The COBACABANA system was introduced by Land (see the first item below) and described in detail in a recent book (second item below). The descriptions in this appendix are based on the details in the book.

- "COBACABANA (Control of Balance by Card-Based Navigation): A Card-Based System for Job Shop Control," by M.J. Land, *International Journal of Production Economics*, Vol. 117, No. 1, 2009, pages 97–103.
- *Card-Based Control Systems for a Lean Work Design*, by M. Thurer, M. Stevenson, and C. Protzman, Taylor & Francis, 2016.

Appendix D: Capacity Clusters Make High-Level MRP Feasible

Guest Authors: Ignace A.C. Vermaelen and Antoon van Nuffel

Most manufacturing companies today use an Enterprise Resource Planning (ERP) system to manage their companywide operations. All ERP systems include in their functionality some form of Material Requirements Planning (MRP) to manage the production and purchasing areas of the operations. This appendix will focus on the MRP portion of such systems.

The Need for a High-Level Unit of Capacity

Classic MRP tries to plan all dependent and independent demand in detail. To do so it needs thorough definitions regarding lead-times, available capacity, and required capacity. To determine this required capacity, an exhaustive description of all products and their processes, operations, and time elements is needed. In most cases it comes down to the fact that every bill of material (BOM) needs its own customized bill of labor (BOL), listing all necessary operations (in the right order) and all accompanying time elements (set up, run, wait, and so on). This means that a lot of data elements have to be defined (and maintained!). Not only is this an enormous effort to start with, but in the case of highly customized products it is a never-ending story when BOMs need to be added or modified. Because of the large number and the high rate of modifications, the time elements might never get calculated or measured correctly, leading to inaccurate schedules and bottlenecks on the shop floor, resulting in rescheduling, expediting, and increasing lead-times.

To make matters worse, unpredictable factors—such as tool breakages or customer changes in request dates—deteriorate the quality of the calculated schedule even further. This explains why most of the time those schedules do not meet reality on the shop floor. (Also see Chapter 3 for more discussion on this point.)

One of the root causes of this sequence of events is that the level of definition required by the traditional MRP system is simply too high to be used practically in an environment with high-mix, low-volume and customized (HMLVC) products. It is already well-recognized in the manufacturing strategy literature that to reduce lead time and improve delivery performance, such companies need to be organized into cells. Further, High-Level MRP (HL/MRP) has been recommended in the literature for companies with a cellular manufacturing structure (for example, see the book *Quick Response Manufacturing* by R. Suri, Productivity Press, 1998). In the HL/MRP approach, instead of planning the progress of work orders at the detailed operation level, you plan the work order at the cell level. However, a cell may include several different operations on a work order. So, this presents the question, how can one reasonably plan capacity at a cell level without accounting for all the details in each of the operations? What is needed is the ability to define capacity at a higher level and in a more universal way. That is the goal of this appendix. An analogy will help to illustrate our goal:

> Consider the issue of sizing the electric supply feeding into a factory. The factory has many different types of electrical machines and appliances, each using varying amounts of voltages and currents. However, there is an established way to aggregate all these needs into a high-level description, and that is through using the measure of kilowatts (kW). Although the machines are very different, each has a kW rating, and by adding these ratings together you can get an aggregate capacity (in kW) that is needed for the feeder cable to the whole factory. You can similarly get aggregate needs for cables going to separate areas of the factory by aggregating only the equipment in those areas. Note that this can be done despite that fact that there are all kinds of different electrical equipment in each area.

Our aim in this appendix is to provide a similar aggregate measure for capacity at a cell level that does not require the exhaustive details in traditional MRP systems.

Introducing a Capacity Cluster as an Approach to this High-Level Unit of Capacity

What is needed is a high-level measure that can be used easily in capacity and lead time calculations. At the same time, this measure needs to be easy to maintain, and, therefore, in contrast with the classical MRP approach, it should not be a function of the underlying operations and time elements (needed capacity), nor of the available labor or machine hours (available capacity). This goal may seem surprising or even impossible. However, we will explain that it is indeed achievable and it is possible to define a macro-level aggregate capacity measure that reasonably reflects reality in an empirical way.

Below we will explain how we can define a *Capacity Cluster* as a collection of multiple resources (machines/labor) in a cell that performs similar processes to manufacture similar products. We will show how Capacity Clusters are suitable for defining the available capacity of cells. Note that typically each cell will have its own Capacity Cluster defined, because of its different constellation of resources. We will go into more details on all these points in the following sections.

The best way to explain how Capacity Clusters can be used is through a detailed example. In the rest of this Appendix we will work through such an example and use it to illustrate all the concepts related to defining and using Capacity Clusters.

The MadTran Example Extended

We will use the MadTran example described in Appendix A of the book *It's About Time* (by R. Suri, Productivity Press, 2010). However, we will extend the example to help explain several of the concepts of Capacity Clusters. MadTran is a company that makes transmissions. In our extension of this example we will describe two end-products made by MadTran:

- A Heavy Truck Transmission (HTT-534)
- A Light Vehicle Transmission (LVT-142)

Each product has its own multi-level BOM structure, as illustrated in Figure D.1. As shown in the figure, each product is assembled using three main parts:

- One purchased drive shaft for transmitting the rotation to the wheels.
- One housing, which is machined in-house from a purchased casting.

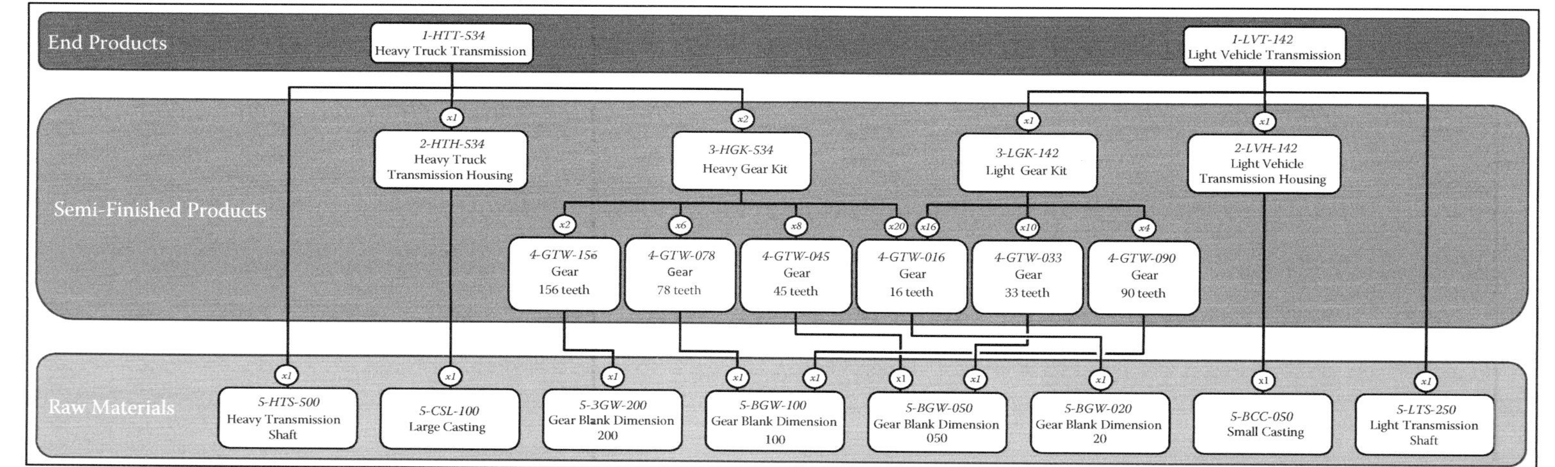

Figure D.1 Overview of Bill-of-Material (BOM) structure for both transmissions.

- Multiple gears machined in-house from purchased blanks, and then organized into kits for the final assembly.

Note that the quantity and type of these parts can differ for each of the two transmissions. Next, for every product (end-product or semi-finished product) we can define its single-level BOM, which shows the manufacturing process needed, the type of components needed, and the quantity of each component (see Figure D.2).

In our example, MadTran has organized its shop floor into cells so that each of the main items at a given level in the multilevel BOM is processed in a particular cell. There are four cells, and the routing of the components through the various cells is shown in Figure D.3 and explained here:

- All gears are machined in the Gear Manufacturing Cell, starting from purchased blanks.
- The Gear Kit Assembly Cell prepares and organizes the various gears into the two types of gear kits needed for the final assembly.
- The housings are machined in the Housing Machining Cell, starting from purchased castings.
- The transmissions are finally assembled in the Transmission Assembly Cell. As can be seen in Figures D.1 and D.2, this assembly requires a housing and a gear kit from the in-house operations, as well as a transmission shaft, which is purchased.

At the next level of detail, we need to describe the operations for all the processes, along with their relevant time elements. These are listed in Figure D.4. Note that the first three processes in the figure are the same for all the products, but the processes for the manufacturing of the Large Gear and Small Gear have different time elements.

In classic MRP, the operations and all related time elements must be defined for all processes involved. In our MadTrans example, the BOMs and the BOLs were kept relatively simple in comparison to many real-world manufacturing scenarios. Even so, we needed to define 26 time elements in 19 operations spread over five processes. In the next section we will show that, with the use of Capacity Clusters, it is possible to define capacity directly at the level of the BOM so that only one definition for every BOM will needed.

Product Definition				Construction (Bill of Material)				
Product		*Type*	*Process*	*# Stock Units Needed*		*Part/Component*		*Type*
1-HTT-534	Heavy Truck Transmission	End Product	Assemble Transmission	1	Piece	2-HTH-534	Heavy Truck Transmission Housing	Semi-Finished Product
				2	Piece	3-HGK-534	Heavy Gear Kit	Semi-Finished Product
				1	Piece	5-HTS-500	Heavy Transmission Shaft	Raw Material
1-LVT-142	Light Vehicle Transmission	End Product	Assemble Transmission	1	Piece	2-LVH-142	Light Vehicle Transmission Housing	Semi-Finished Product
				1	Piece	3-LGK-142	Light Gear Kit	Semi-Finished Product
				1	Piece	5-LTS-250	Light Transmission Shaft	Raw Material
2-HTH-534	Heavy Truck Transmission Housing	Semi-Finished Product	Manufacture Housing	1	Piece	5-CSL-100	Large Casting	Raw Material
2-LVH-142	Light Vehicle Transmission Housing	Semi-Finished Product	Manufacture Housing	1	Piece	5-CSS-050	Small Casting	Raw Material
3-HGK-534	Heavy Gear Kit	Semi-Finished Product	Assemble Gear Kit	20	Piece	4-GTW-016	Gear 16 teeth	Semi-Finished Product
				8	Piece	4-GTW-045	Gear 45 teeth	Semi-Finished Product
				6	Piece	4-GTW-078	Gear 78 teeth	Semi-Finished Product
				2	Piece	4-GTW-156	Gear 156 teeth	Semi-Finished Product
3-LGK-142	Light Gear Kit	Semi-Finished Product	Assemble Gear Kit	16	Piece	4-GTW-016	Gear 16 teeth	Semi-Finished Product
				10	Piece	4-GTW-033	Gear 33 teeth	Semi-Finished Product
				4	Piece	4-GTW-090	Gear 90 teeth	Semi-Finished Product
4-GTW-016	Gear 16 teeth	Semi-Finished Product	Manufacture Small Gear	1	Piece	5-BGW-020	Gear Blank Dimension 20	Raw Material
4-GTW-033	Gear 33 teeth	Semi-Finished Product	Manufacture Small Gear	1	Piece	5-BGW-050	Gear Blank Dimension 50	Raw Material
4-GTW-045	Gear 45 teeth	Semi-Finished Product	Manufacture Small Gear	1	Piece	5-BGW-050	Gear Blank Dimension 50	Raw Material
4-GTW-078	Gear 78 teeth	Semi-Finished Product	Manufacture Large Gear	1	Piece	5-BGW-100	Gear Blank Dimension 100	Raw Material
4-GTW-090	Gear 90 teeth	Semi-Finished Product	Manufacture Large Gear	1	Piece	5-BGW-100	Gear Blank Dimension 100	Raw Material
4-GTW-156	Gear 156 teeth	Semi-Finished Product	Manufacture Large Gear	1	Piece	5-BGW-200	Gear Blank Dimension 200	Raw Material

Figure D.2 Single-level BOM for each item specifying the process, types of components, and quantities needed.

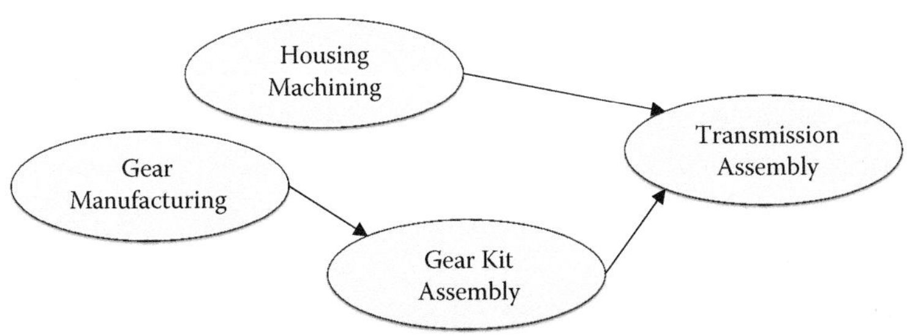

Figure D.3 Cells at MadTran and the product flow.

Process	*Operations*	*Description*	Time Elements (minutes)		
			Run Labor	*Setup Machine*	*Run Labor and Machine*
			Variable	*Fixed*	*Variable*
Assemble Gear Kit	GearCollection	Sort out and collect the appropriate gears	10		
	GearPacking	Pack the gears together	20		
Assemble Transmission	GearAssembly	Set up the jig for the appropriate transmission and put the housing and gears on the jig		40	30
	TransmissionFixing	Fix all parts	50		
	TransmissionGreasing	Grease the assembled gears	20		
	TransmissionClosing	Seal the transmission housing			20
Manufacture Housing	HouseDrilling	Drilling of holes in the Transmission Housing		10	16
	HouseMilling	Milling of openings in the Transmission Housing		20	40
	HouseDeburring	Deburring of the drilled and milled housing			20
Manufacture Large Gear	BlankDegreasing	Cleaning and degreasing of blank	14		
	GearMilling	Milling of the blank to the appropriate gear wheel		20	50
	GearHobbing	Cutting the gear shapes		15	48
	GearDeburring	Deburring of the machined gear			20
	GearQualityCheck	Check if the machined gear meets the quality standards			16
Manufacture Small Gear	BlankDegreasing	Cleaning and degreasing of blank	10		
	GearMilling	Milling of the blank to the appropriate gear wheel		20	40
	GearHobbing	Cutting the gear shapes		10	30
	GearDeburring	Deburring of the machined gear			10
	GearQualityCheck	Check if the machined gear meets the quality standards			16

Figure D.4 List of all processes, operations, and time elements.

Defining Available and Needed Capacity Through Capacity Clusters

The starting point for defining a Capacity Cluster is to think about the minimum typical set of resources that are needed to complete the processes in a cell for a given item. As two examples of this: (i) to manufacture any of the gears at MadTran, one worker is needed along with a milling machine, a gear hobbing machine, and a deburring machine—we can call this one Gear Cell Capacity Cluster; (ii) to assemble the transmission a group of two workers is required along with specialized jigs and a set of power tools—this becomes one Assembly Cell Capacity Cluster.

Next, we define the number of stock units of a given item that can be produced in a cell with one Capacity Cluster over one production period (e.g., an eight-hour shift). Building on the gear example above, let's say one

Capacity Cluster can produce 10 small gears of a given type, or only two large gears of another type. Using the inverse of these numbers, this also tells us how much of a Capacity Cluster is needed to make one unit of a given product. In the preceding example, we need 0.1 Capacity Cluster for one small gear and 0.5 for one large gear. Similarly, such numbers can be calculated for all the other items and cells.

As a result of the above approach, the capacity of a cell can be defined with only one parameter: the number of Capacity Clusters available in a production period. So, for the Assembly Cell Capacity Cluster just described, if there are three workers and additional jigs and tools, we could say that we have an aggregate of 1.5 Gear Cell Capacity Clusters available.

Observe that instead of trying to give values to the diversity of available resources in a production cell, only one aggregate parameter is needed instead of a complex and time-consuming definition of the time elements of all operations of the related process. As an aggregation, this will obviously be an approximation of the details needed in reality. On the other hand, it substantially reduces the amount of necessary data and effort required in managing and maintaining the system. Further, given the errors and inaccuracies that exist in highly detailed BOMs and BOLs, it can be argued that the more detailed descriptions do not necessarily provide a higher level of accuracy anyway.

In general, it is to be expected that various products and cells would have different characteristics. Some operations on products need a lot of labor and only a few machines, while for other operations it could be the opposite. The same is true for cells: some will contain a lot of machines and only a few people to operate them, while others will only mostly people to perform the operations. Therefore, it is likely that different Capacity Clusters will be defined for each cell.

For our MadTran example we will define four Capacity Clusters (Figure D.5). Every cell gets its own Capacity Cluster definition (Figure D.6), which probably will be the case in most situations as just explained. It is, however

Capacity Cluster	*Representing the set of resources needed*
GearKitAssemblyCluster	To Assemble Gear Kits Starting from the in-house manufactured gears.
GearManufacturingCluster	To manufacture gears from a blank. The cluster consists of one person operating a milling, hobbing, and deburring machine.
HousingMachiningCluster	To prepare the housing starting from a casting. The cluster consists of one person using a milling machine and drill.
TransmissionAssemblyCluster	To perform an assembly. The assembly is executed by two people using a set of machinery such as a multi-purpose jig and power tools.

Figure D.5 Four Capacity Clusters defined for MadTran.

Cell (Production Unit)	Available Capacity	
	# Capacity Clusters/Production Period	
Gear Kit Assembly	3	GearKitAssemblyCluster
Gear Manufacturing	5	GearManufacturingCluster
Housing Machining	4	HousingMachiningCluster
Transmission Assembly	6	TransmissionAssemblyCluster

Figure D.6 Capacity Cluster type and amount available for each cell.

possible that different cells could make use of the same Capacity Cluster definition. Even if another cell had different machines and products, but a similar high-level way to define the available capacity made sense in that context, then the same definition could be extended to this other cell.

Also, the available capacity in a cell does not need to be fixed over time. Just as in normal MRP, you can input different capacity levels for different periods, in order to take into account staff vacation time, sickness, machine maintenance or breakdown, and so on. In this appendix, we will keep the capacity fixed over time for MadTran, so as not to complicate the example.

We now complete the details of the data for MadTran. For each of the in-house manufactured items, Figure D.7 shows the capacity needed per product, expressed in Capacity Clusters. Let us go over an example to explain the details. Both the Heavy Truck Transmission and the Light Vehicle Transmission are assembled in the same cell (Transmission Assembly) and they make use of the same Capacity Cluster (TansmissionAssemblyCluster). However, for the former product, only two transmissions can be manufactured with one TransmissionAssemblyCluster, while for the latter, five transmissions can be manufactured. As a reminder of the earlier definition, the capacity needed was

Product		Capacity Needed			
		Stock Units/Capacity Cluster			*# Capacity Clusters/ Stock Unit*
		# Stock Units		*Capacity Cluster*	
1-HTT-534	Heavy Truck Transmission	2	Piece	TransmissionAssemblyCluster	0.500
1-LVT-142	Light Vehicle Transmission	5	Piece	TransmissionAssemblyCluster	0.200
2-HTH-534	Heavy Truck Transmission Housing	8	Piece	HousingMachiningCluster	0.125
2-LVH-142	Light Vehicle Transmission Housing	12	Piece	HousingMachiningCluster	0.083
3-HGK-534	Heavy Gear Kit	5	Piece	GearKitAssemblyCluster	0.200
3-LGK-142	Light Gear Kit	10	Piece	GearKitAssemblyCluster	0.100
4-GTW-016	Gear 16 teeth	20	Piece	GearManufacturingCluster	0.050
4-GTW-033	Gear 33 teeth	15	Piece	GearManufacturingCluster	0.067
4-GTW-045	Gear 45 teeth	10	Piece	GearManufacturingCluster	0.100
4-GTW-078	Gear 78 teeth	6	Piece	GearManufacturingCluster	0.167
4-GTW-090	Gear 90 teeth	4	Piece	GearManufacturingCluster	0.250
4-GTW-156	Gear 156 teeth	2	Piece	GearManufacturingCluster	0.500

Figure D.7 Capacity needed per product, expressed in Capacity Clusters.

defined as the number of stock units that can be produced with one Capacity Cluster, so using the data just presented, one Heavy Truck Transmission needs 0.5 of this Capacity Cluster, and similarly one Light Truck Transmission needs 0.2 of the cluster, as shown in the right-hand column of the figure.

Later sections of this appendix will provide more examples on calculating the number of capacity clusters in a cell as well as the number of stock units that could be produced with such a cluster. If you are concerned about the number of different Capacity Cluster units that you may need to define, note that this not that large: the upper limit is the number of cells, if each has its own unit of capacity definition. If different cells can use the same definition as explained earlier, then this number could be lower.

In summary, as seen from this detailed example, Capacity Clusters define capacity at a high level. Using such a high level might introduce some approximations and errors, but trying to define the capacity on the lowest level of the operations, may not necessarily lead to more reliable planning for several reasons explained earlier. On the other hand, the gain from using the Capacity Cluster approach is that it provides the ability to more easily tune the parameters when evaluating results: because there are far fewer parameters to maintain, tuning becomes more feasible. After a few iterations, the tuned parameters will lead to far more realistic scheduling.

Using Dedicated Throughput Calculation as a Verification

We can use the data in the previous figures to calculate the total throughput of a cell during one period (e.g., a shift) if it is dedicated to one of the products. Production planners and cell teams should already have a good idea of what an achievable throughput should be if the cell is focused on getting one product out (e.g., from past efforts of trying to get rush jobs out). This calculation can then be used to check the degree of realism of the chosen parameters or to fine tune them if needed. The specific calculation to be used is:

$$\begin{aligned} &\textit{Dedicated throughput of } \textit{Product}_x \textit{ in } \textit{Cell}_y \\ &= (\#\textit{Stock Units}/\textit{Capacity Cluster})_{\textit{BOM}_{\textit{Product}_x}} \\ &\quad \times (\#\textit{Capacity Clusters}/\textit{Production Period})_{\textit{Cell}_y} \end{aligned}$$

Applying this calculation to all the in-house manufactured products at MadTran gives us the numbers in the right-hand columns in Figure D.8.

Product		Capacity Needed			Cell (Production Unit)	Available Capacity		Max Throughput	
		Stock Units/Capacity Cluster				*# Capacity Clusters/Production Period*		*# Stock Units/ Production Period*	
		# Stock Units		*Capacity Cluster*					
1-HTT-534	Heavy Truck Transmission	2	Piece	TransmissionAssemblyCluster	Transmission Assembly	6	TransmissionAssemblyCluster	12	Piece
1-LVT-142	Light Vehicle Transmission	5	Piece	TransmissionAssemblyCluster	Transmission Assembly	6	TransmissionAssemblyCluster	30	Piece
2-HTH-534	Heavy Truck Transmission Housing	8	Piece	HousingMachiningCluster	Housing Machining	4	HousingMachiningCluster	32	Piece
2-LVH-142	Light Vehicle Transmission Housing	12	Piece	HousingMachiningCluster	Housing Machining	4	HousingMachiningCluster	48	Piece
3-HGK-534	Heavy Gear Kit	5	Piece	GearKitAssemblyCluster	Gear Kit Assembly	3	GearKitAssemblyCluster	15	Piece
3-LGK-142	Light Gear Kit	10	Piece	GearKitAssemblyCluster	Gear Kit Assembly	3	GearKitAssemblyCluster	30	Piece
4-GTW-016	Gear 16 teeth	20	Piece	GearManufacturingCluster	Gear Manufacturing	5	GearManufacturingCluster	100	Piece
4-GTW-033	Gear 33 teeth	15	Piece	GearManufacturingCluster	Gear Manufacturing	5	GearManufacturingCluster	75	Piece
4-GTW-045	Gear 45 teeth	10	Piece	GearManufacturingCluster	Gear Manufacturing	5	GearManufacturingCluster	50	Piece
4-GTW-078	Gear 78 teeth	6	Piece	GearManufacturingCluster	Gear Manufacturing	5	GearManufacturingCluster	30	Piece
4-GTW-090	Gear 90 teeth	4	Piece	GearManufacturingCluster	Gear Manufacturing	5	GearManufacturingCluster	20	Piece
4-GTW-156	Gear 156 teeth	2	Piece	GearManufacturingCluster	Gear Manufacturing	5	GearManufacturingCluster	10	Piece

Figure D.8 Calculation of dedicated throughput for each product in each cell.

As an example, to help explain these numbers, let us consider the assembly of the Heavy Truck Transmission. Suppose the "Production Period" being used for these data is one shift. The data show that there are six Capacity Clusters available in the Transmission Assembly Cell in one shift. Further, two Heavy Truck Transmissions can be assembled with one Capacity Cluster. Thus, if the whole cell is dedicated to this product, it should be able to assemble 12 Heavy Truck Transmissions in a shift, and this is the number that you find towards the end of the first row of the table. This number (12) now serves as a reality check for the setting of the Capacity Cluster for this cell. Similarly, all the other numbers in this column can be used to verify or refine the definition of the Capacity Cluster for a given cell and given product type.

Defining a Common Capacity Cluster for Companywide Comparisons

In the previous sections, we introduced the concept of multiple Capacity Cluster definitions, each one typically related to a given cell (or occasionally, a few cells). However, it may not be clear how to use these separate definitions to compare the needed and available capacity at a higher aggregate level, such as for areas of a factory, or even a whole factory. If there is a need to do so, we propose here an approach to defining a super-level common Capacity Cluster for a group of cells or an entire factory. This can be done by relating all other Capacity Clusters to this common one using a weighting factor to calculate the equivalent common capacity of a cell (say "a") as follows:

$$\begin{aligned}&\#\mathit{Common\ Capacity\ Clusters}\\&\quad = \#\mathit{Capacity\ Cluster}_a \times \mathit{Weighting\ Factor}_a\end{aligned}$$

Adding up the common Capacity Clusters for a collection of cells provides an even more aggregated capacity number for an area of a factory or a whole factory, as will be shown below.

To determine the weighting factors, and by consequence the proportionality of all cells, common sense must be used, because these factors depend on the company priorities. For example, the relative contribution to the company profit or revenue could be used, or alternatively in a labor-constrained or machine-constrained company also the importance in terms of labor or machine power

Capacity Cluster	Representing the Set of Resources Needed	Weight Factor	Explanation
CommonCluster	Common capacity cluster to which all other capacity clusters could be recalculated, to enable high-level comparison on company level	1.00	–
GearKitAssemblyCluster	To Assemble Gear Kits starting from the in-house manufactured gears	2.00	2 common capacity clusters represent an equivalent capacity
GearManufacturingCluster	To manufacture gears from a blank. The cluster consists of one person operating a milling, hobbing and deburring machine.	0.80	80% of a common capacity cluster represents an equivalent capacity
HousingMachiningCluster	To prepare the housing starting from a casting. The cluster consists of one person using a milling machine and drill.	1.50	150% of a common capacity cluster represents an equivalent capacity
TransmissionAssemblyCluster	To perform an assembly. The assembly is executed by two people using a set of machinery such as a multi-purpose jig and power tools.	0.50	Half of a common capacity cluster represents an equivalent capacity

Figure D.9 Weighting factors chosen by MadTran planners for common capacity calculation.

Cell (Production Unit)	Available Capacity per Production Period					Share
	# Capacity Clusters		Weighting Factor	# Common Capacity Clusters		
Transmission Assembly	6	TransmissionAssemblyCluster	0.50	3	CommonCluster	15.79%
Housing Machining	4	HousingMachiningCluster	1.50	6	CommonCluster	31.58%
Gear Kit Assembly	3	GearKitAssemblyCluster	2.00	6	CommonCluster	31.58%
Gear Manufacturing	5	GearManufacturingCluster	0.80	4	CommonCluster	21.05%
Company Total				**19**	**CommonCluster**	**100.00%**

Figure D.10 Total aggregated capacity for MadTran along with the capacity share for each of the cells.

is another option. Figure D.9 illustrates the choice of these weighting factors for our MadTran example, along with an explanation in the right-hand column. Using these factors, in Figure D.10 we see that at this super-level aggregations, MadTran has a total of 19 common Capacity Clusters. Further, using these common clusters, we can calculate the share of capacity that is contributed by a given cell, simply by looking at the proportion that the cell contributes to the total common clusters (see the last column in Figure D.10). This can also be viewed as a measure of importance of a given production cell or area. For instance, the figure shows that even though the Housing Machining and Gear Kit Assembly are different types of cells, they have exactly the same weight for the company regarding available capacity. Thus, we see that using the construct of common Capacity Clusters makes it possible to compare capacity in an aggregated way across different cells or areas of the factory.

Calculating High-Level Capacity for Work Orders

One of the main goals of our approach, of course, is to enable capacity planning, which requires comparing the available capacity with the needed capacity for all the promised orders. In order to do this, we need to calculate the capacity needed by each order, and then cumulate this across all the orders in the system. In this section, we show how to go from a final order to the specific work orders needed for that job, and then to the capacity needed for each work order.

Suppose that MadTran gets a customer order for five Heavy Truck Transmissions. For the purpose of this numerical example, we will assume that for this low-volume and customized business, all the parts for this job are made to order; specifically, we assume that the intermediate parts are not stocked ahead of time, nor are their work orders combined with other jobs to save on setups.

We will now determine all the necessary work orders for this customer order. Refer to Figures D.1 and D.2 for the multilevel BOM for this transmission. We will start from the end product and work our way back to the raw materials. The last work order (we'll label it WO-001), is for the final assembly of the five transmissions mentioned in the sales. From Figure D.1 we see that each transmission requires three components: two of them (a housing and a gear kit) come from in-house operations, while the third (a shaft) has to be purchased.

The two in-house components need to be processed with two separate work orders, which are:

- WO-002 for machining of five housings, using five purchased castings.
- WO-003 for assembly of 10 Gear Kits (two kits for each transmission).

Next, we need the work order to manufacture all the gears for the Gear Kits. Each type of component will have a separate work order. Referring to the BOM and detailed data in Figures D.1 and D.2 we can calculate the number of pieces needed for each, and this results in the following work orders:

- WO-004 for 200 of the 16-teeth gears, using 200 purchased Dimension 20 blanks.
- WO-005, for 80 of the 45-teeth gears, using 80 purchased Dimension 50 blanks.
- WO-006 for 60 of the 78-teeth gears, using 60 purchased Dimension 100 blanks.
- WO-007 for 20 of the 156-teeth gears, using 20 purchased Dimension 200 blanks.

Of course, all the purchased components listed above need to be already in stock or else they need to be purchased; however, this is done in a separate module of the MRP system and here we are concerned only with capacity planning for the in-house processed items, so we will not consider the purchased components further.

Finally, we will calculate the capacity needed for each work order. This capacity is expressed in the Capacity Cluster of the cell where the appropriate step will be executed (and which must be the same as the unit of capacity used in the BOM to define how many stock units can be manufactured). The capacity needed for processing a given work order (WO) in a cell is simply the number of stock units in the work order multiplied by the number of Capacity Clusters required per stock unit, given in Figure D.7:

$$\#\textit{Capacity Clusters Needed}_{WO}$$
$$= \#\textit{Stock Units}_{WO}$$
$$\times \#\textit{Capacity Clusters per Stock Unit}_{BOM_{Product_{WO}}}$$

Figure D.11 shows the results of this calculation for all the work orders. The last column in the figure shows the number of Capacity Clusters needed for each work order in the appropriate cell. As a result of our approach and the corresponding BOM data, this capacity needed is expressed using the same kind of Capacity Cluster as the cell by virtue of all the definitions made above. Let's go over the calculation for work order WO-003, which assembles 10 Heavy Truck Gear Kits in the Gear Kit Assembly cell. This work order requires 10 stock units to be assembled, and from Figure D.7 we see that each such unit requires 0.2 Capacity Clusters in this cell. Thus, we need

Workorder						Capacity Needed		
						Capacity Clusters/Stock Unit		*Total # Capacity Clusters*
No.	*# Stock Units*		*Product (Bill of Material)*		*Cell (Production Unit)*	*#*	*Capacity Cluster*	
WO-001	5	Piece	1-HTT-534	Heavy Truck Transmission	Transmission Assembly	0.500	TransmissionAssemblyCluster	2.500
WO-002	5	Piece	2-HTH-534	Heavy Truck Transmission Housing	Housing Machining	0.125	HousingMachiningCluster	0.625
WO-003	10	Piece	3-HGK-534	Heavy Gear Kit	Gear Kit Assembly	0.200	GearKitAssemblyCluster	2.000
WO-004	200	Piece	4-GTW-016	Gear 16 teeth	Gear Manufacturing	0.050	GearManufacturingCluster	10.000
WO-005	80	Piece	4-GTW-045	Gear 45 teeth	Gear Manufacturing	0.100	GearManufacturingCluster	8.000
WO-006	60	Piece	4-GTW-078	Gear 78 teeth	Gear Manufacturing	0.167	GearManufacturingCluster	10.000
WO-007	20	Piece	4-GTW-156	Gear 156 teeth	Gear Manufacturing	0.500	GearManufacturingCluster	10.000

Figure D.11 Capacity needed for each work order for the final order for five Heavy Truck Transmissions.

10 × 0.2 = 2.0 total Capacity Clusters, as can be seen in the last column of the figure.

Defining the Quantum for POLCA in Terms of Capacity Clusters

As explained in Chapter 5, in designing your POLCA system an important decision to be made is the choice of *quantum*: the amount of work that can be done with one POLCA card. Chapter 5 gave examples of different ways of defining the quantum, such as number of pieces, hours of work, or material-handling container size. The concept of Capacity Cluster allows for a logical choice of quantum.

Since a POLCA card returning from a downstream cell signals available capacity at this downstream cell, it makes sense for this signal be in units of capacity of that cell. The Capacity Cluster for a cell is designed to be just such a unit. So, it is a rational choice to define the quantum for a given POLCA loop in terms of number of Capacity Clusters of the downstream cell. For example, for a loop going to the Transmission Assembly Cell, the quantum could be set at two TransmissionAssemblyClusters. Then this quantum can be converted to number of stock units for a given product. This number will in general be different for each product, based on the number of units that can be made with one Capacity Cluster; this is exactly how it should be, since we want to use a certain amount of capacity downstream, rather than to send a fixed number of pieces. We will now illustrate these calculations using the MadTran example.

Suppose that MadTran has implemented POLCA on its shop floor. Based on the four cells in place and the product flows already described, it is clear that three POLCA loops will be needed (Figure D.12).

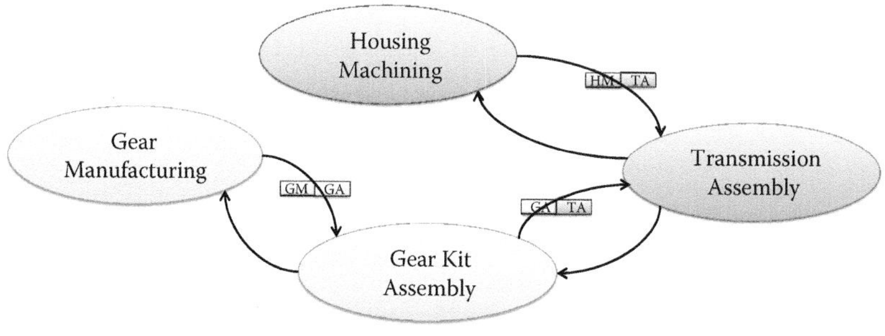

Figure D.12 POLCA loops and cards at MadTran.

We will use the HM/TA loop (from Housing Machining to Transmission Assembly) to illustrate the approach. Based on the design issues laid out in Chapter 5, let's say the MadTran team has decided that the quantum for the HM/TA loop is set to a capacity of two TransmissionAssemblyClusters. If the HM cell is machining the housing for a Heavy Truck Transmission, the product to be manufactured in the downstream cell (TA) will be the Heavy Truck Transmission End Product. Because two pieces of this end product can be manufactured with one Capacity Cluster, and because the POLCA card represents two Capacity Clusters, four pieces of the end product can be made with one POLCA card (quantum). Since each end product needs one housing, this means that HM can work on up to four housings with one POLCA card. Similarly, if HM is machining a Light Vehicle Transmission Housing, five pieces can be manufactured with the same Capacity Cluster, resulting in 10 housings that can be machined with one POLCA card. Although defining the capacity of a POLCA card using a Capacity Cluster makes the quantum variable in terms of pieces, it makes it more accurate in terms of the goal of the POLCA card, which is to send work appropriate to the capacity of the downstream cell.

In general, we can state the calculation needed using this formula:

$$\begin{aligned}&\textit{Downstream stock units per POLCA card}\\&\quad= \textit{Downstream Capacity Clusters per quantum}\\&\quad\quad\times \textit{Downstream stock units per Capacity Cluster}\end{aligned}$$

Applying this formula to all the loops gives us the set of quanta shown in Figure D.13.

Next, we can use the previously calculated capacity needed for a work order (see Figure D.11) to calculate the number of POLCA cards needed for that work order using the formula:

$$\#\textit{Polca Cards Needed}_{Wo} = \frac{\#\textit{Downstream Capacity Clusters Needed}_{Wo}}{\#\textit{Downstream Capacity Clusters per Quantum}}$$

Note that this number will need to be rounded up to the next integer.

Using the previous example of the customer order for five Heavy Truck Transmissions, Figure D.14 calculates the number of POLCA cards needed for all work orders for this customer order. For instance, for WO-003 we need 2.5 Capacity Clusters (downstream). Dividing this by the quantum of

POLCA Card					(Downstream) Product					# Stock Units per POLCA Card	
			Capacity per Quantum				Capacity needed				
							Stock Units/Capacity Cluster				
Loop	*From*	*To (Downstream Cell)*	*# Capacity Clusters*		*Product*		*# Stock Units*		*Capacity Cluster*		
HM/TA	Housing Machining	Transmission Assembly	2	TransmissionAssemblyCluster	1-HTT-534	Heavy Truck Transmission	2	Piece	TransmissionAssemblyCluster	4	Piece
HM/TA	Housing Machining	Transmission Assembly	2	TransmissionAssemblyCluster	1-LVT-142	Light Vehicle Transmission	5	Piece	TransmissionAssemblyCluster	10	Piece
GA/TA	Gear Kit Assembly	Transmission Assembly	2	TransmissionAssemblyCluster	1-HTT-534	Heavy Truck Transmission	2	Piece	TransmissionAssemblyCluster	4	Piece
GA/TA	Gear Kit Assembly	Transmission Assembly	2	TransmissionAssemblyCluster	1-LVT-142	Light Vehicle Transmission	5	Piece	TransmissionAssemblyCluster	10	Piece
GM/GA	Gear Manufacturing	Gear Kit Assembly	3	GearKitAssemblyCluster	3-HGK-534	Heavy Gear Kit	5	Piece	GearKitAssemblyCluster	15	Piece
GM/GA	Gear Manufacturing	Gear Kit Assembly	3	GearKitAssemblyCluster	3-LGK-142	Light Gear Kit	10	Piece	GearKitAssemblyCluster	30	Piece

Figure D.13 Quantum calculated for each POLCA loop at MadTran.

Workorder					Capacity Needed	POLCA Cards Needed				
No.	*# Stock Units*		*Product (Bill of Material)*		*Cell (Production Unit)*	*# Capacity Clusters*	*Loop*	*Downstream Cell*	*# Capacity Clusters/ Quantum*	*Downstream Capacity Clusters Needed*	*# POLCA Cards*
WO-001	5	Piece	1-HTT-534	Heavy Truck Transmission	Transmission Assembly	2.500	–	–	–	–	–
WO-002	5	Piece	2-HTH-534	Heavy Truck Transmission Housing	Housing Machining	0.625	HM/TA	Transmission Assembly	2	2.5	2
WO-003	10	Piece	3-HGK-534	Heavy Gear Kit	Gear Kit Assembly	2.000	GA/TA	Transmission Assembly	2	2.5	2
WO-004	200	Piece	4-GTW-016	Gear 16 teeth	Gear Manufacturing	10.000	GM/GA	Gear Kit Assembly	3	2.0	1
WO-005	80	Piece	4-GTW-045	Gear 45 teeth	Gear Manufacturing	8.000	GM/GA	Gear Kit Assembly	3	2.0	1
WO-006	60	Piece	4-GTW-078	Gear 78 teeth	Gear Manufacturing	10.000	GM/GA	Gear Kit Assembly	3	2.0	1
WO-007	20	Piece	4-GTW-156	Gear 156 teeth	Gear Manufacturing	10.000	GM/GA	Gear Kit Assembly	3	2.0	1

Figure D.14 Number of POLCA cards needed for each work order for the customer order of five Heavy Truck Transmissions.

2.0 Capacity Clusters gives us the result of 1.25 POLCA Cards. Rounding up, this results in the number of 2.0 in the figure. Similar reasoning can be used for the other entries. The only other explanation is that for WO-001 no POLCA cards are needed because there is no downstream cell.

This number of POLCA cards is needed for the decision about launching each work order into the cell. Depending on the decision rules being used, the full set of POLCA cards may be needed before the work order can be launched (see Chapter 5).

This calculation is also important for another important reason. In order to calculate the number of POLCA cards in each loop, we need to estimate the flow of POLCA cards in each loop during a given planning period (Chapter 5). This number of cards, totaled across all anticipated work orders, will be required for this calculation (also explained in Chapter 5).

High-Level Capacity Definitions Enable Easier Adding of Data for Customized Products

As stated at the beginning of this appendix, for companies that make a lot of customized products it can be very time-consuming to first determine and then input into the MRP system the data for each new customized product. The whole concept of high-level Capacity Clusters is aimed at simplifying this task. Even when companies make customized products, there are a lot of similarities with previously made products.

When defining similar products to be manufactured in a similar process and similar cells, the high-level definition of capacity with Capacity Clusters helps to predict the impact of the introduction of such a similar product. In this case the derived bill of material for the new product could simply refer to an existing Capacity Cluster, but the number of stock units and lead time could be different if needed. Even when products are highly customized and different, when designing the processes, it should be possible for engineering to relate the different stages of the production process to existing Capacity Clusters.

We will illustrate this once again using the MadTran example. Let's say that MadTran decides to expand its existing offering of products (Heavy Truck Transmission and Light Vehicle Transmission) with a new transmission for Heavy Vehicles. The engineers start with the design, but it is clear that the general process and the standard parts will not be very different from the existing products. One of the results of the engineering session will be a new BOM. It turns out that for the new product the same housing and shaft can be used as for the Light Vehicle Transmission, but a specific Gear Kit is needed. Still, this kit uses the existing Gears. The resulting BOM structure for this Heavy Vehicle Transmission is shown in Figure D.15 along with the details in Figure D.16.

From the figures, you can see that most items in the multi-level BOM were already defined earlier. Only the new end-product (Heavy Vehicle Transmission) and a new semi-finished product for the Heavy Vehicle Gear Kit need to be defined. Next, in terms of capacity, we only have to define the capacity needed and an appropriate lead time for these two new items in the BOM. For this, we can start with the already-defined Capacity Clusters and lead times, and then more appropriate numbers can be decided for the new items. For example, the engineers could reason that because the transmission for heavy vehicles can be positioned between the transmissions for light vehicles and heavy trucks, it can be assumed that the needed capacity and lead times lie between the numbers for these other two products. The result of the engineers' estimates are shown in Figure D.17. We can also check whether these results are realistic by calculating the maximum throughput (Figure D.18) and verifying this estimate with production staff. For the new Heavy Vehicle Transmission, we get a throughput of 24 pieces a day, which can be compared with the throughput of 30 pieces a day for the existing Light Vehicle Transmission (see Figure D.8). If needed, based on feedback from

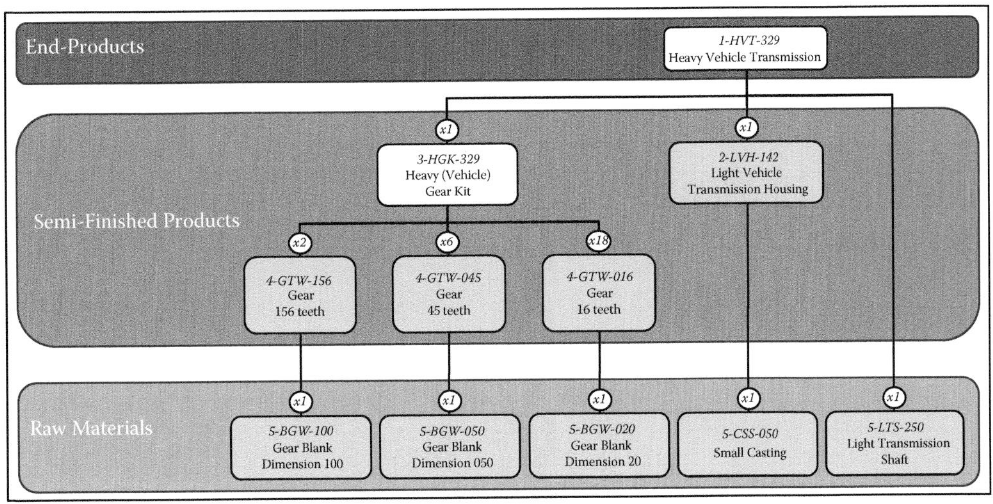

Figure D.15 Overview of BOM structure for the new Heavy Vehicle Transmission.

Product Definition				Construction (*Bill of Material*)				
Product		***Type***	***Process***	***# Stock Units Needed***		***Part/Component***		***Type***
1-HVT-329	Heavy Vehicle Transmission	End Product	Assemble Transmission	1	Piece	2-LVH-142	Light Vehicle Transmission Housing	Semi-Finished Product
				1	Piece	3-HGK-329	Heavy (Vehicle) Gear Kit	Semi-Finished Product
				1	Piece	5-LTS-250	Light Transmission Shaft	Raw Material
2-LVH-142	Light Vehicle Transmission Housing	Semi-Finished Product	Manufacture Housing	1	Piece	5-CSS-050	Small Casting	Raw Material
3-HGK-329	Heavy (Vehicle) Gear Kit	Semi-Finished Product	Assemble Gear Kit	18	Piece	5-GTW-016	Gear 16 teeth	Semi-Finished Product
				6	Piece	4-GTW-045	Gear 45 teeth	Semi-Finished Product
				2	Piece	4-GTW-156	Gear 156 teeth	Semi-Finished Product
4-GTW-016	Gear 16 teeth	Semi-Finished Product	Manufacture Small Gear	1	Piece	5-BGW-020	Gear Blank Dimension 20	Raw Material
4-GTW-045	Gear 45 teeth	Semi-Finished Product	Manufacture Small Gear	1	Piece	5-BGW-050	Gear Blank Dimension 50	Raw Material
4-GTW-156	Gear 156 teeth	Semi-Finished Product	Manufacture Large Gear	1	Piece	5-BGW-200	Gear Blank Dimension 200	Raw Material

Figure D.16 Detailed BOM for the new Heavy Vehicle Transmission.

Product		Lead Time	Capacity Needed			
			Stock Units/Capacity Cluster			# Capacity Clusters/ Stock Unit
			# Stock Units		Capacity Cluster	
1-HTT-534	Heavy Truck Transmission	4.00	2	Piece	TransmissionAssemblyCluster	0.500
1-HVT-329	Heavy Vehicle Transmission	3.00	4	Piece	TransmissionAssemblyCluster	0.250
1-LVT-142	Light Vehicle Transmission	2.00	5	Piece	TransmissionAssemblyCluster	0.200
3-HGK-534	Heavy Gear Kit	5.00	5	Piece	GearKitAssemblyCluster	0.200
3-HGK-329	Heavy (Vehicle) Gear Kit	3.00	8	Piece	GearKitAssemblyCluster	0.125
3-LGK-142	Light Gear Kit	2.00	10	Piece	GearKitAssemblyCluster	0.100

Figure D.17 Lead time and capacity needed for the Heavy Vehicle Transmission items.

production, the numbers for the new products can be fine-tuned based on these types of reality checks.

Using Capacity Clusters for High-Level Planning

We now build on all the preceding concepts to show how Capacity Clusters can be used in the High-Level MRP (HL/MRP) planning process. Note that most of the processes described below are the standard MRP processes; the difference is only in the retention of the BOM structure and use of the higher-level capacity constructs at several places—both of these are explained below.

The HL/MRP process starts in the usual way with the release of sales orders. A sales order is released with a promised delivery date, so a factory due date can be calculated from this to be sure the ordered product can be delivered to the customer on time. The next step is the generation of the dependent demand, so for a sales order the complete underlying work order and purchase order structure is generated. One difference in relation to classic MRP where all needed work orders are generated unrelated to each other, in HL/MRP the relation between the work orders (dictated by the multilevel BOM structure) is kept in a hierarchical work order structure. Every work order in the structure is assigned to a specific cell, using the matching Capacity Cluster (the product to be manufactured by the work order with a given BOM is related to a Capacity Cluster, so the order has to be assigned to a cell related to the same Capacity Cluster).

With backward-scheduling from the due date, the start dates are calculated for the work orders. If the order flow is being managed by a POLCA control system then these start dates are termed Authorization Dates. After

Product		Capacity Needed			Cell (Production Unit)	Available Capacity		Max Throughput	
		Stock Units/Capacity Cluster				# Capacity Clusters/Production Period		# Stock Units/ Production Period	
		# Stock Units		Capacity Cluster					
1-HVT-329	Heavy Vehicle Transmission	4	Piece	TransmissionAssemblyCluster	Transmission Assembly	6	TransmissionAssemblyCluster	24	Piece
2-LVH-142	Light Vehicle Transmission Housing	12	Piece	HousingMachiningCluster	Housing Machining	4	HousingMachiningCluster	48	Piece
3-HGK-329	Heavy (Vehicle) Gear Kit	8	Piece	GearKitAssemblyCluster	Gear Kit Assembly	3	GearKitAssemblyCluster	24	Piece
4-GTW-016	Gear 16 teeth	20	Piece	GearManufacturingCluster	Gear Manufacturing	5	GearManufacturingCluster	100	Piece
4-GTW-045	Gear 45 teeth	10	Piece	GearManufacturingCluster	Gear Manufacturing	5	GearManufacturingCluster	50	Piece
4-GTW-156	Gear 156 teeth	2	Piece	GearManufacturingCluster	Gear Manufacturing	5	GearManufacturingCluster	10	Piece

Figure D.18 Dedicated throughput calculation for the Heavy Vehicle Transmission items.

determining the start date (Authorization Date), the work order is put into particular time buckets based on that date. The time buckets are chosen for the specific MRP implementation (typical choices are shifts, days, weeks, and so on). Next, the number of Capacity Clusters needed per work order is determined as explained earlier in the appendix. This amount can be put in a single time bucket (e.g., the one associated with the start date), or evenly spread over the buckets containing the time needed to execute the work order based on its lead time; again, this is a choice of the MRP implementation at the company. Because all work orders assigned to the same cell have their capacity defined with the same Capacity Cluster, the capacity of all these work orders assigned to the same time bucket can be aggregated. Since the available capacity is also defined for a cell using the same Capacity Cluster, the available capacity and the needed capacity per time bucket can be compared and the relative usage of capacity can be calculated. If the capacity usage for a given bucket at a certain cell exceeds a critical threshold (e.g., 90%), the usual planning processes can be used to deal with this issue. For example, typical solutions involve authorizing overtime, adding temporary workers, outsourcing work, or in some cases changing a promised delivery date.

We will illustrate the use of the above approach for the work orders for the final order for five Heavy Truck Transmissions (see Figure D.11). The details of each of the above steps for this set of work orders, along with the associated formulas and calculations, can be found on the authors' website (www.QRM-University.com); here we present the results of the calculations and the insights obtained.

Using only the work orders generated for the mentioned final order (in other words, without taking into account any other orders), we get the High-Level Capacity Plan shown in Figure D.19. The plan shows, for instance, that for Gear Manufacturing during the first two days, 17% of the available capacity is planned (0.83 clusters planned out of the 5.00 clusters available), for the next two days 37% of capacity is planned (1.83/5.00), and then the capacity planned rises to 82% during the next days. If the company sets its "alarm level" for capacity at 80%, then because the alarm level is exceeded, the schedule shows there is a risk that lead times will be longer than planned because of shop floor congestion. Thus, no more orders should be planned in that period.

However, if other orders need to be planned, or if we wish to remove the alarm, then by drilling down into the details in the MRP system we can see which products and work orders are responsible for the needed capacity. At

Cell	Type	Product	Ticket	Today	Day 1	Day 2	Day 3	Day 4	Day 5	Day 6	Day 7	Week 2	Week 3	Week 4	Week 5	Week 6
Housing Machining	Cap%			0.00	0.00	0.00	0.00	0.00	0.00	0.00	0.00	1.00	1.00	0.00	0.00	0.00
	Available			4.00	4.00	4.00	4.00	4.00	4.00	4.00	4.00	28.00	28.00	28.00	28.00	28.00
	Planned	2-HTH-534	WO-002									0.21	0.42			
Gear Manufacturing	Cap%			17 00	17.00	37.00	37.00	82.00	82.00	82.00	82.00	47.00	0.00	0.00	0.00	0.00
	Available			5 00	5.00	5.00	5.00	5.00	5.00	5.00	5.00	35.00	35.00	35.00	35.00	35.00
	Planned	4-GTW-045	WO-005					1.00	1.00	1.00	1.00	4.00				
		4-GTW-016	WO-004					1.25	1.25	1.25	1.25	5.00				
		4-GTW-078	WO-006			1.00	1.00	1.00	1.00	1.00	1.00	4.00				
		4-GTW-156	WO-007	0 83	0.83	0.83	0.83	0.83	0.83	0.83	0.83	3.33				
Gear Kit Assembly	Cap%			0.00	0.00	0.00	0.00	0.00	0.00	0.00	0.00	6.00	4.00	0.00	0.00	0.00
	Available			3.00	3.00	3.00	3.00	3.00	3.00	3.00	3.00	21.00	21.00	21.00	21.00	21.00
	Planned	3-HGK-534	WO-003									1.20	0.80			
Transmission Assembly	Cap%			0.00	0.00	0.00	0.00	0.00	0.00	0.00	0.00	0.00	6.00	0.00	0.00	0.00
	Available			6.00	6.00	6.00	6.00	6.00	6.00	6.00	6.00	42.00	42.00	42.00	42.00	42.00
	Planned	1-HTT-534	WO-001										2.50			

Figure D.19 High-Level Capacity Plan for the example order.

this point the planners can use any of the usual solutions listed above, such as authorizing overtime, moving delivery dates, or other similar solutions as might be appropriate and typically used at the company.

As you can see from the preceding discussion, apart from a few special considerations the processes involved are the usual ones with which the MRP/Planning staff at any company is already familiar. Hence the implementation of Capacity Clusters will not be difficult; in fact, it should simplify the tasks of the people involved for all the reasons explained in this appendix.

Relation Between Capacity Clusters, Lead Time (MCT), and Gray or White Space

It is also possible to connect Capacity Clusters to lead time, and more specifically to MCT (see Appendix A). As explained in Appendix A, the MCT metric for an order can be decomposed into "gray space" and "white space" components. Capacity Clusters can be used to calculate these components. Readers interested in seeing these details for the MadTrans order for five transmissions can find them on the authors' website (www.QRM-University.com).

Determination of Capacity Clusters and Tuning the Parameters

It can be seen that the concept of the Capacity Cluster is very appealing in regard to getting control over capacity definition at a high level. This is helped in particular by the limited number of parameters that need to be defined to make the system work. A key question, however, is how to define those parameters reasonably. We discussed this briefly with the MadTran example and illustrated the answer with some instances. We now cover this point in more detail.

Analysis of historical data can lead to fine-tuning the relation between BOMs, cells, and Capacity Clusters, but such data are not always available and even when they are, this will only give a first raw setup that needs to be tuned. In many cases it will come down to using common sense when defining the parameters the first time. This makes the possibility of tuning them, based on the first results, even more important. Thus, a possible scenario to set up the parameters could be the following (refer to Figure D.20):

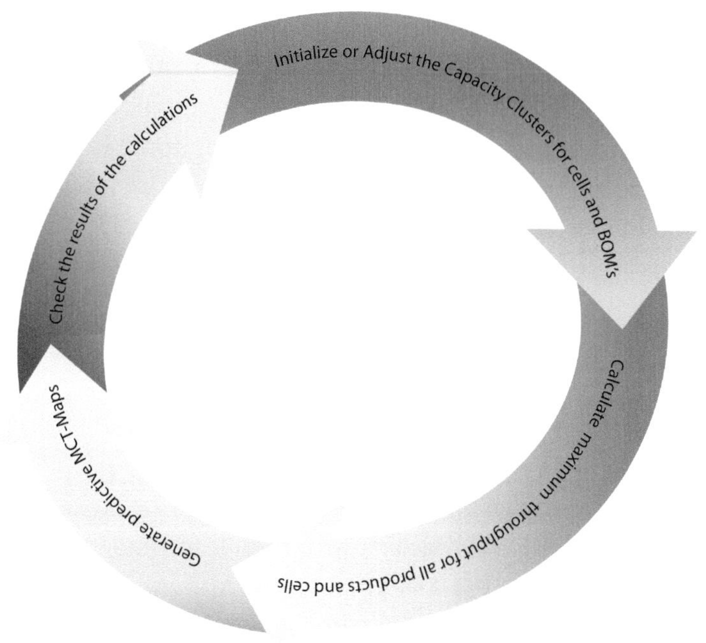

Figure D.20 Tuning the parameters by iteration.

- Set up the parameters using common sense.
- Calculate the dedicated throughput for each product in every cell (how many units of each product can be manufactured, when only that product is made in the cell) as shown in Figure D.8 and the accompanying explanation earlier in this appendix.
- Workers and managers responsible for each cell will help you to evaluate how realistic these numbers are. Based on the feedback the model can be tuned until the numbers for dedicated throughput are viewed as being reasonable.

If, after initial use of this approach, work orders still do not proceed as planned, or when measured lead times show large discrepancies from the planned lead times, then a more major tuning of the parameters can be considered by reviewing the relation between BOMs, cells, and Capacity Clusters using one more of the following additional actions:

- Relate a different, more suited, Capacity Cluster to a BOM or a cell.
- Create a whole new Capacity Cluster because the product is not as similar to the other related products as presumed or the competence of a cell does not match the related Capacity Cluster anymore.

- Modify the number of stock units per Capacity Cluster in the BOM.
- Modify the number of available Capacity Clusters for a cell.
- Modify the number of Capacity Clusters for a POLCA card (i.e., the quantum).
- Modify the lead time of the product.

With some experience, engineers and planners will be able to use a combination of the above approaches to improve the accuracy of the high-level planning.

Concluding Remarks

This appendix has shown how the concept of the Capacity Cluster can bring the task of scheduling to a higher level, making the HL/MRP idea more concrete. A major consequence of defining capacity at such a high level is that far fewer data items need to be initiated and maintained, resulting in several benefits:

- Tuning a manageable number of parameters becomes feasible. While it may seem like a lot of work to manage the tuning steps described in the previous section, note that this will still be much less work than trying to calculate and enter hundreds or even thousands of data (time elements, descriptions of processes and operations) as would be needed in standard MRP.
- When adding variations of products, less information and effort is needed.
- For companies making custom-engineered products, configuration of all the details for new products (such as BOM and capacity requirements for all component items) becomes much easier.
- By also defining the capacity of POLCA cards in the same way, the Capacity Cluster implements more accurate capacity signals from downstream cells to their upstream supplier cells.

Capacity Clusters are the missing link between classic MRP and the newer alternative HL/MRP that has been recommended in the literature for companies with a cellular manufacturing structure. The Capacity Cluster concept enables rational and logical planning for this HL/MRP while at the same time reducing the time demands for such planning. In other words, Capacity Clusters make HL/MRP feasible and practical.

About the Authors

Ignace A.C. Vermaelen is a Senior Partner and co-founder of 3rd Wave (Lede, Belgium). As an expert business consultant with a track-record of nearly 30 years in multiple areas such as discrete manufacturing, logistics, retail, food industry, and healthcare, he has been involved in several large software development projects concerning ERP/MRP, management information and business intelligence. Within 3rd Wave he is the architect of a QRM-tooling set and the ongoing development of the Supply Chain Management and ERP/MRP functionality.

Antoon van Nuffel is Senior Partner, co-founder, and CEO of 3rd Wave. In the mid-1980s, as a pioneer, he brought the theoretical concepts of relational databases (a la Codd and Date) to a useable level with the introduction of a powerful code generator. As manager of a team of professionals, he implemented many ERP projects in multiple areas. Today, he is the driving force in 3rd Wave dealing with the marriage between QRM and ERP.

Appendix E: The Sociotechnical Success Factors of POLCA

Guest author: Jannes Slomp

The operational benefits of POLCA have been discussed earlier in this book. In particular, the people side of the system was emphasized in Chapter 3 as well as in several other places in this book, and also illustrated by the case studies in Part III. Readers with a deeper interest in human aspects of technological systems may be wondering: "Are there specific scientific principles that explain the success of POLCA relative to other shop floor control systems, in regards to the human interactions with the system?" The answer is as follows: indeed there are such scientific principles, and this appendix will describe how POLCA is linked to *sociotechnical* success factors of system design and implementation. (*Sociotechnical systems* is the term for the formal approach to complex organizational work design that recognizes the interaction between people and technology in workplaces.) The material here will be helpful for managers at companies that are planning to adopt POLCA, and who would like to have a deeper understanding of how sociotechnical factors are properly addressed in POLCA specification and implementation.

The POLCA system was first introduced in the book *Quick Response Manufacturing* by Rajan Suri (Productivity Press, 1998), and since then there have been many other publications and case studies on this system (see Appendix H). Since POLCA has already been described in detail in the main part of this book, we will assume the reader is familiar with the workings of POLCA and related terminology, and we will not provide any details on POLCA here.

Bridging the Gap Between Theory and Practice

The scientific literature includes more than fifty years of formal publications on scheduling theory and scheduling algorithms, and hundreds—if not thousands—of algorithms can be found in the collection of leading international journals. And yet, one finds few applications of most of these algorithms in practice. There are many reasons for this practicality gap. Scheduling algorithms presented by scientists are mostly complex, and practitioners prefer solutions that are simple, even if inferior, over-complex and (supposedly) superior solutions from algorithms that they do not understand. It is often costly to implement the complex algorithms. Also, a lot of data is needed by these algorithms, which is not easily available. Furthermore, it is doubtful to what extent the scheduling algorithms actually do come up with good solutions. When proving the optimality of the algorithms, scientists assume that the available data is correct. They also assume that their model does grasp the most important elements of the real system. This is seldom true. Lack of accurate data on processing times, further exacerbated by the variability in processing times needed by different workers, variability in the number of available workers on a given day, and the presence of sequence-based setup times are, amongst other things, issues that are rarely incorporated in scheduling algorithms. Also, most scheduling algorithms schedule around a single class of constrained resource (e.g., only machines), while reality has to cope simultaneously with multiple constrained resources (machines, workers, fixtures, and so on). On top of all this is the real-world fact that the whole environment is itself dynamic, with customer order changes, unexpected "drop-in" orders, and real-time system changes (machine failures, absent workers, rework or scrap, missing parts), which raises doubts about whether a schedule that was calculated earlier can still be considered "optimum" in any reasonable sense.

This appendix helps to explain a significant aspect of how POLCA bridges the gap between theory and practice. POLCA is a transparent and robust planning and control system and has proven its value in practice. The system supports the realization of short lead times in a high-mix, low-volume and custom (HMLVC) environment. POLCA has clear logic and has the robustness to cope with many types of variabilities. The technical abilities of POLCA are one aspect of the practical success of the system. However, we will see here that human issues, furthermore, play a major role in the acceptance of the system by the people in a company.

Need for a Sociotechnical Approach

In practice, there is an overemphasis on the technical aspects of planning and control systems. It is essential to consider human and organizational aspects when specifying and implementing a new planning and control system. The relative failure of many shop floor planning and control systems can be explained, at least partially, in terms of the lack of a true sociotechnical approach to the design and implementation of these systems.

A well-known list of ten sociotechnical factors to be used in the design and implementation of sociotechnical systems is given by Cherns in two articles published in 1976 and 1987 (see "For Further Reading" at the end of this appendix). This set of factors has been used in several studies and provides a handy checklist for the assessment of technical systems to be used in an organizational context. Here we will divide Cherns' factors into two categories: *process-oriented* factors and *design-oriented* factors. Process-oriented factors relate to the process that should be followed in order to gain acceptance of the system by those who will work with the system. Design-oriented factors are important to link the social and the technical needs while working with the system.

In this appendix, we will explain how these factors relate to the activities that occur during the adapting and implementing of POLCA to a specific situation. To illustrate these points in practice, we use the case study of Variass Electronics BV, the Netherlands, where POLCA has been successfully implemented. Variass is a system supplier and EMS (Electronics Manufacturing Services) specialist in electronic and mechatronic high-tech products and systems. The company manufactures a huge variety of customer-specific products. About six years ago, Variass implemented POLCA in its manufacturing department. An ERP system is used to generate manufacturing orders, and then the POLCA card system manages the flow of jobs on the work floor. More details on this case study will follow as we discuss the success factors below.

Process-Oriented Factors for Adapting, Implementing, and Improving a System

Our category of process-oriented factors consists of three out of Cherns' list of sociotechnical factors that we believe are important in the process of adapting, implementing and improving the application of a system such as POLCA. As indicated in Figure E.1, these factors are (i) compatibility, (ii) transitional organization, and (iii) incompletion.

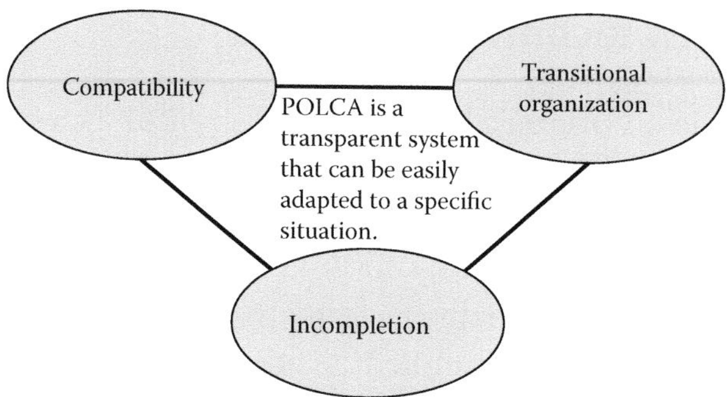

Figure E.1 Three process-oriented sociotechnical success factors and a key characteristic of POLCA for these success factors.

The first factor, *compatibility*, states that the way in which design is done should be compatible with the design's objective. It is the objective of POLCA to balance the workload on the shop floor while realizing productivity and due date targets. The system explicitly asks for cooperation from the workers. They have to accept the rules of the system with respect to the flow of jobs between cells. Furthermore, they have to self-organize the work within the cells. Therefore, the self-organization of the workers and the production manager in the POLCA system should be reflected in the process of adapting and implementing POLCA. The sociotechnical success factor of compatibility demands appropriate participation of the workers and the production managers in the adaptation and implementation of POLCA. In the case of Variass, the improvement manager developed a simulation game for the workers and the production manager. Playing this game led to some adaptations of POLCA. (Appendix F describes similar experiences with other organizations using games to help teach and train people about POLCA.)

Since the shop floor area was not too large and cells were quite close to each other, the teams at Variass decided to replace the POLCA Boards at each cell with one board centrally located and accessible to all the cells. They also made a minor modification in the way that cards connect cells, which will be explained in a later section. Figure E.2 shows the central board. There are two main points to note about this board. First, if a row (which represents a cell) has a lot of cards, it means the cell needs work from supplying cells. If there are no cards on a particular row, then the cell has no capacity and supplying cells have to perform operations for other cells. This information on the board helps to make sequencing and dispatching decisions in the cells. Second, the board provides staffing information for each cell, as also shown in Figure E.2. The combination of these pieces of

Figure E.2 Bulletin board with staffing information along with POLCA cards indicating the need for jobs. The manager is pointing to the number of workers in a particular cell.

information helps the production manager decide whether to move workers from one cell to another, if needed. The cell teams also helped management to develop rules for the re-assignment of workers to cells in order to cope with such temporal imbalances.

By doing the above activities, the adaptation and implementation process of POLCA at Variass has been compatible with the self-organizing element within POLCA. POLCA is a relative simple, but smart, system which can be simulated easily: it does not need computer support. These elements of POLCA further enabled the sociotechnical factor of compatibility.

The success factor of *transitional organization* requires that those who are responsible for adapting POLCA to the specific situation should also be responsible for the implementation of the system. As explained with many examples in Chapter 5, the design of a POLCA system needs to accommodate many company-specific situations and requires detailed knowledge of the production system. This knowledge exists among people on the shop floor. In the case of Variass, the workers made important design decisions under guidance of the improvement manager. They also implemented the operational details of the system in their areas. This was helped by the fact that implementation activities needed on the shop floor are not complex in the case of POLCA. The improvement manager also communicated with the planning department to create an appropriate connection of POLCA with the planning system of the company. As noted in Chapter 4 and Appendix C, a High-Level MRP (HL/MRP) approach is desirable to support the operation of POLCA. In the case of Variass, this was not problematic: the planning system only had to control the total required capacity. Using the central board as an indicator and the worker reallocation mechanisms as a solution, it was easy to create flexibility on the shop floor to cope with fluctuations in demand.

The sociotechnical factor of *incompletion* recognizes the fact that systems need to change in the course of time, because of a changing context and/or improvement ideas. There should be room for improvement. POLCA is not a complex system and does not need much computer support. This creates the possibility to change the system easily, if needed. It is, for instance, not complex to add or remove cells. Similarly, it is easy to add or remove cards. Furthermore, the cards themselves provide clear improvement signals, as explained in Chapters 3, 7, and 9.

Design-Oriented Factors to Link the Social and Technical Needs

As mentioned, POLCA is a system that can be easily adapted to a specific situation. Adaptation of POLCA should also obey sociotechnical design factors. Cherns (1976, 1987) present seven design-oriented factors: (i) minimal critical specification, (ii) variance control, (iii) boundary control, (iv) information flow, (v) power and authority, (vi) multi-functionality, and (vii) support congruence. Table E.1 gives a short explanation of these factors, and then we discuss how POLCA relates to each factor.

The success factor of *minimal critical specification* demands a careful specification of the essential elements of decision support needed by the workers. Only these elements should be dealt with by the system. No more, no less. A planning and control system should not unnecessarily

Table E.1 Design-Oriented Sociotechnical Success Factors for the Adaptation of a System

Principle	*Explanation*
i. Minimal critical specification	No more should be specified than is absolutely essential. What is essential should be specified.
ii. Variance control	Variances should not be exported across unit, departmental, or other organizational boundaries.
iii. Boundary location	Boundaries should not be drawn so as to impede the sharing of information, knowledge, and learning.
iv. Information flow	Information for action should be directed first to those whose task it is to act.
v. Power and authority	Those who need equipment, materials, or other resources to carry out their responsibilities should have access to them and authority to command them.
vi. The multi-functional principle	If the environmental demands vary, it then becomes more adaptive and less wasteful for each element to possess more than one function.
vii. Support congruence	Systems of social support (systems of selection, training, conflict resolution, work measurement, etc.) should be designed so as to reinforce the behaviors that the organization structure is designed to elicit.

limit the freedom of the workers to take planning and control decisions. In the adaptation of POLCA toward a particular situation, the factor of minimal critical specification needs to play a role. POLCA gives some minimal rules to be obeyed: a limited number of semi-autonomous cells are needed along with load-oriented rules for linking these cells. The production manager at Variass did have experience with another card-based control system called CONWIP (see "For Further Reading," and Appendix C and H). In his opinion CONWIP as a control mechanism running globally across the entire shop floor gave workers too much freedom (this is explained further in Appendix H). The production manager's experience with this system was that workers collected and stored jobs in order to keep busy at their preferred machines. There was no collective responsibility. In contrast, he was enthusiastic about POLCA because it served the setting of both local and global accountability. Hence, in this situation POLCA served the sociotechnical factor of minimal critical specification.

In the case of Variass, cells were defined such that the number of different routings was limited, while keeping the within-cell flows controllable by the autonomous teams. The design team experimented in a game situation with POLCA and decided to link the POLCA cards to cells instead of loops. This is an option in case of simple routings and a transparent load situation, and works as follows. Each cell team looks at the rows for all its customer (downstream) cells. If a card is available on a row, it means that particular customer has capacity to receive more work. Then, if the upstream cell decides to work on a job for this customer, it takes the customer cell's card back to its own cell, completes a job for this customer cell, and when it delivers the job it puts the card back on the row belonging to the customer. The workers adopted the FIFO (First-In-First-Out) dispatching rule, combined with the requirement of an available customer POLCA card. If a customer card is not available, then the next job in the FIFO row is chosen. Imbalance between cells is solved by moving workers. This was an easy way to avoid queues on the work floor. The central board and the row of jobs provide the necessary information for this dispatching rule. This example shows that POLCA can easily be adapted to a particular situation and how the design team searched for minimal critical specification.

The success factor of *variance control* states that variances should not be exported across unit, departmental, or other organizational boundaries. This means that each organizational level in a production control system should be able to cope with the variances that may arise at that level. In other words, decision-making tasks at each level should reflect the variances that may arise at that level. The concept of POLCA serves the success factor of variance control to a certain extent. POLCA defines clear decision-making tasks connected with the organizational levels. Control at the unit level is the cell control, guided by the availability of cards and Authorization Dates (which typically come from High-Level MRP control). Within the cells, there is room to cope with variances in processing time and some fluctuation in demand. Important for this is the setting of cell lead times. The coordination between cells is accomplished through the part of POLCA that creates pairs of cells with overlapping loops of cards. This is a self-organizing mechanism, able to cope with the varying output of the connected cells. High-level MRP control takes care of loading the production cells. This level is able to cope with variation in the importance of the various jobs. Authorization Dates and lead times are set by the High-Level MRP. The division of responsibilities are clear and offer flexibility to cope with variances at each level without frustrating other levels. In the case of Variass, workers are assigned daily to the various cells based upon the predicted workload. Temporary imbalances are solved by the production manager, who may move workers to other cells.

The success factor of *boundary control* demands that boundaries should not be drawn so as to impede the sharing of information, knowledge, and learning. In a functional layout of a factory, each specific group of machines has its own characteristics and may make suboptimal decisions with respect to sequencing and dispatching. In a cellular layout, each cell may focus on its own interests and may not be willing to share information and resources. POLCA links the various resources in a company such that all value streams can be processed on one system consisting of connected cells. According to the Quick Response Manufacturing (QRM) philosophy, the focus of POLCA is on reducing lead time. This sole focus on the lead time needed to finish manufacturing jobs reduces suboptimal decisions. As a conclusion: the design of POLCA-connected cells and the sole focus supports the boundary control success factor.

With respect to boundary control, it is interesting to compare the old manufacturing situation of Variass with the current POLCA system. Originally, the company had a functional layout. Each functional group optimized its own performance based on local efficiency metrics. However, there was substantial sub-optimization of the factory as a whole, visible in the buffers between the various processes. Then the company decided to implement Lean. Based on the advice of a consultant, the company implemented a flow line for the high-volume, fast-moving products. Other products, partly needing the same machines, were seen as exceptions. As a result, the lead times of the fast-moving products decreased, but the lead times of the large number of remaining products increased substantially. Boundaries were not correctly drawn. This led the company to investigate other methods and they eventually arrived at POLCA. The company replaced the flow line by a number of cells. Although these cells have mainly a functional character (i.e., not multidisciplinary) the new boundaries, together with the accompanying POLCA coordination system, work well for both fast-moving as well as slow-moving products.

The principle of *information flow* stresses the need that information for action should be directed first to those whose task it is to act. POLCA keeps all information gathering and processing on the work floor. At Variass, workers and the production manager use the information on the central board for decision-making. There is no intermediate planning manager needed to make assignment decisions and to inform workers.

The success factor of *power and authority* stresses the importance that workers, who have the responsibility for realizing lead times and shortening throughput times, should be able to deal with supporting means. In the cells used as building blocks of the POLCA system, the workers have complete ownership of their resources and decisions. They get the means and support to improve cell functioning. The workers of various cells are together the owners of the coordination system. They work with the system and are able to make appropriate decisions. This aspect of POLCA is important in comparison to an automated scheduling system, which can only be dealt with by a planner.

The success factor of *multi-functionality* is important for realizing system robustness. In the cells used with POLCA, workers are cross-trained. This multi-functionality of workers supports mix flexibility and gives the system robustness. Robustness is also realized by POLCA's card-based coordination system. Decision support tools that can only be used by one employee are not usable if the employee is absent. There are many small companies where the planner literally cannot go on holiday, being the only person able to make appropriate release and dispatching decisions! POLCA simplifies the actions to be performed by the planner. The planner just has to release orders based on the Authorization Date at the Planning Cell. Dispatching is organized on the shop floor, through the cards and POLCA rules. There is no additional software needed. The division of responsibilities and the transparency of the cards support the multi-functionality rule. The system is not dependent on a planning specialist. Hence, POLCA is a robust system.

The success factor of *support congruence* is related to the context of the system. If workers become responsible for the scheduling and control of their tasks using POLCA, then it is a natural progression that they should also become responsible for other decision-making tasks such as hiring temporary personnel, organizing preventive maintenance, prioritizing improvement projects, and so on. (Chapter 5 gives a practical example of a cell team being authorized to hire temporary workers.) Without support congruence, other systems in the company may frustrate the proper functioning of POLCA. Hence companies implementing POLCA should keep in mind this principle, and, in turn, POLCA helps to reinforce its application.

Conclusion: POLCA Obeys Sociotechnical Success Factors

This appendix has linked POLCA to the accepted sociotechnical success factors of system design and implementation. POLCA has characteristics that support these success factors. POLCA is a transparent system, it doesn't need advanced tools, it provides a clear division of planning and control responsibilities, and it can easily be adapted to a particular situation. All these factors, together with the system rules and coordination logic of POLCA, explain its success in practice, as can be seen by the numerous case studies in this book.

For Further Reading

"The Principles of Sociotechnical Design," by A. Cherns, *Human Relations*, Vol. 29 (8), 1976, pages 783–792.

"Principles of Sociotechnical Design Revisited," by A. Cherns, *Human Relations*, Vol. 40 (3), 1987, pages 153–162.

"CONWIP: A pull alternative to Kanban," by Spearman, M.L., D.L. Woodruff, and W.J. Hopp, *International Journal of Production Research*, Vol. 28 (5), 1990, pages 879–894.

Quick Response Manufacturing: A Companywide Approach to Reducing Lead Times, by R. Suri, Productivity Press, 1998.

About the Author

Jannes Slomp is Full Professor of World Class Performance at the HAN University of Applied Sciences, the Netherlands. He was Professor of Operations Management at the Faculty of Economy and Business at the University of Groningen until 2012. He has published over 100 papers in academic journals, professional journals, scientific conferences, and scientific books. His main expertise is on flexible automation, cellular manufacturing, cross-training, layout design, lean manufacturing, QRM, and planning and control for low-volume, high-variety production. He has been involved in many industrial projects. Currently, he is also director of the HAN Lean QRM Center, which links research, higher education, and practice, and has around 50 industrial partners.

Appendix F: Experiences with Using a POLCA Simulation Game

Guest author: Hans Gerrese

Leanteam is a consulting firm formed in the 1990s with experience in assisting many companies in the Netherlands in implementing Lean. Around 2006, we noticed that the Lean concepts were no longer a good fit for the problems that our customers were encountering as a result of the market trend towards lower and lower volumes and higher variety. Specifically, we saw that companies in the high-mix, low-volume and custom (HMLVC) environment were struggling to apply many of the Lean tools. Using the internet, we began searching for some more relevant tools and came across POLCA as one of the tools developed as part of the Quick Response Manufacturing (QRM) strategy. Both POLCA and QRM were completely new to us. It also seemed at the time that they were new to Europe as a whole, and we did not have local resources knowledgeable on these subjects. However, POLCA seemed to fit what we were looking for, and so several of our principals decided to invest in some trips to the U.S. during 2006–2007, to attend workshops conducted by Professors Ananth Krishnamurthy and Rajan Suri at the Center for Quick Response Manufacturing, University of Wisconsin–Madison.

Challenges in Communicating the Right Message

We returned from our trips to the U.S. with some excitement about the potential for QRM and POLCA in the Netherlands. However, as we tried to roll out these concepts to our clients, we found a lack of appreciation of the need for lead time reduction. Managers had not heard about QRM and also had no understanding of how long lead times harmed their businesses, so they did not absorb the QRM message. Instead, when we made our presentations, they tried to compare QRM with Lean and so they looked for tools within QRM which they could use in their organization to complement their existing Lean programs. With the shift to HMLVC manufacturing, these companies were experiencing increasing problems of managing their orders and workload, and were looking for tools that could help them in taking back the control over their production. So, as they looked at the tools within QRM, they latched onto POLCA, since this system was seen as the tool that would help them resolve their planning and control problems. Thus it was that during our early efforts we received a lot of inquiries from manufacturing companies that wanted to implement POLCA in their organization.

As is usual with these kind of requests, as a consultant you first arrange a meeting with the customer to hear and see what their problems are. What we observed were disorganized shop floors with very high work-in-process (WIP), too many open orders on the shop floor, lots of rework, and people working hard, including a lot of overtime. But at the same time, a lot of the hard work included running around and chasing down jobs, and dealing with supervisors, who were changing priorities every hour but still yelling that jobs had to be done quickly. The end result was employees getting demoralized and losing interest in doing a good job. Our analysis was that the solution of this problem was not just implementing POLCA, or, as they saw it, a better "planning tool." The root cause of the problems lay much deeper and the solutions required management to first create cells and simplify the work flow, and then to focus on lead time reduction as a goal, including such strategies as planning for spare capacity. POLCA could then help as a planning and control method to make the whole system function properly.

In summary, our observations showed that for many of these companies, implementing POLCA was not the first step. But to drive this point home, we had to teach these customers what POLCA was and when and how you could use it. This was not easy—we found it very difficult to explain just through presentations what POLCA really is, what it could offer, and particularly the prerequisites for implementing POLCA. Managers in these companies were only focused on "planning" and felt they just needed a good planning tool to solve their problems.

We decided that we needed a hands-on simulation game to demonstrate what POLCA was and how it worked. We felt that, through the game, participants would also understand the pre-conditions for implementing POLCA.

As a starting point for our game, we looked at the POLCA game developed by the Center for Quick Response Manufacturing (see Appendix I), which we had experienced when we attended the workshop in 2006. This is a production game in which participants have to fabricate and assemble custom partitions with doors, windows, or both. The participants actually use thick and thin paper stock to simulate the various items being produced, along with scissors and scotch tape to do the fabrication and assembly operations. This game is played in two rounds, the first one using Kanban, and the second using POLCA, so the game also helps to point out the differences between these two systems, as well as the advantages of POLCA in HMLVC production.

Using our experience with this game as a base, we decided to develop our own game. Since a lot of the companies that we dealt with were already familiar with Lean and Kanban, it was important to maintain the goal in the U.S. game of demonstrating the difference between Kanban and POLCA. Beyond this, we developed our game with two additional goals. First, we wanted the game to involve a product that people could relate to in their everyday life, and second, we wanted to increase the impact of dynamical changes during the game. The reason for this latter goal was that in our initial explanations of POLCA, people were not clear about the added value of the POLCA cards, given that there was already an Authorization List with a sequence of dates for each cell. "Since the jobs are already sequenced at each cell," people asked, "the team already knows its priorities, so why do we also need the POLCA cards?" Hence, we felt that our game should clearly show the benefits of having both the Authorization Dates and the POLCA cards. Later, we discovered one more goal for the game we had developed: it also ended up being very useful as part of the training for employees at companies that wanted to implement POLCA.

Details of Our POLCA Game

To satisfy the first of our additional goals for our POLCA game we decided that participants would build cars with Lego blocks—both cars and Lego blocks are very familiar to people! For the second goal, we decided on a large variety of cars. As will be seen from the figures below, there are six basic configurations of cars based on size (narrow/wide, regular/long, and so on), and in addition, each can be ordered with special options, including more features on the body and a spoiler in the rear. Altogether, this results in around 10 different cars that could be ordered. Figure F.1 shows examples of a standard and a special car.

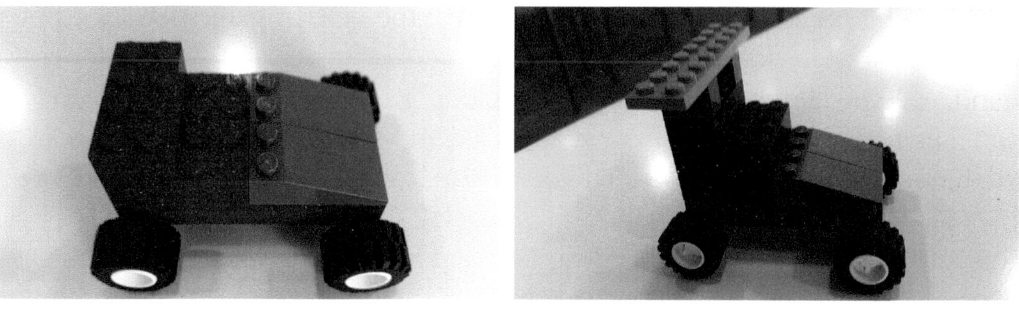

Figure F.1 Standard car with wide body (left) and a special car with a higher body and a spoiler (right).

The simulated car factory consists of seven assembly cells and one shipping cell (Figure F.2). There are two cells for assembling the base of the car, depending on the size. Here the participants put together the chassis and the axles. Next are three cells for the body assembly, again with different operations based on the car type. For example, bodies have to be made bigger with more Lego bricks, or spoilers have to be installed for the specials, and all body stations also have to mount the wheels. The third set of assembly cells attach the nose to the car. Finally, the cars are sent to the shipping area where they are staged for shipping on trucks. Since this is just a staging area, it is not included in the POLCA loops as no capacity signals are needed (see Chapter 5 for a discussion on this point). On the shipping date for any order, that car is removed from the staging area and put in another area, which represents that it has been shipped out on a truck.

As seen from Figure F.2, products have several possible routings. In particular, a given cell could have more than one upstream cell and more than

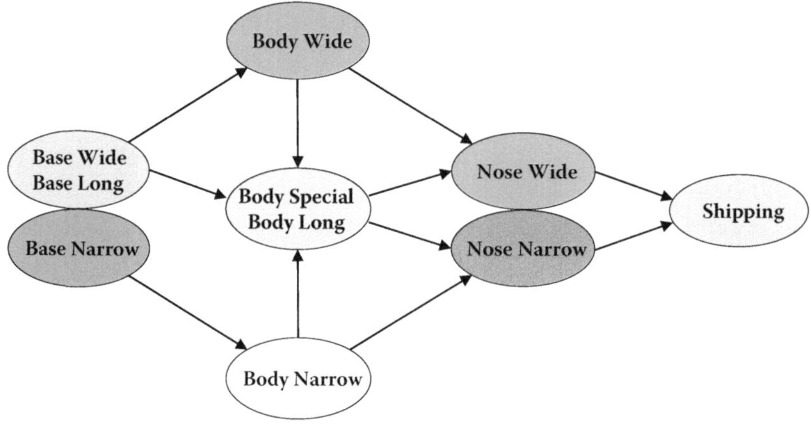

Figure F.2 Layout of the simulated factory.

one downstream cell. In addition, the assembly times vary significantly: some of the special options can take twice as long to assemble compared with a standard option at the same station. So, the variation in products, routings, and assembly times, combined with the demand variations (below), provide the high variability needed to demonstrate the strengths of the POLCA system.

We normally play the game with eight participants—one working at each cell—and one observer. The role of the observer is to watch over the whole operation, make notes about any remarkable situations, and also to observe and calculate the metrics described below.

The game is played in two rounds, and each round is for a month's worth of orders. These orders have been prepared by us ahead of time to help illustrate the main dynamics of the game. In the first round, we use the Kanban system and in the second round we use POLCA. To make the game realistic with pressure on people to get products out in time, we have a Powerpoint animation in which people see the date, hear the start of the day by the sound of a rooster, and then they hear music, during which everyone is allowed to assemble. Just like in musical chairs, when the music stops they have to stop too! For each round, we record some key metrics, specifically the delivery performance, WIP, and finished goods (the amount of products in the staging area). In order to get multiple samples of the metrics, we have several "counting moments" during the game. At such moments, we stop and "freeze" the game, and record all the key metrics.

For the first round with the Kanban system we start with some initial products at each workstation. This is necessary because Kanban is a replenishment system (see Chapter 3 and Appendix C for more details). There are no Authorization Lists—cells make a new product if they receive a Kanban card from a downstream cell. The whole process starts with shipping as the first trigger. In the case of Kanban, there is a Kanban loop from shipping to the preceding cells. Every time a product is shipped, it generates a signal for the upstream cell, which then creates a signal for its upstream cell, and so on.

In the second round, we play the game with the POLCA system managing the production, so we have POLCA cards and Authorization Lists. Based on the possible routings between the cells, Figure F.3 shows that there are nine POLCA loops along with the corresponding two-colored cards. (Earlier we explained why there are no loops to Shipping.) The Authorization List for each cell is made in the usual way, by back-scheduling from the shipping dates (see Chapter 4), allowing for a one-day lead time at each cell. Also, in the POLCA situation we start with no initial WIP at all, as everything will be made based on the Decision Time rules.

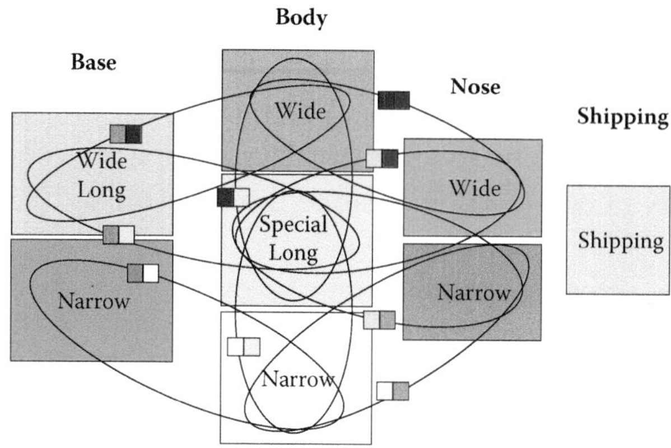

Figure F.3 Set of nine POLCA loops for the simulated factory along with color-coded cards for each loop.

Typical Outcomes from Playing the Game

In Figure F.4 we show the actual outcomes of one of our games, which is representative of the results that we have seen. Let's start with the outcomes for the game using the Kanban system, on the top half of the figure. In the first column, you see the products (we have aggregated the car options into four main categories for the purposes of data-gathering). The next set of columns, under the general heading of "WIP," show the WIP and finished goods (FG) numbers for three counting moments along with their totals. The last column shows the average of these totals (the shaded cells). The next set of data, below the above numbers, shows the delivery performance. Again, we have the data for the three counting moments—note that for the on-time delivery data this is the cumulative performance up to this moment—and then we have the cumulative performance at the end of the simulation (shaded cell).

The lower half of the figure shows the same data for the game when the POLCA system is used. The results of these two phases are representative of what we see in our games, and summarized as follows:

Results with Kanban		*Results with POLCA*	
Average WIP:	17	Average WIP:	12
Average Finished Goods:	4	Average Finished goods:	0
Delivery performance:	85%	Delivery performance:	98%

LEANTEAM	KANBAN								
	WIP									
	Start		Count #1		Count #2		Count #3		Average	
	WIP	FG	WIP	FG	WIP	FG	WIP	FG	WIP	FG
Wide Red	3	1	2	1	6	2	7	2		
Wide Blue	3	1	2	0	5	0	6	1		
Narrow Green	3	1	4	2	6	1	4	0		
Narrow Yellow	3	1	3	0	5	1	6	2		
Total	12	4	11	3	22	4	23	5	17	4

	Delivery Performance							
	Count #1		Count #2		Count #3		Count end	
	On Time	Total	On Time	Total	On Time	Total	On Time	Total
	16	19	31	39	45	56	68	80
On Time (%)	84%		79%		80%		85%	

LEANTEAM	POLCA								
	WIP									
	Start		Count #1		Count #2		Count #3		Average	
	WIP	FG	WIP	FG	WIP	FG	WIP	FG	WIP	FG
Wide Red	4	0	2		3		4			
Wide Blue	2	0	4		3		2			
Narrow Green	4	0	2		3		4			
Narrow Yellow	2	0	4		3		2			
Total	12	0	12	0	12	0	12	0	12	0

	Delivery Performance							
	Count #1		Count #2		Count #3		Count end	
	On Time	Total	On Time	Total	On Time	Total	On Time	Total
	20	20	39	40	58	60	78	80
On Time (%)	100%		98%		97%		98%	

Figure F.4 Example of actual outcomes from a simulation game.

In general, what the participants see through experiencing our games is that in this high-variability situation, as compared with using Kanban, with POLCA the WIP is reduced, finished goods are unnecessary and reduced to zero, and the delivery performance improves to an almost-perfect number. The results are always almost the same every time we play this game: all the metrics in the second round with POLCA are much better than in the first round with Kanban.

An interesting side result is that people realize that the POLCA system reduces the stress during the game. If you are behind at a certain moment, the combination of the Authorization Dates and POLCA cards helps to resolve the backlog. So, people calm down and are more relaxed in their activities.

Results of the Game

The participants in our games have included a mix of people from all levels of the organization, including machine operators, supervisors, planners, managers, and even directors of companies. We have found that the strength of our simulation game is that POLCA can be taught to all levels of a company and that experiencing the operation of POLCA during the game is critical to understanding of POLCA.

To make the game more fun and give it a competitive character, we often play the game with two groups simultaneously. This makes each group alert right from the beginning, and of course they want to see if they can achieve better results. This provides a healthy stress level and makes the game more like real life. At the same time the game clearly demonstrates the limitations and benefits of each of the systems, since each team is trying to do its best, but must operate within the parameters of each system.

We have also found that the POLCA simulation game acts as a springboard for discussion of the prerequisites for successful POLCA implementation. After the game, discussions naturally focus on the importance of actions such as planning for spare capacity, creating the Authorization Lists, and minimizing component shortages, before moving ahead with the POLCA implementation. The feedback from our participants is always very positive and acknowledges that the game helped them to really understand how POLCA works and its benefits.

About the Author

Hans Gerrese is a senior consultant with 17 years of experience in implementing Lean and QRM in a wide variety of manufacturing companies. After completing his study at the automotive university, he started working at several first- and second-tier manufacturers in the automotive world. There, he came in contact with Lean and after a while decided to start his own consultancy firm in Lean manufacturing. After almost seven years, he came in contact with QRM and was one of the first people in Europe to implement QRM at his customers' locations. More recently, his implementation work has been supplemented with training activities at an institute at which people are certified to three levels of QRM expertise.

Appendix G: Explanation of Formula for Number of POLCA Cards

For readers interested in technical details behind the formula for the number of POLCA cards in a loop, we provide here a mathematical basis for the formula. If you are familiar with queueing theory, the formula is based on estimates of various elements of lead time, followed by application of Little's Law. (If you are not familiar with this concept, you don't need to read this section; it is provided for readers who are more mathematically inclined and have some knowledge of queueing theory.)

We will use the A/B POLCA loop example that was presented in Chapter 5 in the section "Calculate the Initial Number of POLCA Cards in Each Loop." It is important that you review that section before reading this appendix, since the explanation below assumes you are familiar with that example and the terminology used there. The approach used here is to first estimate the total sojourn time for an A/B card to make its way all around the loop. This sojourn time along with the throughput rate of the loop allows us to use Little's Law to calculate the number of cards.

Let's follow an A/B card around the loop, starting from when it arrives at Cell A's POLCA Board, through the events that take it to Cell B, and then back to Cell A's POLCA Board. The total time for this sojourn can be broken down into the following segments. We use a symbol to denote the value of each segment, and more detailed explanations of some of the segments are provided after the definitions:

WAuthA = Wait to be assigned to an Authorized job present at Cell A

QA = Wait in queue for Cell A

TA = Time spent in Cell A
MovAB = Time to move the job and card to Cell B
WAuthB = Wait for this job to be Authorized in B
WBC = Wait for B/C card to be available
QB = Wait in queue for Cell B
TB = Time spent in Cell B
RetBA = Time to return the card from B to A

Let's use the symbol SAB to represent the total time for the sojourn of an A/B POLCA card. Then SAB is given by the sum of the above times, namely:

$$\text{SAB} = \text{WAuthA} + \text{QA} + \text{TA} + \text{MovAB} + \text{WAuthB} + \text{WBC} + \text{QB} + \text{TB} + \text{RetBA}$$

Note that the issue of estimating times for MovAB and RetBA was already discussed in Chapter 5. In our analysis below we will also use the following symbols, which were introduced and explained in Chapter 5:

LA = Planned lead time for Cell A
LB = Planned lead time for Cell B
D = Number of working days in planning period
FlowAB = Number of times an A/B card flows from A to B during the planning period
S = Safety margin, expressed as a decimal

As also explained in Chapter 5, we assume here that the move time MovAB is *not* included in either LA or LB. If the move time is already incorporated in one of these lead times then you can just set MovAB = 0 in the following discussion.

Now we will use all the above symbols to explain the derivation of the formula. Let's start by explaining WAuthA and QA. There are several processes taking place in parallel and influencing the POLCA decisions for Cell A. One is that A/B cards are returning from Cell B and arriving onto A's POLCA board. The second is that jobs are arriving from upstream cells. Third, as the clock keeps ticking, jobs at Cell A move from a status of "Not Authorized" to "Authorized." And the last one is that when the team at Cell A arrives at its next Decision Time, it goes through the POLCA logic to see if it can launch a job. Note that these processes are occurring asynchronously, which means they can occur in any order. For example, a job could be Authorized, but have to wait for a POLCA card. Since we are taking the viewpoint of an A/B POLCA card starting from when it arrives at Cell A's POLCA Board, this means that the card is available, so we do not need to account for this wait in the analysis that follows. Next, although our POLCA card has arrived, it may need to wait for a job to be Authorized. Or else, a job could be Authorized, but the material has not yet arrived from upstream. The waits for these last two processes (Authorization and arrival of material) have been combined into the quantity WAuthA above, which represents the average wait regardless of the order in which these events occur. Finally, a job could be Authorized, the material has arrived, and our POLCA card is available, but Cell A could be busy with other work, and so the job (and POLCA card) would wait in a (virtual) queue until the team is ready to launch the next job. The wait for this process is represented by the quantity QA.

The average values for both these waiting times (WAuthA and QA) depend on factors such as the variability in the demand and processing times, the number of machines and people in the cell and their workloads, the actual number of cards currently in use, and many other factors. It would take a complex dynamical simulation along with representative demand and production data to estimate all these quantities precisely. However, we will use an approach often taken by practitioners solving such problems to enable us to get a reasonable estimate for the final result, which is the number of cards needed. The particular approach we use here is to estimate a *lower bound* for the number of cards, and then to use this to make our initial decision. You will see how we do this as our explanation unfolds.

We will use a *best-case* scenario, and then justify its use at the end of the discussion. Let's say the system is fine-tuned and working well so that a returning A/B card is placed on the POLCA board just in time to be matched up with a job that has also arrived and been Authorized. Then WAuthA is effectively zero. Also, if the system is working as planned, then the planned lead time for Cell A should be its queue time plus the time in the cell. Hence, QA + TA can be replaced by LA.

Let's use similar best-case thinking at Cell B. If the system is working as planned, then our job should arrive at B just in time to be Authorized. Also, if the system is well synchronized, a returning B/C card will be placed on B's POLCA Board just in time to be matched up with this job. So, both WAuthB and WBC are effectively zero. (We should point out that there is some degree of self-correction in these assumptions because of the way that POLCA works. If the job is finished earlier than planned at A, and so it has to wait at B to be Authorized, then this waiting time has already been included in the value of LA, which was used in setting the Authorization Date at B, so the assumption that WAuthB = 0 is still valid.) For the last portions of the time in Cell B, if the system is working as expected, then the planned lead time for Cell B should be the queue time plus the time spent in Cell B. Hence, QB + TB can be replaced by LB.

If we put these values in the formula above for the sojourn time SAB (while removing all the items whose values are zero), we get the result that the best-case sojourn time for an A/B card is:

$$\text{SAB} = \text{LA} + \text{MovAB} + \text{LB} + \text{RetBA}$$

Next, note that the flow rate of A/B cards must equal the throughput rate of the system, which is FlowAB/D. We can now apply Little's Law based on the flow rate and sojourn time of A/B cards to get the number of A/B cards that are consistent with this flow rate and sojourn time. We call this number IdealAB, since it is based on the idealized (best-case) analysis. Using Little's Law, this is given by:

$$\text{IdealAB} = \text{SAB} \times (\text{FlowAB/D})$$

Since we have assumed values of zero for some of the waiting times in the preceding discussion, the actual value of SAB will always be greater than the above value. Therefore, the actual number of cards needed will be higher than IdealAB. Thus, we see that IdealAB provides us with the desired lower bound for the number of cards needed—in other words, it represents the minimum number of cards required.

As already mentioned, it would be difficult to get more accurate estimates for all the waiting times above without performing some complex simulation, and even then, such simulations could be quite sensitive to detailed assumptions about the actual demand and other production data. So, a simpler and more practical approach is as follows. Keep in mind that we don't want to add too much slack in the system compared with the best-case scenario, and at the same time, we can fine-tune the number of cards easily after the system is launched. So a simple and practical approach is to add a small safety margin (S), and then round up to the next integer. If we put all these points together, we get the value for NumAB, the recommended number of POLCA cards:

$$\text{NumAB} = (\text{LA} + \text{MovAB} + \text{LB} + \text{RetAB}) \times (\text{FlowAB/D}) \times (1 + \text{S})$$

This is the formula that was presented in Chapter 5. While based on some simplifications and pragmatic approaches, this formula has resulted in good initial estimates for POLCA implementations, such as those described in Chapters 10 through 15.

Appendix H: POLCA: A State-of-the-Art Overview of Research Contributions

Guest Author: Jan Riezebos

Introduction

Production control is an important part of the planning and control system that has to organize the various flows in production in a company. These include not only the material flow, but also the flow of resources (such as tools, capacity of work stations, and employees) and information (what to produce, how to produce, where and when to deliver).

In the academic literature, a lot of attention is given to various methods of planning and control (Orlicky, 1975; Silver, Pyke, and Peterson, 1998; Jacobs, 2003; Umble, 2003; Schönsleben, 2004; Vollmann et al., 2005; Riezebos, 2013).* Academic contributions describe, for example, the sequence and hierarchical layers of decision-making that should be used in typical configurations of production systems, such as make-to-stock and make-to-order environments (e.g., Bertrand et al., 1990). Or they analyze how much capacity is required given the order book and forecasts of future demand (Jodlbauer and Reitner, 2012; Bahl and Ritzman, 1987; Plenert, 1999). Or they analyze the impact of uncertainty on the design of a planning and control system and provide methods

* Since this appendix is aimed at a research-oriented audience, we will use a formal journal style for the references to publications, which are listed at the end of the appendix.

that are able to take into account the existing uncertainty in, for example, material availability or processing time length (van der Vaart, de Vries, and Wijngaard, 1996; Wu et al., 2012; Brennan and Gupta, 1993; Enns, 2002; Mula et al., 2006; Lin and Parlaktürk, 2012; Inderfurth, 2009; Dolgui and Prodhon, 2007; Humair et al., 2013). Finally, the literature gives a lot of attention to the question of whether specific methods can be developed that are able to find optimal solutions to specific problem configurations (Bensoussan et al., 2009; Wang and Sarker, 2006; Louly, Dolgui, and Hnaien, 2008; Iravani, Liu, and Simchi-Levi, 2012; Homem-De-Mello, Shapiro, and Spearman, 1999; Serel, 2009; Tadj, Bounkhel, and Benhadid, 2006). For other methods that do not guarantee such optimal solutions (i.e., heuristics), the quality of the results are analyzed and worst-case or average performance metrics are provided (Roundy, 1993; Chand, Hsu, and Sethi, 2002; Pochet and Wolsey, 1988; Riezebos, 2011; Kaminsky and Kaya, 2009).

Production control literature (Slomp, Bokhorst, and Germs, 2009; Fernandes and Filho, 2011; Framinan, González, and Ruiz-Usano, 2003; Geraghty and Heavey, 2005; Powell et al., 2013; González and Framinan, 2009; Gaury, Kleijnen, and Pierreval, 2001; Riezebos, 2010b; Powell, Riezebos, and Strandhagen, 2013; Riezebos, 2010a; Johnson, 2003; Khojasteh, 2016; McKay, 2000) takes a different perspective. Production control is not a preparatory activity that is done several days or weeks before the actual production starts. Instead, production control focuses on questions such as: how to regulate and organize the actual usage of resources; what jobs should be released to (another stage of) the shop floor; what jobs should get priority or be postponed; and so on. These control decisions are still necessary notwithstanding the previous planning decisions that have been made. The planned availability of resources and materials may differ from the actual availability at a specific moment in time, and a production control system helps to gather the relevant information and supports the decision-making process in the short term in order to realize the overall objectives set during the planning phase for the production system as a whole. Production control can hence be denoted as the collection of all the regulatory activities during the execution of the plan.

Production control systems are often defined in terms of Pull or Push systems. In this chapter we follow the Pull and Push definitions of Hopp and Spearman (2004, p. 142): "A pull production system is one that explicitly limits the amount of work in process that can be in the system. By default, this implies that a push production system is one that has no explicit limit on the amount of work in process that can be in the system." Hence, Pull systems will limit the workload at the production system (which includes the shop floor) by controlling one or more of the following decisions: order allocation, order release, order progress, and/or outsourcing. Order release and outsourcing directly affect the amount of workload that is within the system. Order allocation (the assignment of orders to routings with specific work stations) determines what stations will be loaded and might experience limits. Order progress management (i.e., dispatching), determines the timing of the arrival of workload at various work stations and may prohibit the delivery to another stage of production due to workload limits.

This chapter focuses on production control and more specifically on the POLCA system. The next section, "Where is a POLCA Production Control System Applicable?" provides an overview of insights from literature on the POLCA system. The literature review mainly addresses the suitability of POLCA for typical production environments. It also discusses the attention given in the literature to POLCA's ability to cope with variety and uncertainty. The aim of this section is to provide the reader access to relevant academic literature in this area. Hence, you may find many references to journal papers that provide an in-depth analysis of sometimes very detailed aspects of production control. These papers are not always accessible to non-academic readers, so through this overview I hope to make the main insights accessible and help both academic and non-academic readers find their way in the enormous amount of academic literature on important aspects of relatively simple production control systems such as POLCA. Moreover, it might help academic readers to identify under-researched areas and further developments in this field. This literature review doesn't aim to provide readers with an overview of alternatives to POLCA. Systems such as CONWIP (Spearman, Woodruff, and Hopp, 1990), Production Authorization Cards (Buzacott and Shantikumar, 1992), and Workload Control (Breithaupt, Land, and Nyhuis, 2002) are not discussed in full detail, as our focus is just on helping the reader to understand the background of POLCA, not of every alternative to POLCA.

After the extensive discussion in the section "Where is a POLCA Production Control System Applicable?" on the areas where POLCA might be applicable, we move on to discuss literature related to the two important production control decisions that POLCA performs. The section "Order Release Decision" addresses order release, and the section "Shop Floor Control and Workload Balancing" addresses shop floor control and workload balancing. Finally, in the last section we provide some conclusions.

Where Is a POLCA Production Control System Applicable?

This section aims to describe for which production environments POLCA might be a good production control system. Such a description does not imply that POLCA is the only or the best system to apply in these environments. It implies that characteristics of POLCA match with characteristics of these production environments. Production systems that have characteristics that do not match with POLCA may require additional support tools or even a different approach to production control.

Most academic papers that discuss the applicability of production control systems actually compare various production control approaches and trade the benefits of one approach off with the other. We take a different perspective by focusing on the characteristics of the production environments for which production control is needed and see whether there is a match with characteristics of POLCA.

A typical distinction in literature on production environments is what production and preparatory activities take place even before an actual order of a customer has been received. If we view the total sequence of activities to be performed for fulfillment of the order as a process, part of this process might be performed when the actual customer that will receive this specific item is not yet known. In the next part of the process, the actual customer is known. The point in the process where the actual customer of the item that is being produced has become known is denoted as the Decoupling Point (Reichhart and Holweg, 2007; Storey et al., 2006; Lehtonen, Holmström, and Slotte, 1999; Fogliatto, da Silveira, and Borenstein, 2012; Demeter, 2013) or Customer Order Decoupling Point (Hedenstierna and Ng, 2011; Tavares Thomé et al., 2012; Olhager, 2013). A similar concept in literature is known as the Make-to-Order/Make-to-Stock (MTO/MTS) interface (Liu, Parlar, and Zhu, 2007; Adan and van der Wal, 1998; Akkerman, van der Meer, and van Donk, 2010; Zhang et al., 2013; Iravani, Liu, and Simchi-Levi, 2012; Su et al., 2010), which introduces various production environments based on the primary control decision: make/engineer-to-order versus make-to-stock or even make-to-forecast. Assemble-to-order is a hybrid configuration that makes components to stock and then builds final products to order. By making products (partly) to stock before the decoupling point, the customer lead time is being reduced, but the total throughput time of the material is increased.

This distinction on what production activities are being performed while the actual customer order has not (yet) been received is important from a control perspective. In make-to-stock environments, production orders for items are initiated by a planning department and are not associated with strict delivery dates that have been promised to an external customer. In make-to-order environments, these due dates are set and communicated to the customer before the activities for the specific order starts. In engineer-to-order situations, delivery dates are negotiated and set while there might still be some uncertainties on what production activities and materials are required and whether they will be available in time.

Table H.1 provides an overview of literature on POLCA based on the characteristics of the paper (theory, case study, optimization approach) or of the production situation in which it is been tested or applied. The table summarizes what aspects the various papers that discuss POLCA have either included or ignored within their conceptualization of POLCA.

The first characteristic of the production environment is the decoupling point. The decoupling point has impact on many control aspects. From a control perspective, delivery dates are sometimes exogenous and in other situations endogenous, process plans (i.e., routings, setup times, and processing times) are known or not at the moment of order acceptance, etc. Some production control systems assume that specific information is available at the time of the control decision, while other systems take into account that even when estimates are available, this information is not always reliable for decision-making, given that the state of the system changes very rapidly. For example, production control using detailed Gantt charts (such as used in many advanced planning and scheduling systems) assumes that all information on the amount to be produced, processing times, setup times, etc. is available before a schedule can be generated. Hence, we see that production control systems have to be characterized based on their suitability for the stage before the decoupling point and the stage after the decoupling point.

Another characteristic is how the system copes with dynamics and uncertainties. Dynamics are for example differences between the orders in terms of routing length, routing sequence, etc. They are related to observing variety between orders: similar orders might obtain different due dates or other characteristics due to this variety. Uncertainties may refer to material availability, time uncertainty (setup, processing), demand changes, due date changes, changes in machine allocation (routing), and machine availability (breakdowns, maintenance). The main issue to identify is whether the production control system provides support for handling variety and/or uncertainty, or does it assume that all information is known and reliable before the production starts? In other words, how does the control system cope with dynamics and uncertainties at the shop floor?

Table H.1 Overview of Research on POLCA

Characteristic	Approach		Environment		Coping with Dynamics: Variety		Coping with Uncertainties				
Sub-Characteristic / Source	Theory/ Cases	Optimization/ Simulation	Engineer-To-Order/ Make-To-Order	Make-To-Stock/ Build-To-Forecast	Routing Length	Routing Sequence	Time Uncertainty	Demand Uncertainty	Due Date Changes	Breakdowns	Material Availability
Suri, 1998	+	–	+	–	+	+	+	+	+	+	+
Krishnamurthy, Suri, and Vernon, 2000	–	+	+	–	–	+		+		+	
Luh, Zhou, and Tomastik, 2000	–	+	+	–	+	+	–	+	–	–	–
Suri and Krishnamurthy, 2003	+	–	+	–	+	+	+	+	–	+	+
Suri, 2003	+	–	+	–	+	+	+	+	–	+	+
Lödding, Yu, and Wiendahl, 2003	+	–	+	–	+	+	+	+	–	–	–
Stevenson, Hendry, and Kingsman, 2005	+	–	+	+	+	–	+	+	–	–	–
Fernandes et al., 2006		+	+	+	+	+	+	+	–	–	–
Vandaele et al., 2008	+	+	+	–	+	+	+	+	+	+	+

(*Continued*)

Table H.1 (Continued) Overview of Research on POLCA

Characteristic	Approach		Environment		Coping with Dynamics: Variety		Coping with Uncertainties				
Sub-Characteristic / Source	Theory/ Cases	Optimization/ Simulation	Engineer-To-Order/ Make-To-Order	Make-To-Stock/ Build-To-Forecast	Routing Length	Routing Sequence	Time Uncertainty	Demand Uncertainty	Due Date Changes	Breakdowns	Material Availability
Krishnamurthy and Suri, 2009	+	–	+	–	+	+	+	+	+	+	+
Riezebos, 2010a	+	–	+	–	+	+	+	+	–	+	+
Germs and Riezebos, 2010	–	+	+	–	–	+	+	+	–	–	–
Ziengs, Riezebos, and Germs, 2012	–	+	+	–	–	+	+	+	–	–	–
Aziz et al., 2013	–	+	–	+	–	+	–	+	–	–	–
Farnoush and Wiktorsson, 2013	+	+	+	–	–	+	+	+	–	–	+
Harrod and Kanet, 2013	–	+	+	–	+	+	+	+	–	–	–
Lödding, 2013b	+	+	+	–	–	+	–	–	–	–	–
M. Braglia, Castellano, and Frosolini, 2014	–	+	+	–	+	+	+	+	–	–	–

(*Continued*)

Table H.1 (Continued) Overview of Research on POLCA

Characteristic	Approach		Environment		Coping with Dynamics: Variety		Coping with Uncertainties				
Sub-Characteristic / Source	Theory/ Cases	Optimization/ Simulation	Engineer-To-Order/ Make-To-Order	Make-To-Stock/ Build-To-Forecast	Routing Length	Routing Sequence	Time Uncertainty	Demand Uncertainty	Due Date Changes	Breakdowns	Material Availability
Marcello Braglia, Castellano, and Frosolini, 2015	–	+	+	–	+	+	+	+	–	–	–
Eng, Ching, and Siong, 2015	+	+	+	–	+	+	+	+	–	–	–
Eng and Ching, 2016	+	+	+	–	+	+	+	+	–	–	–
Thürer, Stevenson, and Protzman, 2016c	+	–	–	+	–	+	–	–	–	–	–
Thürer, Stevenson, and Protzman, 2016b	+	–	–	+	–	+	–	–	–	–	–
Frazee and Standridge, 2016	+	+	+	–	–	–	+	+	–	–	–

It appears that POLCA literature addresses some of these issues in specific papers, e.g., by discussing a specific case or setup of a simulation study, but not all have been addressed in the books that have been published so far. This causes confusion in other studies that aim to compare POLCA with other approaches for production control, such as workload control, Kanban, CONWIP, etc. Examples of such comparison studies that have used a rather limited view of the POLCA system are Lödding et al., 2003, Stevenson et al., 2005, and Thürer et al., 2016b. We have included them in the overview of Table H.1 in order to demonstrate what aspects these papers have either included (+) or ignored (–) within their conceptualization of POLCA. If the cell is left empty, it was not clear to the author whether the issue has been considered or ignored.

Let us take a look of the results of the overview in Table H.1. The first characteristic of the papers or books that discuss POLCA is the approach taken towards studying POLCA. We classify the various contributions based on whether theoretical or practical insights are being described at a conceptual level, either based on actual case studies or implementations or on comparisons with other production control approaches. Next, we identify whether the contribution is based on optimization or simulation studies, which typically aim to investigate the design choices of a POLCA system that would result in an effective control of the material flow. Design choices that are being studied are, e.g., the number of POLCA cards in a loop, priority rules at cells, and the size of a quantum (i.e., the maximum workload associated with a single POLCA card). This type of study often tests these design choices in various specific shop floor configurations. These configurations may differ with respect to the existence and location of a bottleneck, divergent or convergent material flows, and whether there are differences in routing lengths or routing sequence. The optimization and simulation studies provide important insights for theory on POLCA, but we characterize such a study only as a theory/case approach if it includes a strong conceptual component as well. In Table H.1 we observe that many recent contributions from the last ten years focus on optimization and simulation research, i.e., on identifying design choices for POLCA that result in effective control in either practical or hypothetical shop floor configurations. This has resulted in several important insights, such as:

- the effect of using load based cards (quantums) (Vandaele et al., 2008);
- the existence of workload balancing capability of POLCA and related systems (GPOLCA, m-CONWIP) (Fernandes et al., 2006; Germs and

Riezebos, 2010; Ziengs, Riezebos, and Germs, 2012; Farnoush and Wiktorsson, 2013);
- the possibility of heavy blockings (Lödding, Yu, and Wiendahl, 2003), lockups (Harrod and Kanet, 2013), or deadlocks (Lödding, 2013b) in case very low numbers of POLCA cards are used to control loops that connect both directions of the material flow (i.e., A→B as well as B→A);
- the possibility to use mixed integer linear programming and genetic algorithms to configure the POLCA system and identify the number of cards per loop (M. Braglia, Castellano, and Frosolini, 2014; Aziz et al., 2013);
- the role of priority rules in combination with a POLCA system (Marcello Braglia, Castellano, and Frosolini, 2015; Eng and Ching, 2016; Barros et al., 2016; Thürer, Stevenson, and Protzman, 2016b).

The second characteristic that has been used in Table H.1 to classify the various books and papers on POLCA is the production environment. Most contributions acknowledge that POLCA has unique properties that allow for application in custom-made production, such as make-to-order and engineer-to-order situations. In such production environments, jobs will not be released at the shop floor before the actual customer has issued the order. As the order specification might be different for each order (process plan, operations, materials, machines, quality control, packaging materials, etc.), there is no possibility to produce in advance. In such production environments, traditional Pull systems that are based on refilling inventories of components through the process cannot be used. Some contributions (e.g., Stevenson, Hendry, and Kingsman, 2005; Fernandes et al., 2006; Aziz et al., 2013; Thürer, Stevenson, and Protzman, 2016b; Thürer, Stevenson, and Protzman, 2016c) discuss the effectiveness of POLCA in situations where it still is possible to refill stocks in the process. They conclude that in such circumstances it might be better to provide more precise control using traditional Pull systems, such as Kanban. Note that Suri (2010, p. 131) argues against this by stating that these traditional Pull systems use intermediate storage, which creates waste in cases when demand is very low, such as for infrequently ordered spare parts. POLCA is more efficient in such cases, as it prevents the buildup of stocks that barely flow at all. For more complex material flows, such as in traditional job shops, the number of different routings might become too high for relatively simple visual control systems such as POLCA, although cases are reported of successful use of POLCA in situations with thousands of low-volume complex parts, 44 loops, and

845 cards (Pelto and Mueller, 2012). However, other authors in the field of workload control (e.g., Lödding, 2013a; Thürer, Stevenson, Protzman, et al., 2016) advise using more central planning and a central release authority and investing in information systems that gather detailed information on work order progress at the shop floor. Note that such an approach takes responsibility from the shop floor and involves a huge investment in information systems.

The third characteristic is the ability to cope with dynamics on the shop floor. We distinguish between variety and uncertainty. Variety is known beforehand, it can be anticipated upon. Uncertainty means that the outcome is not known until the activity has been completed or effectuated. An example might help to understand the difference. If half of the orders have a processing time equal to one hour and the other half has a processing time of 11 hours, there exists variety, but no uncertainty with respect to processing time. If all orders would have the same expected processing time equal to five hours, but the actual processing times would differ from the expected or planned processing times, uncertainty exists, but (initially) no variety.

POLCA allows every order to be different. If the differences become more elaborate, variety increases. We distinguish between variety in routing length and routing sequence. Routing length can be defined as the number of production steps required for the order. Often, a cell is counted as one comprehensive production step, although within the cell several machines might be required for the necessary operations of the order. If the number of cells that should be visited in the sequence of production steps differs per order, the production control system should be able to cope with these differences. POLCA is quite effective in that respect, while other systems, such as CONWIP and Workload control, are more sensitive to differences in routing length. It appears that many studies that examine the effectiveness of POLCA do not take this aspect into consideration. That is, they experiment with many shop characteristics, such as different processing times, setup times, routings, but leave the length of the routings the same for all orders. Sometimes a follow-up paper studies the effect of a different routing length (e.g., Germs & Riezebos, 2010 and Ziengs et al., 2012), but further research might be needed in this area. Another type of variety is routing sequence variety. Every order can be characterized by the sequence of its routing, i.e., the sequence of cells or operations that need to be followed in order to produce the product. Several orders may use the same routing, even if the actual processing required at the cells or machines may be different. In case the routing sequence is equal among all orders, there is no routing variety.

But if there are differences in routing amongst orders, routing sequence variety increases. Routing sequence variety is considered in almost all papers. Stevenson, Hendry, and Kingsman's paper (2005) is the only one to suggest that POLCA is applicable only in flow systems with fixed sequences, which is a bit strange given the flexible chaining characteristic of the various POLCA cards that allows for all directions of routings (back and forth, from any cell to any other cell).

The last characteristic is the ability to cope with uncertainties. Uncertainty means that the outcome is not known until the activity has been completed or effectuated. Earlier in this section, we distinguished this from variety. Uncertainty may concern the processing times, setup times, routing, demand, due date, etc. Demand uncertainty occurs even if it is known that demand will change over time; for example, if it is not known beforehand when it will change or how much, demand is considered to be uncertain. Due date uncertainty may be apparent in case due dates may still change after order acceptance/release. Breakdowns and material availability might also be unpredictable and affect the operation of the POLCA system as an uncertainty component. We have included them as sub-characteristics of coping with different uncertainty types in Table H.1.

It appears that literature on POLCA addresses time and demand uncertainty, but most simulation/optimization studies model this uncertainty as a type of variety, i.e., the outcomes of demand size, processing times, and routings, are known from the moment the order is known and will not change anymore during processing. POLCA, however, is able to cope with a lot of uncertainties even during the shop floor operations. For example, processing time uncertainty is incorporated in its operation, as it waits upon the actual completion of all processing times in the two cells listed on the POLCA card before a signal is sent back to the original cell. This feedback, which is also present in workload control and CONWIP, is essential for production control systems, but is not considered when comparing the effectiveness of the different production control methods. As the delay in the information flow will be different for the various systems (e.g., CONWIP postpones the feedback signal till the whole product loop has been completed), this might have an impact on the effectiveness of the production control systems in case of uncertainties.

With respect to breakdowns and uncertainty about material availability, it is surprising that this is only considered in conceptual studies as well as simulation/optimization studies that involve practical applications. Most simulation and comparison studies that use hypothetical shop configurations do not include the occurrence of issues with machines or materials in their conclusions on the effectiveness of POLCA compared to other systems, such as m-CONWIP, GPOLCA, workload control, and Kanban. This is clearly a factor to consider in interpreting the outcomes of their research. For example, Farnoush and Wiktorsson (2013) did include uncertainty factors that were omitted in the original work of Ziengs et al. (2012) and concluded that the effectiveness of POLCA compared to m-CONWIP improved. Future research on comparing production control approaches should include these important uncertainty sub-characteristics in order to support decision-making on what system to use in what circumstances.

To conclude this overview of research on POLCA, it is clear from the length of this section that research on POLCA is very active. Many characteristics have been included in order to study its applicability in various production environments. POLCA appears to be a robust method that has been examined in very different circumstances, which is good news for companies that have plans to grow in size, product variety, explore new markets, or would like to change their decoupling point. POLCA has been examined on its ability to cope with variety as well as uncertainty and appears to be capable to cope with these shop floor dynamics. Other methods might do so as well (this section was not aimed at comparing POLCA with other methods or demonstrating which system would be better), so it is good to explore in more detail how POLCA supports two of the most important production control decisions. The next section explores the order release decision, while the section "Shop Floor Control and Workload Balancing" examines the operation of this control system after orders are released to shop floor.

Order Release Decision

This section focuses on the release decision. In order to be able to start production, an order first has to be released to the shop floor. Between order acceptance/arrival and order release, several preparatory activities need to be completed. These activities may concern the acquisition of the right raw materials or components (sometimes they still have to be ordered from an external supplier), the composition of order information details (process plans, bills of material, delivery details, drawings, tolerances, quality assurance forms, etc.), or even other information that is not needed for the shop floor, but still required before an order can be started (insurance procedures, advance payment guarantees, etc.).

Timing of Order Release

The timing of the order release decision is itself an important decision in production planning and control systems. Some preparatory activities may not yet be completed at the time of the release decision. As long as the activities are completed before the actual start of the processing stage in which they are required, there seems to be no problem. However, it inserts uncertainty during the shop floor production phase. If the decision for order release would be postponed till all preparatory activities have been completed, this might increase the lead time of the order, even though the activity might not have any direct relation to the shop floor processes. For example, the only information that might be lacking is where to deliver the final product. In such a case, shop floor processes can be started and completed while the activity is still pending. Longer lead times lead to more uncertainty as well. The information might get outdated soon if lead times increase. Materials that seemed available might in the meantime have been used for other products/customers, customers might have lost their financial credibility, etc.

This discussion shows that the timing of the order release decision is a relevant strategic planning decision. In the academic literature, we distinguish between systems that allow for time-phased order release decisions and systems that simply assume that all preparations should have been completed before the actual (and initial) order release takes place. Examples of time-phased systems are as follows:

- Period Batch Control (Benders and Riezebos, 2002; Burbidge, 1996; Kaku and Krajewski, 1995),
- (High-level) Material Requirements Planning (Krishnamurthy, Suri, and Vernon, 2004; Milne, Mahapatra, and Wang, 2015; Hopp and Spearman, 2000),
- POLCA (Suri, 1998), Synchronous Manufacturing (M. Umble, Umble, and Murakami, 2006; Riezebos, 2011; Vairaktarakis, 2002), and
- Advanced Resource Planning (Van Nieuwenhuyse et al., 2011).

This is not to say that these systems advocate the release of orders in an incomplete state, but they allow initial release and have built-in authorization decisions at the shop floor level that enable them to withhold orders that are not yet ready to be released to the next stage of production. Other systems, such as Just-In-Time planning/level scheduling (Miltenburg, 2007), workload control (Stevenson et al., 2011; Thürer et al., 2012; Land, 2009), and Theory of Constraints (Mabin and Balderstone, 2003; Takahashi, Morikawa, and Chen, 2007; Tibben-Lembke, 2009; Riezebos, Korte, and Land, 2003; Jodlbauer and Huber, 2008) normally do not consider the option of time-phased release, although in practice variants are found that use these methods in several subsequent stages of production.

Time-phased release decisions may have a positive or negative effect on the total throughput time. The negative effect (increase of throughput time) is caused by the inclusion of waiting times till the next release decision. The positive effect (reduction of throughput time) is caused by starting already with a parallel activity while still waiting on the not-yet-available activity or component. The final throughput time will only be reduced if the waiting time for the not-yet-available activity or component exceeds the induced waiting times to advance to the next stages. Many comparisons of the effectiveness of time-phased release methods often only show the throughput-time-increasing effect of time-phased release, but the need for time phasing is only apparent in case of synchronizing parallel processes that otherwise would have to be made sequential. My own Ph.D. thesis on production planning for cellular manufacturing addressed this issue in detail (Riezebos, 2001).

Decision-Making Based on the State of the Shop Floor

According to Hopp and Spearman, two influential scholars in the field of Push/Pull systems, the decision to release orders to the shop floor may be considered a Push decision if the state of the shop floor is not included in the decision to release. For example, a planner who would release an order as soon as all documents and materials are available will push the job to the floor, irrespective of the actual capacity available at the shop floor. A planner that would prioritize orders to release only based on characteristics of the orders, such as due date, routing, and processing times, is still considered to push the orders to the shop floor, notwithstanding that he or she might postpone some orders to avoid unnecessary work in progress. The state of the system hence does not refer to the size of the order book waiting to be released; it refers to the situation after release, identified through work-in-progress inventories and available capacity. As soon as the state of the shop floor is included in the release decision, not the planned (expected) state, but the actual state (although there might be a time delay in updating the information), the release mechanism is denoted as a Pull system. In a Pull system, the progress of orders to the shop floor is initiated or affected by the state of the system, i.e., the progress of already-released orders at the shop floor. This definition of Pull versus Push is due to Hopp and Spearman (2004), who introduced the concept of Pull being related to the (possibly constrained) state of the system that needs to handle the order.

In light of this discussion, it is now time to take a look at order release in the POLCA system. First of all, order release in POLCA is only possible as long as

a) a capacity signal is available from the first cell that the order has to visit (POLCA card from planning cell to initial manufacturing cell),
b) orders are available (necessary preparatory activities for the first cell have been completed, so the order is waiting to be released), and
c) the planner has provided approval to start the order at the shop floor (earliest release date signal).

The first characteristic of order release in POLCA (a) makes the release decision reactive on the state of the shop floor, as the POLCA card will only be returned—and hence the signal provided for a new release—after completion of a previous order in the same routing segment. Hence, POLCA may be denoted as containing Pull characteristics. Note that the signal upon which POLCA reacts is not perfect, but may be somewhat compromised due to information delays. For example, at the time of receiving the signal at the order release stage, the first cell might be out of operation due to a breakdown. Information delays are caused by waiting and travelling of the card back to the originating cell in the POLCA loop. In case of a release decision, this is the time needed to return to the planning cell. The travelling might be organized using so called milk runs (i.e., intermittently), or continuously (as soon as the card becomes available it is being returned). The issue of information delays has been examined in the literature (e.g., Weitzman and Rabinowitz, 2003), but its effect on the POLCA system has, as far as I know, not yet been addressed.

Quality of Capacity Signal

An issue that has received ample attention in literature is the quality of the capacity signal provided by (a) above. Several authors (e.g., Fernandes et al., 2006; Germs and Riezebos, 2010; Ziengs, Riezebos, and Germs, 2012; Riezebos, 2006; Vandaele et al., 2008) have considered ways to improve the quality of the capacity signal to the release authority that POLCA cards provide. The problem is that this signal is of limited quality as it only refers to (1) time needed for an already-completed order, irrespective of the workload of the new order, and (2) the first cell to be visited, ignoring whether subsequent cells needed in the routing of the newly released order will have capacity available to process that order. To address the variety in workload, research has experimented with quantum cards (already suggested in Suri, 1998), or load-based POLCA cards (Vandaele et al., 2008), which resolves the first issue, or suggested to split orders in order to reduce the workload variance per order (Suri and Krishnamurthy, 2003; Riezebos, 2010a; Powell, Riezebos, and Strandhagen, 2013). Recent research has investigated the effect of order splitting in detail (Barros et al., 2016; Riezebos, 2004).

To extend the capacity signal to the whole routing, Fernandes et al. (2006) suggested the requirement that all POLCA cards for the whole routing should be available upon release. They denoted this modification of POLCA as GPOLCA (Generic POLCA). In this system, more POLCA cards are needed for especially the latter pairs of cells in the routings, as the cards travel along with the order. It is similar to the m-CONWIP system that originates from Hopp and Spearman (2000) and is compared with POLCA in papers by Germs and Riezebos (2010), Ziengs, Riezebos, and Germs (2012), and Farnoush and Wiktorsson (2013). However, GPOLCA differs from m-CONWIP in the speed of returning capacity signals from a pair of cells to the originating cell, as the signal is already provided by returning the POLCA card as soon as the two cells have completed their operations on this order. Although the simulation research of Fernandes et al. (2006), Germs and Riezebos (2010), and Ziengs, Riezebos, and Germs (2012) shows that these approaches slightly improve the results of the original POLCA system, Farnoush and Wiktorsson (2013) show that these differences deteriorate quickly if the simulations are done for an actual implementation in the automotive industry. Moreover, the drawbacks of the extended routing information is a less precise control of the workload capacity reservation (denoted by indirect workload in the workload control literature: Bechte, 1988; Stevenson et al., 2011; Breithaupt, Land, and Nyhuis, 2002). The total number of cards that is needed to operate the production control system should be increased, both for the GPOLCA and the m-CONWIP system, which leads to more control cards waiting on boards, and less focus on the visual signals of the cards that are at the shop floor. Finally, the extensions require more administrative efforts from the operators and/or planner, which is not very attractive. Hence, the research has shown that it is possible to include this level of detail and make the POLCA system more effective in case of high workload or routing variety, but it comes at a cost for the practical issues of operating the control system. We suggest that this should be included in future research that compares several production control approaches.

Readiness Assurance

Until now, we only have focused on the first signal that should have become available upon order release (capacity available at the shop floor), so it is time to discuss the second (orders prepared and materials available) and third ones as well (authority signal of planner available). The availability of orders that have been prepared (i.e., materials available, tools and auxiliary resources available, process plans, drawings, testing protocols, etc.) and are waiting before release is an obvious condition for order release. But, in practice, this is an often overseen one. Planners might base their release decision on other information than can be found in reality. Errors in the availability of materials, tools, drawings, etc. could lead to the release of orders that cannot proceed at the shop floor where all things come together. At the shop floor, such orders have been assigned a POLCA card that prevents the signaling of available capacity, while at the same time the machines and employees cannot start working on the incomplete order. Hence, the importance of complete and correct information before order release is essential. POLCA literature has addressed several issues and suggested solutions such as safety cards, on-hold cards, material reservation policies, etc. (Krishnamurthy and Suri, 2009; Riezebos, 2010a). Note that these solutions are meant to act as a safety valve against unforeseen events, errors, or mistakes, but not for lack of discipline. POLCA sets out rules that should be respected by both employees, planners, and other authorities in a company. However, in cases of unforeseen events, solutions are available, at least within the post-release phase—not yet in the pre-release phase. More general books on production control, such as (Thürer, Stevenson, and Protzman, 2016a; Lödding, 2013a) often simply suggest guaranteeing that orders are complete before they are being released to the shop floor, which is a typical way of ignoring the inherent complexity of organizing this process. Future research on POLCA could aim to support the pre-release process better in order to avoid releasing orders that should have been postponed.

Authority Division through Release Lists

The third signal (authority signal of planner available) is provided through a release list from the planner. Such a release list can often easily be generated through an ERP system. It specifies for each cell the second cell in the routing of the orders, in order to also inform the originating cell which POLCA card will have to be attached before they can actually start. Next, it specifies an earliest release date of this order in the first cell. By providing this date, the planner prevents an early release of orders that might fit in the capacity profile of the first cell in the routing, but will lead to the unnecessary build-up of work-in-progress in remaining parts of the routing. It is a way to assure short throughput times by reducing the total amount of work in progress at the shop floor. Researchers have examined how to calculate the earliest release dates. In the typical production environment for which POLCA has been developed (engineer/make-to-order), due dates of orders have been set and communicated with the customer. Hence, it will be possible to apply backward-calculation, taking into account the planned lead times in the various cells (setups, processing, testing, and some minor waiting times) and between the cells (transport and waiting times for capacity and next POLCA signal).

Backward-calculation is available in ERP systems through the Material Requirements Planning (MRP) algorithm. See Lödding (2013b) for a description of this method in the context of POLCA. However, there are several drawbacks to apply this algorithm directly to the processing plans in order to calculate the earliest release date of the order, due to MRP's fundamental problems. First, MRP focuses on processing steps instead of cells and includes a predetermined waiting time allowance between each step. Next, MRP allows the use of time phasing, which will lead to the inclusion of additional waiting times before the next release takes place, as discussed in subsection "Timing of order release." See Riezebos (2001) and Krishnamurthy and Suri (2009) for a detailed discussion of this phenomenon.

The literature provides two major approaches to avoid the first fundamental problem of MRP (predetermined waiting times between processing steps). The first one, suggested by Vandaele et al. (2008) and Van Nieuwenhuyse et al. (2011), focuses on the improved calculation of the planned lead times. Instead of using fixed waiting time allowances, this method calculates the planned lead times based on the expected queuing behavior at the time the order is being processed. Hence, the planned lead times are obtained through Advanced Resource Planning instead of the MRP-algorithm. The deterministic MRP approach is replaced by a stochastic approach that might result in different planned lead time estimates if the due date of the order would change.

The second approach to avoid this fundamental problem of MRP is less calculation-intensive and has already been suggested in the first publication on POLCA (Suri, 1998). It is denoted as High-Level MRP (HL/MRP), which is a less detailed, more aggregated use of the MRP algorithm at the cell level, where waiting time allowances are no longer calculated for every processing step, but rather, for the cell as a whole, mainly to accommodate for the between-cell waiting times. In case the variance of the total estimated processing time within the cell is small, the processing times might be replaced by the average cell throughput time. The focus of HL/MRP is to stimulate the alignment of activities within the cells in order to have them realize short throughput times. Prior to the publication of the current book (with its Appendix D, "Capacity Clusters Make High Level MRP Feasible"), there have been no journal publications or books that discuss in detail the modifications needed to operate MRP as a HL/MRP system. However, important parallels can be seen with the design of Period Batch Control systems, which can be configured through MRP systems by avoiding certain choices of parameter settings. In my contribution to the 2016 World QRM conference (Riezebos, 2016), I discussed several of these modifications and how to calculate the allowances at cell level, based on a practical case study at a company in the Netherlands. In my view, these modifications still need further and more fundamental research. I would suggest this as a fruitful area for future research with practical relevance.

POLCA uses a unique authorization mechanism that respects the division of autonomy between planner and shop floor. This mechanism has to cope with the second fundamental problem of MRP (inclusion of waiting times due to time phased release authorization). POLCA allows planners to specify earliest release times per cell, but provides autonomy to the cells on the actual release time. This includes the choice of order in case more orders apply for the selection criteria (i.e., all signals are available). It is not required in POLCA that a planner specifies earliest release dates for every cell. Some planners leave this to the shop floor and set the same earliest release date for every cell listed in the routing of the order. In that case, all cells except the first one receive complete autonomy of the planner. This avoids the inclusion of waiting times on the authority signal of the planner during the post-release phase. In fact, this is the configuration that has been implemented initially in the POLCA systems that I am aware of in practice, although some of them have emerged into the more complex system including authorization release dates per cell (see Chapter 12 on Bosch and Chapter 13 on Provan). The Release and Flow POLCA (RF/POLCA) that has been described in Chapter 5 along with the POLCA rules for this option does in fact allow for setting a single earliest release date per order, while the regular POLCA system allows for setting earliest release dates per cell. Hence, it is a fundamental choice within POLCA to also allow the planner to specify different earliest release times for the different cells that an order needs to visit. In simulation and optimization studies so far, the effect of authorization per cell has been ignored. Several authors have criticized POLCA on the possibility it offers to include this type of waiting times (e.g., Thürer, Stevenson, and Protzman, 2016; Lödding, 2013a) as it increases total throughput times and might lead to an increase of tardiness, but Riezebos (2010a) argued that it is unclear how this affects the average shop floor throughput time of all orders, and that the main reasons for introducing this mechanism focus on achieving other performance indicators. For example, a planner might aim to prohibit cannibalization of components inventory, which might occur if cells start too early with orders that have enough time before they are due. The same holds true for using scarce capacity. In both cases, the planner may prefer some jobs to wait in order to prioritize the progress of other jobs. This might improve due-date related performance measures (e.g., tardiness or lateness), but it does not need to have a positive influence on average shop floor throughput time. Future research might further investigate this issue. I would recommend that such studies would include one or more of the above-mentioned characteristics of the production environments (variety, uncertainty, time-phased availability of materials) that might make the use of such intermittent authorization behavior of planners with the shop floor logical.

The release list functions as an order backlog or order book for the cells. It shows them what orders are in their backlog, i.e., waiting upon release or part of their indirect workload, allows them to think ahead on the impact of these orders for their internal processes and task division within the cell, identify in advance possible issues that would prevent short throughput times for these orders, helps them to plan auxiliary tasks (e.g., cleaning, training, supporting other cells) in case no new orders are expected to become available within this cell, and allows for some influence of the planner on the behavior of the cell (by changing the earliest release moment of an order in the cell).

The release list might be restricted to orders that are not yet released, but are made available for release within the time horizon at the time of submitting the release list to the shop floor. In that case, the release list only functions as a division of authority between planner and shop floor with respect to the release decision, and does not provide a complete overview of orders in progress at the shop floor, i.e., the backlog per cell.

If the release list is available per cell and includes orders that already have been released to other cells and are expected to arrive at some moment in time at the current cell, it is a matter of choice whether the latter category of orders is a complete list (the total indirect workload) or would be limited to orders that are in process at the directly preceding cell (i.e., the orders to which a POLCA card with the current cell as destination cell has been attached). Implementations of ERP support of POLCA release lists have in general limited the view to the orders with such a POLCA card attached, as this provides a concise overview of the relevant part of the indirect workload (Powell, Riezebos, and Strandhagen, 2013), while the COBACABANA system that is based on workload control theory tends to depict the total indirect workload per cell/work station to give the planner a complete overview for future release decisions (Lödding, 2013a; Thürer, Land, and Stevenson, 2014; Thürer, Stevenson, and Protzman, 2015). Future research might reveal what level of abstraction is more effective for the cells and/or planner.

Let us now focus on the remaining parameter design choices for the release list. There are still three main parameters to be decided upon for a release list:

- its time bucket granularity (time bucket length expressed in shifts, days, or weeks);
- time horizon (how many periods to look ahead); and
- frequency of being updated.

As a general rule, the time between updates is longer than the time bucket length used in the release list, and shorter than or equal to the time horizon. So, if the release list specifies for a set of orders its earliest release times in terms of either the morning or afternoon shift at the current or next day, the time period used is a day shift, the time horizon is four shifts (two days), and the frequency of updating the release list is probably daily or every two days. The choices for these three parameters are important, as they have an impact on the total throughput times as well as on the tasks and division of autonomy between planner and shop floor.

The time bucket granularity (period length) in which the earliest release times are specified has an impact on the waiting time before release. If the granularity is too high (weeks instead of days or hours), the system becomes non-responsive and starvation might occur. This has been identified by (Lödding, 2013b; Thürer, Stevenson, and Protzman, 2016b). Similar effects could be seen in workload control literature and intermittent release. New insights in this field suggest using continuous release (such as in POLCA) and starvation avoidance (Fredendall, Ojha, and Patterson, 2010; Thürer et al., 2014). An appropriate choice of time period length allows for a responsive system.

The time horizon length should be chosen such that the shop floor has the possibility to anticipate expected occurrences in the near future. It is not necessary to include all orders known to the planner to be ready for release if some of these orders are beyond the scope of cells and do not need to be anticipated. Typically, this type of information on orders that are not yet available for release but will become available within some time periods is being used by cells as a type of advanced demand information (González et al., 2013) to influence the selection of an order for release. But a time horizon that is too long will blur the cell with information it does not need to include in its current decision making-process. Hence, it is a parameter to simplify and enhance decision-making within the cell.

The time between updates should be chosen such that the system becomes adaptable, responsive, but not nervous. An updated release list allows planners to modify the information on available orders for release and specify or change earliest release times. The update frequency should allow the cells to look ahead for some amount of time and base their decisions on the information provided in the time window. Research on nervousness has been published in the context of MRP (Van Nieuwenhuyse et al., 2011; Simpson and Erenguc, 1996; Pahl, Voss, and Woodruff, 2007; Jonsson and Mattsson, 2006; González et al., 2013), but research on nervousness and update frequencies in POLCA systems has not yet been published, as far as I know. I strongly recommend future research in this area, as it clearly affects the operation of a POLCA system in practice.

Conclusion on Order Release Decision

This completes our discussion on the contribution of research into POLCA in the area of order release. I would suggest further research in this area (as specified above), and especially would welcome research toward modifications/extensions of POLCA that would focus on a better representation of the indirect workload and starvation avoidance. POLCA could benefit from recent insights developed in workload control literature in order to make the order release decision more effective. On the other hand, I welcome the strong focus of POLCA on providing (at least partial) autonomy to cells instead of giving full autonomy to the central planner for the release decision. Most of the simulation studies ignore the impact of local information on the realized throughput times and do not include the benefits of using the information available at cell level in their models. In practice, local or decentralized control is very important, as noticed by Lödding (2013a); also see Appendix E for more on this point.

Shop Floor Control and Workload Balancing

This section covers issues related to the operation of the POLCA control system after orders are released to the shop floor.

Product-Specific, Product-Anonymous, and Route-Specific Control

POLCA does not provide product-specific, but route-specific control of the material flow at the shop floor. Product-specific control is, for example, provided by Toyota's Kanban system. Product-specific control systems signal the requirement for a refill of a stock position for a specific component or stock-keeping unit. Krishnamurthy, Suri, and Vernon (2004) have shown that such systems are not effective in cases of make-to-order companies. The main reason is that the number of products these firms can offer is generally much higher than make-to-stock companies offer, while at the same time the frequency of demand of these products is much lower. Together, these effects lead to significant inefficiency of product-specific control systems in the case of make-to-order companies.

An alternative for product-specific control is product-anonymous control. Several systems have been developed that provide product-anonymous control, such as CONWIP (Hopp and Spearman, 2000) and Generic Kanban (Chang and Yih, 1994). The idea of product-anonymous control is that it only signals the need for a new release or next step of the order irrespective of the actual contents of the processing step or order.

POLCA further elaborates on this type of signal by introducing route-specific control of the material flow (Riezebos, 2010a). Route-specific control is a special kind of product-anonymous control. However, there is an important difference, as the control system takes into account the availability of capacity in the next parts of the routings of the orders that are available for release. Hence, essential information on the remaining routing of the order is taken into account when deciding on the release or progress of an order at the shop floor.

Visual Signal

The visual signal that a POLCA card represents should provide information to the shop floor employees on route-specific characteristics of the order to which it is or should be attached. Suri (1998) suggested providing the following information:

- Originating cell (From);
- Destination cell (To);
- Card identifiers (Company name, serial number of card, etc.).

Riezebos and Pieffers (2006) suggested increasing the visual contents of these cards by using color-coding to identify the cells. If every cell has a unique color, each card consists of two colors: one for the from-cell and one for the to-cell. Employees of a cell can easily identify the direction of the card flow. If their color is listed in the to-column, the card should be returned to the originating cell after completing the operations in the current cell. If their color is listed in the from-column, the card should remain attached to the order and be sent to the to-cell after completing the operations in the current cell.

Vandaele et al. (2008) choose to use an electronic signal, mainly because of the shop floor layout in their case study. Electronic signals are shown on screens at the work stations and operators can attach and detach these signals to orders in their system. The main advantage of electronic signals is that there is no information delay nor manual handling caused by the transportation of the visual signal if it is detached from an order. Other advantages are the avoidance of card damage and loss, mistakes when attaching or detaching cards, search time in order to localize cards, and ease of changing the number of cards in the system by the shop floor manager or planner. The main disadvantage is the loss of connection between the flow of material and information, which might lead to less confidence in the system by the employees. Insights from lean manufacturing indicate that the strength of that connection often determines the success of a control policy.

Based on four case studies of implementations of Pull systems, Powell, Riezebos, and Strandhagen (2013) identified what support ERP systems (could) provide to implement and support a Pull system such as POLCA. Moreover, they discuss the use of manufacturing execution systems that allow for a visual representation of POLCA, so-called E-POLCA or Digital POLCA. Chapters 12, 13, and 15 describe several electronic POLCA systems that have been developed and are commercially available.

In practical applications of POLCA, I observe variants that have not yet been analyzed using optimization/simulation studies. For example, companies have decided to split the cards, so they are no longer route-specific (see Appendix E). The cell that would like to start an order first needs a capacity signal from the next cell, but this signal is no longer specific for this routing segment in cases of split cards. This solution avoids the dedication of cards to routing segments and makes it easier to calculate the number of cards, as this number directly reflects the available capacity in the cell for the near future. Each card is seen as a capacity reservation for the near future. Employees did understand the principle and visual signal of these cards better. Moreover, in case of a temporary increase in available capacity at a cell, no information on the mix of orders that will need this capacity is needed. However, there might also be some drawbacks of this approach; for example, whether the predefined mix of route-specific cards would prevent imbalance in prioritizing one type of orders (e.g., orders with small workload). I would strongly suggest future research towards the effectiveness of this modification of POLCA.

Control Loops and Workload Balancing

The second issue that has to be addressed is the design of the control loops. POLCA uses overlapping loops and introduces a (virtual) "assembly waiting time" before a job can start in a cell, in the following way. First, a suitable POLCA card needs to be attached to the job in order to signal the authorization to start processing in the next cell. The rationale of this procedure is that a job will only be allowed to ask for capacity in a cell upstream (e.g., Cell A) if it is expected that the required downstream cell (Cell B) will be able to continue processing afterwards. If Cell B would face a breakdown, no A/B POLCA cards will be returned to Cell A and no authorization is given to start new jobs that need to be processed in Cell B after completing Cell A. If Cell B is a bottleneck, a lot of A/B POLCA cards are probably waiting before Cell B, connected to orders that are queueing before this bottleneck cell. Hence, no new orders will be released from Cell A that have Cell B in their routing until an order in Cell B has finished and the A/B POLCA card is detached and returned to Cell A. As Cell A cannot proceed with the jobs that have to visit Cell B afterwards, but don't already have an A/B POLCA card attached, it will stay idle or it might select another order that has a different cell than Cell B as immediate follower in its routing. Orders selected in Cell A will therefore for the moment bypass Cell B, which has important benefits for shortening average throughput times. We denote this control behavior as workload balancing. There has been ample attention in literature, e.g., Germs and Riezebos (2010), that POLCA is able to provide this type of workload balancing, which is very important in many make-to-order situations that face shifting bottlenecks over time. Initially, it was thought that Pull systems could not provide effective workload-balancing capability. Early literature on workload control (e.g., Melnyk and Ragatz (1988); Kanet (1988)) even suggests the existence of a paradox related to the absence of this workload-balancing capability in Pull systems. While practical implementations show significant reductions in the total throughput time of orders, simulation studies show that constraining the workload on the shop floor leads to both shorter shop floor throughput times and longer total throughput times. There are some studies, such as those of Land and Gaalman (1998) and Breithaupt, Land, and Nyhuis (2002) that have shown the existence of an effective workload-balancing capability in load-based Pull systems. Germs and Riezebos (2010) were the first that showed that the unit-based version of POLCA has effective workload-balancing capability (i.e., it reduces the total throughput time, not just the shop floor throughput time). Some recent comparisons of production control systems tend to overlook this important capability of POLCA, as they focus on load balancing capability only (i.e., balancing based on processing time differences) while ignoring the workload balancing capability of a more simple unit-based system such as POLCA; for example, see Thürer, Stevenson, and Protzman (2016b).

Number and Workload of Cards

Literature has addressed the issue of determining the number of POLCA cards in a single loop in extension. Suri (1998, p. 256) provides a simple formula based on Little's law (Little, 2011). Little's law states that the average throughput rate of a system equals the quotient of the average number of elements in the system and the average time an element is in the system. (Riezebos, 2010a) provided a correction and extension of this formula in cases where earliest release times are specified per cell. More recently, Lödding (2013b) provided formulas for logistic positioning of the manufacturing cells and advanced release windows. In Appendix G, Suri now provides a more rigorous explanation and derivation of an improved formula to calculate the number of POLCA cards.

Suri (1998, p. 256) also introduced the notion of a quantum, i.e., the maximum workload (amount of hours) that should be associated with a single card or the amount of material that should accompany a single card. Jobs that exceed this limit need more than one card in order to be authorized to start processing in a cell. Using multiple POLCA cards per order based on quantum rules is better than using a limit on the number of jobs in a loop, as can be concluded from workload control theory, e.g., Bergamaschi et al. (1997) and Land and Gaalman (1998). In make-to-order job shop configurations, the correlation between workload and throughput time is much stronger than the correlation between number of orders and throughput time. Still, there may be valid reasons to not use a quantum rule. For example, ease of use and transparency of the system might lead to the use of a POLCA system with no quantum rule at all.

Deadlocks and Priority Rules

Now I would like to make some remarks on the occurrence of deadlocks in a POLCA system. This issue has been identified first by Lödding, Yu, and Wiendahl (2003), who used a simulation study with a known set of orders and noticed that some jobs were not being completed due to heavy blockings. Afterward, Harrod and Kanet (2013) encountered this issue as well in their simulation studies. They denoted this phenomenon as lockups and showed that in 100 experiments, in which 20,000 orders per experiment had to be processed, this behavior could be observed in 30% of the experiments when the number of cards per loop was limited to two. Kanban systems showed even worse performance (50%). They suggested the use of safety cards—a concept that had already been introduced in POLCA in order to facilitate rare and unforeseen circumstances that would otherwise hold up the system, see Suri and Krishnamurthy (2003) and also Chapter 6—which could be sufficient to solve the rare issue that has been described.

Lödding (2013b) and Thürer, Stevenson, and (2016) also discuss the issue of deadlocks in POLCA, but do not provide new solutions. Future research might contribute to a more fundamental solution (see Chapter 6 for Suri's contribution in this respect), but as far as I can see this is primarily relevant for simulation studies that generate such large numbers of orders, while in practice simple solutions are quite effective.

Finally, I would like to address the issue of priority rules. POLCA has been criticized for not prescribing the use of priority rules. Simulation studies have shown that such rules can make a difference in terms of throughput time performance. See, e.g., Harrod and Kanet (2013), Barros et al. (2016), and Mortágua, Fernandes, and Carmo-Silva (2014). However, these studies have not tested the effectiveness of using such priority rules in practice and compared it with POLCA's policy to give autonomy to the cells to decide on their own based on local information. Release lists might be sorted based on the preferred priority rule—in fact, this has been implemented in some of the ERP software add-ons that support the implementation of POLCA in practice (see Powell, Riezebos, and Strandhagen, 2013). But autonomy at the shop floor is a strong element of POLCA. Studies that examine the effectiveness of priority rules should not only compare random, non-intelligent, priority rules as a benchmark, compared to the intelligent rules that they test, but also aim to model the benefits of autonomy of decision-making at the cell level. That would provide more realistic outcomes that are useful for practical implementations.

Concluding Remarks on Shop Floor Control

To conclude, we see that the POLCA system enables control of work order progress based on information of routings between cells in a loop and a measure of work-in-progress that already has been released in that loop. The design of a loop does have to give attention to the cells that are included in a single loop, as this affects the information content on routings in the system. Moreover, it should give attention to the number of POLCA cards that will circulate in that loop and the workload associated with a single POLCA card.

Conclusions

This appendix has discussed a variety of issues with respect to POLCA that have received attention from the research community. I have restricted my attention to books, journal papers, and some conference papers. This selection did not include the many theses (at bachelors, masters, and Ph.D. levels) that have appeared on POLCA. I have been involved in supervising almost 20 student theses myself and am aware of many more that have been written (and sometimes published) over the world. In this overview, I have not included these studies, as they are generally not easily available for other readers. The published academic contributions in books, journals, and presented at conferences have led to many more detailed insights in the effectiveness of POLCA, its applicability, and the mechanisms behind order release and order progress.

I have provided many suggestions for further research in the main text. In the section "Where is a POLCA Production Control System Applicable?", I addressed the issue of the applicability of POLCA and how it has been assessed in the literature thus far. This will allow researchers to contribute by exploring the applicability of POLCA in case studies that have different combinations of characteristics. It will also allow a better comparison of several production control systems in a specific production environment, as I have found that coping with dynamics, especially uncertainties, is an under-researched area in the currently published reviews and comparisons.

In the section "Order Release Decision", the order release decision within POLCA has been investigated. This brings forward many questions on the design parameters of the release system (i.e., frequency of release, the contents of the release list, HL/MRP, time-phased release, etc.). But perhaps even more important, order release should load the shop floor in a realistic and respectful manner, aiming to realize short throughput times. It should be a shared responsibility of the planner (who is able to oversee the boundary activities at both sides of the shop floor, i.e., preparatory activities and customer perspective), and the shop floor employees. POLCA involves the state of the shop floor in the release decision, which is fundamental for a Pull system.

In the section "Shop Floor Control and Workload Balancing", the operation of POLCA on the shop floor has been examined. For any production control system, including POLCA, it is important to consider decisions on various important concepts, such as the type of control used. We discussed product-specific, product anonymous, and routing-specific control. POLCA appears to be one of the only systems that applies route-specific control, which has several advantages in cases where the variety of products and routing sequences increases. Another decision concerns design of the visual signals, i.e., the card (digital or physical, colors, predefined routing segments, or split cards that are paired on the fly). Control loop design has an effect on the workload-balancing capability of production control systems. Literature has demonstrated that POLCA is a production control system that has this capability, i.e., is able to reduce total throughput time by limiting the workload at the shop floor in case of an appropriate choice of the number of cards per loop. Many other popular systems, such as CONWIP and Kanban, do not have this capability, while for others (e.g., workload control) it has been cumbersome to demonstrate this effect, leading to the so-called workload control paradox. The last topic that has been discussed is the determination of the number of cards and the setting of quantum rules. All these areas need further research, as there are many new developments that have to be investigated, both in theory and in practice.

Instead of repeating the suggestions for further research that have been included in the sections, I would like to conclude with an open invitation to researchers all over the world to submit and publish their work, whether it be conceptual/theoretical, practical (implementations, experiences, observations, longitudinal studies), design-oriented (optimization, simulation, experimental), or action-based (issues encountered when training employees, planners, suppliers, etc.). Research could bring forward new insights, extensions, or modifications to the known set of POLCA systems that are being used in practice or described in the literature, and enhance the application of visual control systems in the important field of custom-based manufacturing.

About the Author

Jan Riezebos is a full professor of educational innovation in the Department of Operations, University of Groningen, the Netherlands. He is president of the European QRM network and fellow of the research schools SOM and TRAIL. As academic director Career Services and Corporate Relations of the Faculty of Economics and Business, he is launching many initiatives to incorporate the practice of business and economics in higher education and research at the University of Groningen. Jan's current research interests involve process improvements through lean and quick response approaches and their applications in manufacturing and service, with more recently a focus on the application in higher education. The research approach includes design of methods and algorithms as well as behavioral aspects of operations management. He has conducted important studies on POLCA and team-based production planning approaches, and was involved in the implementation of POLCA and related Pull systems in several companies in Europe.

References

Adan, Ivo J.B.F. and Jan van der Wal. 1998. "Combining Make to Order and Make to Stock." *OR Spectrum* 20: 73–81.

Akkerman, Renzo, Dirk van der Meer, and Dirk Pieter Van Donk. 2010. "Make to Stock and Mix to Order: Choosing Intermediate Products in the Food-Processing Industry." *International Journal of Production Research* 48 (12): 3475–3492. doi: 10.1080/00207540902810569.

Aziz, Muhammad Haris, Erik L.J. Bohez, Roongrat Pisuchpen, and Manukid Parnichkun. 2013. "Petri Net Model of Repetitive Push Manufacturing with POLCA to Minimise Value-Added WIP." *International Journal of Production Research* 51 (15): 4464–4483. doi: 10.1080/00207543.2013.765073.

Bahl, Harish C. and Larry P. Ritzman. 1987. "Determining Lot Sizes and Resource Requirements A Review." *Operations Research* 35 (3): 329–345.

Barros, Cláudia, Cristiana Silva, Sandra Martins, Luís Dias, Guilherme Pereir, Nuno Octávio Fernandes, and Sílvio Carmo-Silva. 2016. "Are Card-Based Systems Effective for Make-to-Order Production?" *Romanian Review Precision Mechanics, Optics and Mechatronics* 2016 (49): 5–9.

Bechte, Wolfgang. 1988. "Theory and Practice of Load-Oriented Manufacturing Control." *International Journal of Production Research* 26: 375–395. doi: 10.1080/00207548808947871.

Benders, Jos and Jan Riezebos. 2002. "Period Batch Control: Classic, Not Outdated." *Production Planning & Control* 13 (6): 497–506. doi: 10.1080/0953728021016294.

Bensoussan, Alain, Metin Çakanyıldırım, Qi Feng, and Suresh P. Sethi. 2009. "Optimal Ordering Policies for Stochastic Inventory Problems with Observed Information Delays." *Production and Operations Management* 18 (5): 546–559. doi: 10.1111/j.1937-5956.2009.01028.x.

Bergamaschi, D., R. Cigolini, M. Perona, and A. Portioli. 1997. "Order Review and Release Strategies in a Job Shop Environment: A Review and a Classification." *International Journal of Production Research* 35 (2): 399–420. doi: 10.1080/002075497195821.

Bertrand, J.W.M., J.C. Wortmann, and Jacob Wijngaard. 1990. *Production Control: A Structural and Design Oriented Approach*. Amsterdam: Elsevier.

Braglia, M., Davide Castellano, and Marco Frosolini. 2014. "Optimization of POLCA-Controlled Production Systems with a Simulation-Driven Genetic Algorithm." *International Journal of Advanced Manufacturing Technology* 70 (1–4): 385–395. doi: 10.1007/s00170-013-5282-5.

Braglia, Marcello, Davide Castellano, and Marco Frosolini. 2015. "A Study on the Importance of Selection Rules within Unbalanced MTO POLCA-Controlled Production Systems." *International Journal of Industrial and Systems Engineering* 20 (4): 457. doi: 10.1504/IJISE.2015.070182.

Breithaupt, Jan-Wilhelm, Martin J. Land, and Peter Nyhuis. 2002. "The Workload Control Concept: Theory and Practical Extensions of Load Oriented Order Release." *Production Planning & Control* 13 (7): 625–638. doi: 10.1080/09537280210000026230.

Brennan, L. and S.M. Gupta. 1993. "A Structured Analysis of Material Requirements Planning Systems under Combined Demand and Supply Uncertainty." *International Journal of Production Research* 31 (7): 1689–1707.

Burbidge, John L. 1996. *Period Batch Control.* Oxford: Clarendon Press.

Buzacott, John A. and J. George Shantikumar. 1992. "A General Approach for Coordinating Production in Multiple-Cell Manufacturing Systems." *Production and Operations Management* 1 (1): 34–52.

Chand, Suresh, Vernon Ning Hsu, and Suresh P. Sethi. 2002. "Forecast, Solution and Rolling Horizons in Operations Management Problems: A Classified Bibliography." *Manufacturing & Service Operations Management* 4 (1): 25–43.

Chang, T.M. and Y. Yih. 1994. "Determining the Number of Kanbans and Lot sizes in a Generic Kanban System: A Simulated Annealing Approach." *International Journal of Production Research* 32 (8). Taylor & Francis: 1991–2004. http://cat.inist.fr/?aModele=afficheN&cpsidt=3306134.

Demeter, Krisztina. 2013. "Time-Based Competition—the Aspect of Partner Proximity." *Decision Support Systems* 54 (4): 1533–1540. doi: 10.1016/j.dss.2012.05.055.

Dolgui, Alexandre and Caroline Prodhon. 2007. "Supply Planning under Uncertainties in MRP Environments: A State of the Art." *Annual Reviews in Control* 31 (2): 269–279. doi: 10.1016/j.arcontrol.2007.02.007.

Eng, Chong Kuan and How Whee Ching. 2016. "Hybrid-POLCA in Job Shop Manufacturing." *Indian Journal of Science and Technology* 9 (36): 1–5. doi: 10.17485/ijst/2016/v9i36/102600.

Eng, Chong Kuan, How Whee Ching, and Bong Cheng Siong. 2015. "Paired-Cell Overlapping Loops of Cards with Authorization Simulation In Job Shop Environment." *International Journal of Mechanical and Mechatronics Engineering* 15 (3): 68–73.

Enns, S.T. 2002. "MRP Performance Effects due to Forecast Bias and Demand Uncertainty." *European Journal of Operational Research* 138: 87–102.

Faria Fernandes, Flavio Cesar, and Moacir Godinho Filho. 2011. "Production Control Systems: Literature Review, Classification, and Insights Regarding Practical Application." *African Journal of Business Management* 5 (14): 5573–5582.

Farnoush, Alireza and Magnus Wiktorsson. 2013. "POLCA and CONWIP Performance in a Divergent Production Line: An Automotive Case Study." *Journal of Management Control* 24 (2): 159–186. doi: 10.1007/s00187-013-0177-z.

Fernandes, Nuno Octávio, Sílvio do Carmo-Silva, Nicole Suclla Fernandez, and S. Docarmosilva. 2006. "Generic POLCA—A Production and Materials Flow Control Mechanism for Quick Response Manufacturing." *International Journal of Production Economics* 104 (1): 74–84. doi: 10.1016/j.ijpe.2005.07.003.

Fogliatto, Flávio Sanson, Giovani J.C. da Silveira, and Denis Borenstein. 2012. "The Mass Customization Decade: An Updated Review of the Literature." *International Journal of Production Economics* 138 (1): 14–25. doi: 10.1016/j .ijpe.2012.03.002.

Framinan, Jose M., Pedro L. González, and Rafael Ruiz-Usano. 2003. "The CONWIP Production Control System: Review and Research Issues." *Production Planning & Control* 14 (3): 255–265. doi: 10.1080/0953728031000102595.

Frazee, Todd and Charles Standridge. 2016. "CONWIP versus POLCA: A Comparative Analysis in a High-Mix, Low-Volume (HMLV) Manufacturing Environment." *Journal of Industrial Engineering and Management* 9 (2): 432. doi: 10.3926/jiem.1248.

Fredendall, Lawrence D., Divesh Ojha, and J. Wayne Patterson. 2010. "Concerning the Theory of Workload Control." *European Journal of Operational Research* 201 (1): 99–111. doi: 10.1016/j.ejor.2009.02.003.

Gaury, E.G., J.P.C. Kleijnen, and H. Pierreval. 2001. "A Methodology to Customize Pull Control Systems." *Journal of the Operational Research Society* 52 (7): 789–799. doi: 10.1057/palgrave.jors.2601153.

Geraghty, John and Cathal Heavey. 2005. "A Review and Comparison of Hybrid and Pull-Type Production Control Strategies." *OR Spectrum* 27 (2–3): 435–457. doi: 10.1007/s00291-005-0204-z.

Germs, Remco and Jan Riezebos. 2010. "Workload Balancing Capability of Pull Systems in MTO Production." *International Journal of Production Research* 48 (8): 2345–2360. doi: 10.1080/00207540902814314.

González-R., Pedro L., and Jose M. Framinan. 2009. "The Pull Evolution: From Kanban to Customised Token-Based Systems." *Production Planning & Control* 20 (3): 276–287. doi: 10.1080/09537280902875393.

González-R., Pedro L., Jose M. Framinan, and Rafael Ruiz-Usano. 2013. "A Methodology for the Design and Operation of Pull-Based Supply Chains." *Journal of Manufacturing Technology Management* 24 (3): 307–330. doi: 10.1108/17410381311318855.

Harrod, Steven and John J. Kanet. 2013. "Applying Work Flow Control in Make-to-Order Job Shops." *International Journal of Production Economics* 143 (2). Elsevier: 620–626. doi: 10.1016/j.ijpe.2012.02.017.

Hedenstierna, Philip and Amos H.C. Ng. 2011. "Dynamic Implications of Customer Order Decoupling Point Positioning." *Journal of Manufacturing Technology Management* 22 (8): 1032–1042. doi: 10.1108/17410381111177476.

Homem-De-Mello, T., A. Shapiro, and Mark L. Spearman. 1999. "Finding Optimal Material Release Times Using Simulation-Based Optimization." *Management Science*. JSTOR, 86–102. http://www.jstor.org/stable/2634924.

Hopp, Wallace J. and Mark L. Spearman. 2000. *Factory Physics: Foundations of Manufacturing Management*. McGraw-Hill Education.

———. 2004. "To Pull or Not to Pull: What Is the Question?" *Manufacturing & Service Operations Management* 6 (2): 133–148. doi: 10.1287/msom.1030.0028.

Humair, Salal, John D. Ruark, Brian T. Tomlin, and Sean P. Willems. 2013. "Incorporating Stochastic Lead Times Into the Guaranteed Service Model of Safety Stock Optimization." *Interfaces* 43 (5): 421–434.

Inderfurth, Karl. 2009. "How to Protect against Demand and Yield Risks in MRP Systems." *International Journal of Production Economics* 121 (2): 474–481. doi: 10.1016/j.ijpe.2007.02.005.

Iravani, Seyed M.R., Tieming Liu, and David Simchi-Levi. 2012. "Optimal Production and Admission Policies in Make-to-Stock/Make-to-Order Manufacturing Systems." *Production and Operations Management* 21 (2): 224–235. doi: 10.1111/j.1937-5956.2011.01260.x.

Jacobs, F.R. 2003. "Enterprise Resource Planning: Developments and Directions for Operations Management Research." *European Journal of Operational Research* 146 (2): 233–240. doi: 10.1016/S0377-2217(02)00546-5.

Jodlbauer, Herbert and A. Huber. 2008. "Service-Level Performance of MRP, Kanban, CONWIP and DBR due to Parameter Stability and Environmental Robustness." *International Journal of Production Research* 46 (8): 2179–2195. doi: 10.1080/00207540600609297.

Jodlbauer, Herbert and Sonja Reitner. 2012. "Material and Capacity Requirements Planning with Dynamic Lead Times." *International Journal of Production Research* 50 (16): 4477–4492.

Johnson, Danny J. 2003. "A Framework for Reducing Manufacturing Throughput Time." *Journal of Manufacturing Systems* 22 (4): 283–298.

Jonsson, Patrik and Stig-Arne Mattsson. 2006. "A Longitudinal Study of Material Planning Applications in Manufacturing Companies." *International Journal of Operations & Production Management* 26 (9): 971–995. doi: 10.1108/01443570610682599.

Kaku, B.K. and L.J. Krajewski. 1995. "Period Batch Control in Group Technology." *International Journal of Production Research* 33 (1). Taylor & Francis: 79–99. doi: 10.1080/00207549508930139.

Kaminsky, Philip and Onur Kaya. 2009. "Combined Make-to-Order/ Make-to-Stock Supply Chains." *IIE Transactions* 41 (2): 103–119. doi: 10.1080/07408170801975065.

Kanet, John J. 1988. "Load-Limited Order Release in Job Shop Scheduling Systems." *Journal of Operations Management* 7 (3–4): 44–58. doi: 10.1016/0272-6963(81)90003-6.

Khojasteh, Yacob. 2016. *Production Control Systems, A Guide to Enhance Performance of Pull Systems*. Tokyo: Springer. doi: 10.1007/978-4-431-55197-3.

Krishnamurthy, Ananth and Rajan Suri. 2009. "Planning and Implementing POLCA: A Card-Based Control System for High Variety or Custom Engineered Products." *Production Planning & Control* 20 (7): 596–610. doi: 10.1080/09537280903034297.

Krishnamurthy, Ananth, Rajan Suri, and Mary Vernon. 2000. "Push Can Perform Better than Pull for Flexible Manufacturing Systems with Multiple Products."

In *Industrial Engineering Research Conference Proceedings*, 1–7. http://citeseerx.ist.psu.edu/viewdoc/download?doi=10.1.1.75.5973&rep=rep1&type=pdf.

———. 2004. "Re-Examining the Performance of MRP and Kanban Material Control Strategies for Multi-Product Flexible Manufacturing Systems." *International Journal of Flexible Manufacturing Systems* 16 (2): 123–150. doi: 10.1023/B:FLEX.0000044837.86194.19.

Land, Martin J. 2009. "Cobacabana (Control of Balance by Card-Based Navigation): A Card-Based System for Job Shop Control." *International Journal of Production Economics* 117 (1): 97–103. doi: 10.1016/j.ijpe.2008.08.057.

Land, Martin J. and Gerard J.C. Gaalman. 1998. "The Performance of Workload Control Concepts in Job Shops: Improving the Release Method." *International Journal of Production Economics* 56–57: 347–364. http://www.sciencedirect.com/science/article/pii/S0925527398000528.

Lehtonen, Juha-Matti, Jan Holmström, and Jonas Slotte. 1999. "Constraints to Quick Response Systems in the Implosive Industries." *Supply Chain Management: An International Journal* 4 (1): 51–57. doi: 10.1108/13598549910255095.

Lin, Yen Ting and Ali Parlaktürk. 2012. "Quick Response under Competition." *Production and Operations Management* 21 (3): 518–533. doi: 10.1111/j.1937-5956.2011.01269.x.

Little, J.D.C. 2011. "OR FORUM—Little's Law as Viewed on Its 50th Anniversary." *Operations Research* 59 (3): 536–549. doi: 10.1287/opre.1110.0940.

Liu, L., Mahmut Parlar, and Stuart X. Zhu. 2007. "Pricing and Lead Time Decisions in Decentralized Supply Chains." *Management Science* 53 (5): 713–725. doi: 10.1287/mnsc.1060.0653.

Lödding, Hermann. 2013a. *Handbook of Manufacturing Control*. Berlin, Heidelberg: Springer Berlin Heidelberg. doi: 10.1007/978-3-642-24458-2.

———. 2013b. "POLCA Control." In *Handbook of Manufacturing Control*, 419–433. Berlin, Heidelberg: Springer Berlin Heidelberg. doi: 10.1007/978-3-642-24458-2_23.

Lödding, Hermann, K.-W. Yu, and H.-P. Wiendahl. 2003. "Decentralized WIP-Oriented Manufacturing Control (DEWIP)." *Production Planning & Control* 14 (1): 42–54. doi: 10.1080/0953728021000078701.

Louly, Mohamed-Aly, Alexandre Dolgui, and F. Hnaien. 2008. "Optimal Supply Planning in MRP Environments for Assembly Systems with Random Component Procurement Times." *International Journal of Production Research* 46 (19): 5441–5467. doi: 10.1080/00207540802273827.

Luh, Peter B., Xiaohui Zhou, and Robert N. Tomastik. 2000. "An Effective Method to Reduce Inventory in Job Shops." *Ieee Transactions on Robotics and Automation* 16 (4): 420–424.

Mabin, Victoria J. and Steven J. Balderstone. 2003. "The Performance of the Theory of Constraints Methodology: Analysis and Discussion of Successful TOC Applications." *International Journal of Operations & Production Management* 23 (6): 568–595. doi: 10.1108/01443570310476636.

McKay, K. 2000. "The Application of Computerized Production Control Systems in Job Shop Environments." *Computers in Industry* 42 (2–3): 79–97. doi: 10.1016/S0166-3615(99)00063-9.

Melnyk, Steven A., and Gary L. Ragatz. 1988. "Order Review/Release and Its Impact on the Shop Floor." *Production and Inventory Management Journal* 29 (1): 13.

Milne, R. John, Santosh Mahapatra, and Chi-Tai Wang. 2015. "Optimizing Planned Lead Times for Enhancing Performance of MRP Systems." *International Journal of Production Economics* 167. Elsevier: 220–231. doi: 10.1016/j.ijpe.2015.05.013.

Miltenburg, John. 2007. "Level Schedules for Mixed-Model JIT Production Lines: Characteristics of the Largest Instances That Can Be Solved Optimally." *International Journal of Production Research* 45 (16): 3555–3577. doi: 10.1080/00207540701223394.

Mortágua, João, Nuno Octávio Fernandes, and Sílvio Carmo-Silva. 2014. "Card-Based versus Workload-Based Production Order Release Mechanisms: Performance Assessment via Simulation." *Romanian Review Precision Mechanics, Optics and Mechatronics*, no. 45: 13–17.

Mula, J., R. Poler, J.P. García-Sabater, and F.C. Lario. 2006. "Models for Production Planning under Uncertainty: A Review." *International Journal of Production Economics* 103 (1): 271–285. doi: 10.1016/j.ijpe.2005.09.001.

Olhager, Jan. 2013. "Evolution of Operations Planning and Control: From Production to Supply Chains." *International Journal of Production Research*, no. March (March): 1–8. doi: 10.1080/00207543.2012.761363.

Orlicky, Joseph. 1975. *Material Requirements Planning: The New Way of Life in Production and Inventory Management*. New York: McGraw-Hill.

Pahl, Julia, Stefan Voss, and David L. Woodruff. 2007. *Production Planning with Load Dependent Lead Times: An Update of Research. Annals of Operations Research*. Vol. 153. doi: 10.1007/s10479-007-0173-5.

Pelto, C. and B. Mueller. 2012. "It's in the Cards: Using POLCA Signals for Better Material Flow." *Apics Magazine*, May/June issue.

Plenert, Gerhard. 1999. "Focusing Material Requirements Planning (MRP) towards Performance." *European Journal of Operational Research* 119 (1): 91–99. doi: 10.1016/S0377-2217(98)00339-7.

Pochet, Yves and Laurence A. Wolsey. 1988. "Lot-Size Models with Backlogging: Strong Reformulations and Cutting Planes." *Mathematical Programming* 40: 317–335.

Powell, Daryl, Erlend Alfnes, Jan Ola Strandhagen, and Heidi Dreyer. 2013. "The Concurrent Application of Lean Production and ERP: Towards an ERP-Based Lean Implementation Process." *Computers in Industry* 64 (3): 324–335. doi: 10.1016/j.compind.2012.12.002.

Powell, Daryl, Jan Riezebos, and Jan Ola Strandhagen. 2013. "Lean Production and ERP Systems in Small- and Medium-Sized Enterprises: ERP Support for Pull Production." *International Journal of Production Research* 51 (2): 395–409. doi: 10.1080/00207543.2011.645954.

Reichhart, Andreas and Matthias Holweg. 2007. "Lean Distribution: Concepts, Contributions, Conflicts." *International Journal of Production Research* 45 (16): 3699–3722. doi: 10.1080/00207540701223576.

Riezebos, Jan. 2001. *Design of a Period Batch Control Planning System for Cellular Manufacturing*. Groningen: Ph.D. Thesis, University of Groningen. http://dissertations.ub.rug.nl/faculties/management/2001/j.riezebos/?pLanguage=en&pFullItemRecord=ON.

———. 2004. "Time Bucket Length and Lot-Splitting Approach." *International Journal of Production Research* 42 (12): 2325–2338. doi: 10.1080/00207540410001666242.

———. 2006. "POLCA Simulation of a Unidirectional Flow System." In *Proceedings of the Third International Conference on Group Technology/Cellular Manufacturing*, edited by Jan Riezebos and Jannes Slomp. Groningen: University of Groningen.

———. 2010a. "Design of POLCA Material Control Systems." *International Journal of Production Research* 48 (5). Kitakyushu, Japan: Waseda University: 1455–1477. doi: 10.1080/00207540802570677.

———. 2010b. "Order Release in Synchronous Manufacturing." *Production Planning & Control* 21 (4): 347–358. doi: 10.1080/09537280903453877.

———. 2011. "Order Sequencing and Capacity Balancing in Synchronous Manufacturing." *International Journal of Production Research* 49 (2): 531–552. doi: 10.1080/00207540903067185.

———. 2013. "Shop Floor Planning and Control in Team-Based Work Processes." *International Journal of Industrial Engineering and Management* 4 (2): 51–56. http://www.iim.ftn.uns.ac.rs/casopis/ijiem_vol4_issue2.php.

———. 2016. "Making Planning More Effective Experiences with High-Level MRP." In *2016 QRM World Conference*. Warsaw.

Riezebos, Jan, Gosse J. Korte, and Martin J. Land. 2003. "Improving a Practical DBR Buffering Approach Using Workload Control." *International Journal of Production Research* 41 (4): 699–712. doi: 10.1080/0020754031000065485.

Riezebos, Jan and Jacob Pieffers. 2006. *POLCA Als Innovatief Materiaal beheersingssysteem*. Groningen: University of Groningen.

Roundy, R.O. 1993. "Efficient, Effective Lot Sizing for Multistage Production Systems." *Operations Research* 41 (2): 371–385. doi: 10.1287/opre.41.2.371.

Schönsleben, Paul. 2004. *Integral Logistics Management: Planning & Control of Comprehensive Supply Chains*. 2nd ed. Boca Raton: CRC Press.

Serel, Doğan A. 2009. "Optimal Ordering and Pricing in a Quick Response System." *International Journal of Production Economics* 121 (2): 700–714. doi: 10.1016/j.ijpe.2009.04.020.

Silver, Edward A., David F. Pyke, and Rein Peterson. 1998. *Inventory Management and Production Planning and Scheduling*. 3rd ed. New York: John Wiley & Sons.

Simpson, Natalie C., and S. Selcuk Erenguc. 1996. "Multiple-Stage Production Planning Research: History and Opportunities." *International Journal of Operations & Production Management* 16 (6): 25–40. doi: http://dx.doi.org/10.1108/01443579610119072.

Slomp, Jannes, Jos A.C. Bokhorst, and Remco Germs. 2009. "A Lean Production Control System for High-Variety/low-Volume Environments: A Case Study Implementation." *Production Planning & Control* 20 (7): 586–595. doi: 10.1080/09537280903086164.

Spearman, Mark L., David L. Woodruff, and Wallace J. Hopp. 1990. "CONWIP, a Pull Alternative to Kanban." *International Journal of Production Research*. doi: 10.1080/00207549008942761.

Stevenson, Mark, Linda C. Hendry, and Brian G. Kingsman. 2005. "A Review of Production Planning and Control: The Applicability of Key Concepts to the Make-to-Order Industry." *International Journal of Production Research* 43 (5): 869–898. doi: 10.1080/0020754042000298520.

Stevenson, Mark, Yuan Huang, Linda C. Hendry, and G.D. Soepenberg. 2011. "The Theory and Practice of Workload Control: A Research Agenda and Implementation Strategy." *International Journal of Production Economics* 131 (2): 689–700. doi: 10.1016/j.ijpe.2011.02.018.

Storey, John, Caroline Emberson, Janet Godsell, and Alan Harrison. 2006. "Supply Chain Management: Theory, Practice and Future Challenges." *International Journal of Operations & Production Management* 26 (7): 754–774. doi: 10.1108/01443570610672220.

Su, Jack C.P., Yih-Long Chang, Mark E. Ferguson, and Johnny C. Ho. 2010. "The Impact of Delayed Differentiation in Make-to-Order Environments." *International Journal of Production Research* 48 (19): 5809–5829. doi: 10.1080/00207540903241970.

Suri, Rajan. 1998. *Quick Response Manufacturing: A Companywide Approach to Reducing Lead Times*. Productivity Press.

———. 2003. "QRM and POLCA: A Winning Combination for Manufacturing Enterprises in the 21st Century." In *Proceedings of the 2003 QRM Conference*, Center for Quick Response Manufacturing. www.qrmcenter.org.

———. 2010. *It's About Time: The Competitive Advantage of Quick Response Manufacturing*. Productivity Press.

Suri, Rajan, and Ananth Krishnamurthy. 2003. "How to Plan and Implement POLCA: A Material Control System for High-Variety or Custom-Engineered Products." In *Proceedings of the 2003 QRM Conference*, Center for Quick Response Manufacturing. www.qrmcenter.org.

Tadj, L., M. Bounkhel, and Y. Benhadid. 2006. "Optimal Control of a Production Inventory System with Deteriorating Items." *International Journal of Systems Science* 37 (15): 1111–1121. doi: 10.1080/00207720601014123.

Takahashi, Katsuhiko, Katsumi Morikawa, and Ying-Chuan Chen. 2007. "Comparing Kanban Control with the Theory of Constraints Using Markov Chains." *International Journal of Production Research* 45 (16): 3599–3617. doi: 10.1080/00207540701228153.

Tavares Thomé, Antônio Márcio, Luiz Felipe Scavarda, Nicole Suclla Fernandez, and Annibal José Scavarda. 2012. "Sales and Operations Planning: A Research Synthesis." *International Journal of Production Economics* 138 (1): 1–13. doi: 10.1016/j.ijpe.2011.11.027.

Thürer, Matthias, Martin J. Land, and Mark Stevenson. 2014. "Card-Based Workload Control for Job Shops: Improving COBACABANA." *International Journal of Production Economics* 147 (PART A). Elsevier: 180–188. doi: 10.1016/j.ijpe.2013.09.015.

Thürer, Matthias, Mark Stevenson, and Charles W. Protzman. 2015. "COBACABANA (Control of Balance by Card Based Navigation): An Alternative to Kanban in the Pure Flow Shop?" *International Journal of Production Economics* 166: 143–151. doi: 10.1016/j.ijpe.2015.05.010.

———. 2016a. *Card-Based Control Systems for a Lean Work Design: The Fundamentals of Kanban, CONWIP, POLCA, and COBACABANA*. Boca Raton: CRC Press/Taylor & Francis, Ltd.

———. 2016b. "Card-Based Production Control: A Review of the Control Mechanisms Underpinning Kanban, CONWIP, POLCA and COBACABANA Systems." *Production Planning & Control* 27 (14): 1–15. doi: 10.1080/09537287.2016.1188224.

———. 2016c. "Inventory Control Plus Material Requirements Planning for the Order Control Problem: POLCA." In *Card-Based Control Systems for a Lean Work Design: The Fundamentals of Kanban, CONWIP, POLCA, and COBACABANA*, 115–126. Boca Raton: CRC Press/Taylor & Francis, Ltd. doi: 10.1016/B978-1-4377-0651-2.10007-4.

Thürer, Matthias, Mark Stevenson, Cristovao Silva, Martin J. Land, and Lawrence D. Fredendall. 2012. "Workload Control and Order Release: A Lean Solution for Make-to-Order Companies." *Production and Operations Management* 21 (5): 939–953. doi: 10.1111/j.1937-5956.2011.01307.x.

Thürer, Matthias, Mark Stevenson, Cristovao Silva, Martin J. Land, Lawrence D. Fredendall, and Steven A. Melnyk. 2014. "Lean Control for Make-to-Order Companies: Integrating Customer Enquiry Management and Order Release." *Production & Operations Management* 23 (3): 463–476.

Tibben-Lembke, Ronald S. 2009. "Theory of Constraints at UniCo: Analysing The Goal as a Fictional Case Study." *International Journal of Production Research* 47 (7): 1815–1834. doi: 10.1080/00207540802624003.

Umble, E. 2003. "Enterprise Resource Planning: Implementation Procedures and Critical Success Factors." *European Journal of Operational Research* 146 (2): 241–257. doi: 10.1016/S0377-2217(02)00547-7.

Umble, M., E. Umble, and S. Murakami. 2006. "Implementing Theory of Constraints in a Traditional Japanese Manufacturing Environment: The Case of Hitachi Tool Engineering." *International Journal of Production Research* 44 (10): 1863–1880. doi: 10.1080/00207540500381393.

Vairaktarakis, George L. 2002. "Workforce Planning in Synchronous Production Systems." *European Journal of Operational Research* 136 (3): 551–572. doi: 10.1016/S0377-2217(01)00056-X.

van der Vaart, J. Taco, J. de Vries, and Jacob Wijngaard. 1996. "Complexity and Uncertainty of Materials Procurement in Assembly Situations." *International Journal of Production Economics* 46–47: 137–152.

Van Nieuwenhuyse, I., Liesje De Boeck, Marc Lambrecht, and Nico J. Vandaele. 2011. "Advanced Resource Planning as a Decision Support Module for ERP." *Computers in Industry* 62 (1): 1–8. doi: 10.1016/j.compind.2010.05.017.

Vandaele, Nico J., I. Van Nieuwenhuyse, D. Claerhout, and R. Cremmery. 2008. "Load-Based POLCA: An Integrated Material Control System for Multiproduct, Multimachine Job Shops." *Manufacturing & Service Operations Management* 10 (2): 181–197. doi: 10.1287/msom.1070.0174.

Vollmann, T.E., L.B. Berry, D. Clay Whybark, and F.R. Jacobs. 2005. *Manufacturing Planning and Control Systems for Supply Chain Management.* 5th ed. New York: McGraw-Hill.

Wang, S. and B. Sarker. 2006. "Optimal Models for a Multi-Stage Supply Chain System Controlled by Kanban under Just-in-Time Philosophy." *European Journal of Operational Research* 172 (1): 179–200. doi: 10.1016/j .ejor.2004.10.001.

Weitzman, R. and G. Rabinowitz. 2003. "Sensitivity of 'Push' and 'Pull' Strategies to Information Updating Rate." *International Journal of Production Research* 41 (9): 2057–2074. doi: 10.1080/0020754031000075862.

Wu, Zhengping, Burak Kazaz, Scott Webster, and Kum Khiong Yang. 2012. "Ordering, Pricing, and Lead-Time Quotation under Lead-Time and Demand Uncertainty." *Production and Operations Management* 21 (3): 576–589. doi: 10.1111/j.1937-5956.2011.01289.x.

Zhang, Zhe George, Ilhyung Kim, Mark Springer, Gangshu (George) Cai, and Yugang Yu. 2013. "Dynamic Pooling of Make-to-Stock and Make-to-Order Operations." *International Journal of Production Economics* 144 (1): 44–56. doi: 10.1016/j.ijpe.2013.01.012.

Ziengs, Nick, Jan Riezebos, and Remco Germs. 2012. "Placement of Effective Work-in-Progress Limits in Route-Specific Unit-Based Pull Systems." *International Journal of Production Research* 50 (16): 4358–4371. doi: 10.1080/00207543.2011.590537.

Appendix I: Additional Resources for POLCA Training and Games

At the time of going to press, the author was aware of the following resources around the world that provide training on POLCA and/or games that illustrate the operation of POLCA.

North America:

- Center for Quick Response Manufacturing at the University of Wisconsin–Madison, www.qrmcenter.org.
- Alexandria Technical & Community College, www.alextech.edu.
- Tempus Institute, www.tempusinstitute.com.

South America:

- GPI Ingenieros Consultores, www.gpi-consultores.cl.

Europe:

- QRM Institute, www.qrminstitute.org. QRM Institute is a pan-European organization with representatives in several European countries.

Belgium:

- Sirris, www.sirris.be.

Denmark:

- QRM Danmark, www.qrmdanmark.dk.

France:

- Quick Response Enterprise, www.quickresponse-enterprise.com.

Germany:

- Axxelia, www.axxelia.com.

Poland:

- 4Results, www.4results.pl.

The Netherlands:

- Censor, www.censor.nl.
- HAN Lean-QRM Centrum, www.han.nl/lean-qrmcentrum.
- Kienologic, www.kienologic.nl.
- LeanTeam, www.leanteam.nl.
- QRM Management Center, www.qrm-managementcenter.nl.
- University of Groningen, www.rug.nl/staff/j.riezebos.

Middle East:

- AVIV AMCG, www.avivamcg.com.

Index

Page numbers followed by f and t indicate figures and tables, respectively.

A

Adaptive system, POLCA as, 6–7, 33
Alexandria Industries, 3–4
  company overview, 183–185
  MCT reduction, 275
  POLCA system, 185–187
    design and training, 189–192
    extension to whole factory, 193
    impact on organization, 196–197
    lessons learned, 194–196
    small-scale implementation, 187–189
    startup and early results, 192
Aluminum extrusion operation, 183–197; *see also* Alexandria Industries
Anticipated shortages, 121
Audits
  periodic, 175–177
  random, 178
Authorization Date, 15
  creation/generation
    in absence of ERP system, 62–63
    through ERP system, 64–66
  ensuring availability of, 61–67
Authorization List, 15–16, 16f
  creation/generation
    in absence of ERP system, 63
    through ERP system, 66–67
  ensuring availability of, 61–67

B

Bachman, Jason, 185
Barros, Cláudia, 409
Bergamaschi, D., R., 408
Bosch Hinges, 3–4, 211–230
  company overview, 211
  initial improvement efforts, 212–213
  POLCA system, 216–230
    color-coded visual system, 219–222, 220f
    digital system, 223–230
    initial results from, 222–223
    Load-Based POLCA option, 227–229
  Rhineland model, 215–216
  situation prior to POLCA, 213–214
Boundary control, as sociotechnical success factor, 361
Bullet Batch Card, 204–205
Bullet Card, 130–131, 130f, 165, 194–195
  processes to be followed for proper use of, 131
  schedule changes and, 132–233

C

Campaign Cards, 204
Canadian facility, POLCA implementation in, 157–167
  checking prerequisites, 159–161
  Decision Time rules, 164–165

decision to launch, 165–166
exceptions, 164–165
POLCA card, 162–163
POLCA loops, 161
quantum, 161–162
Capacity
effective use of, 31–32
high-level unit of, 324–325; *see also* Capacity Clusters
Capacity Clusters, 101, 308, 325–351
calculating capacity for work orders, 335–338
capacity available and needed, 329–332
companywide comparisons, 334–335
customized products, 342–345
determination of, 349–351
high-level planning, 345–349
MadTran example, 325–329
quantum for, 338–342
throughput calculation, 332–334
Cards, *see* POLCA cards
Carlson, Todd, 185
Carmo-Silva, Sílvio, 409
Case studies, 181
Alexandria Industries (AI), 183–197
BOSCH Hinges, 211–230
Patheon, 199–210
Preter, 255–261
Provan, 233–243
Szklo, 245–254
Center for QRM at University of Wisconsin–Madison, 157, 286–287
COBACABANA system, 316–321
Color-coded visual work management, 235–237
Commitment of management, 68–69
Compatibility, as sociotechnical success factor, 356–357
Component, unexpected shortage of, 120–126
CONWIP system, 314–316
Costs for POLCA, 29
CPM, *see* Critical Path Method (CPM)
Critical Path Method (CPM), 271
Custom-engineered hinges, 211–230; *see also* Bosch Hinges
Customer-supplier relationships, 40–41
Custom products, 39–40
Cycle Card, 137–140, 138f
purpose of, 139
rules for use of, 140
Cypher, Jeff, 185

D

Deadlocks, in POLCA system, 408–409
Decentralized decision-making, 6, 44
Decision-making process, 17–21
Decision Time, 19, 24
flowchart, 19f, 110f
identifying triggers for, 109–112
Designing POLCA System, 71–117
finishing touches, 109–117
implementation team and, 72
overview, 71
POLCA cards, 99–109
POLCA Chains, 83–98
POLCA loops, 72–83
Design-oriented sociotechnical success factors, 358–362, 359t
boundary control, 361
information flow, 361
minimal critical specification, 358–360
multi-functionality, 362
power and authority, 361–362
support congruence, 362
variance control, 360–361
Digital POLCA, 105, 226–227, 239–241; *see also* PROPOS
Downstream operations, 37

E

Ending points, for POLCA Chains, 89–92
Energy Recovery Ventilation Systems, 274
Enterprise Resource Planning (ERP) system, 6, 30–31
Authorization Dates and List creation
through system, 64–67
without using system, 62–63
HMLVC production and, 31
Enterprise-wide waste, 266–267, 272, 288–289, 290; *see also* Manufacturing Critical-path Time (MCT)

Exceptional situations, dealing with, 119–140
  expediting rush job for customer need, 129–131
  gridlock, 135–140
  holdups in assembly due to non-synchronized arrivals of components, 133–134
  in-process quality problem, 127–128
  machine downtimes, 128–129
  overview, 119–120
  schedule changes, 131–133
  unexpected shortage of component part, 120–126
Expediting rush job, for urgent customer need, 129–131

F

Feedbacks, for qualitative benefits assessment, 173–174
Fernandes, Nuno Octávio, 397, 409
First-Come First-Serve (FCFS) discipline, 96
Flexibility in product routings, 39–40
Formula for number of POLCA cards, 375–379
Fredrick, Scott, 274

G

Gaalman, Gerard J., C., 407, 408
Gantt Chart, 271
Gates, Chuck, 274, 287–288
Generic POLCA (GPOLCA), 397
Germs, Remco, 397, 407
Glass factory, POLCA in, *see* Szklo, case study
Gray Space, 269–270
Gridlock, dealing with, 135–140; *see also* Cycle Card

H

Harrod, Steven, 408, 409
Healthcare operations, MCT in, 279–280
High-Level MRP (HL/MRP), 400
High-level unit of capacity
  Capacity Cluster for, *see* Capacity Clusters
  need for, 323–324
High-mix, low-volume and custom (HMLVC), *see* HMLVC production
HMLVC production, 5, 27–28
  COBACABANA system, 316–321
  CONWIP system, 314–316
  ERP system and, 31
  Kanban system, 309–314
  MRP system and, 302–309
  shift from mass production to, 28
Hopp, Wallace J., 382, 395, 396, 397
Human capabilities, 42–44

I

Implementation of POLCA; *see also* Designing POLCA System; Roadmap, for POLCA launch
  champion for, 72
  cross-functional team for, 72
  operational environment review for, 51–52
  organizational goals and, 52–53
  phases of, 47–48
  plan for rolling out adjustments/corrections, 151–152
  preparation for, 61–69
    commitment of management, 68–69
    ensuring availability of Authorization Dates and List, 61–67
    manufacturing metrics do not conflict with POLCA rules, 67–68
  prerequisites for, 53–61
    effective planning before control, 57–59
    POLCA-enabling cells, 55–57
    resolving issue of missing material or part shortages, 59–61
    shop floor organizational structure, 54–55
  as sociotechnical success factor, 358
  symptoms leading to, 50–51
Information flow, as sociotechnical success factor, 361
In-process quality problem, dealing with, 127–128
Insurance Industry, MCT in, 280

Inventory control, for low-volume products, 37–39
It's About Time (Suri), 57, 287, 293, 294, 298, 309

J

Job
  conceptual illustration, 53–54
  defined, 53
  expediting for urgent customer need, 129–131
  holding off on, 33–36

K

Kanban card system, 201
  function/operation, 37
  in HMLVC context, 309–314
  inventory proliferation for low-volume products, 37–38
  POLCA *vs.*, 22
  shop floor control, 297
Kanet, John J., 408, 409
Krishnamurthy, Ananth, 365

L

Land, Martin J., 316, 407, 408
Larson, Brian, 185
Lead time, 266
  Capacity Clusters and, 349
  challenges to reducing, 286–287
  enterprise-wide waste and, 288–289, 289f
  hedging with, 113
  metrics, 153–154
  for Planning Cell, 85–86
  QRM for, 287–290
  safety through, 113–114
Lean, 285–286
Leanteam, 365
Lödding, Hermann, 399, 403, 408, 409
Loops, *see* Stocking points, in POLCA loops
Low-volume products, inventory control for, 37–39

M

Machine downtimes, dealing with, 128–129
Management, commitment of, 68–69
Manufacturing Critical-path Time (MCT), 67, 265–282
  application
    in manufacturing and non-manufacturing, 265–266
    in various business contexts, 268–269
  business case for, 272–275
    cost, 273–274
    on-time performance, 273
    productivity and market share, 274
    profitability, 274
    quality, 273
    space and office productivity, 275
  definition
    key terms, 267–268
    reasons for, 266–267
  Gray Space, 269–271
  Map, 275–277
    description, 269
    example, 269f
    insights, 270–271
    VSM *vs.*, 277–279, 278f
  in other industries, 279–280
  overview, 265
  QRM Number, 280–282
  in supply management, 279–280
  White Space, 270
Materials requirements planning (MRP), 12, 13, 58
  approaches to avoid problems of, 399–400
  "Dispatch List," 15, 23
  ERP system, 29, 30
  in HMLVC context, 302–309
  POLCA *vs.*, 23
  QRM theory, 297
  time phasing, 399
M-CONWIP system, 397
MCT, *see* Manufacturing Critical-path Time (MCT)
MCT Quick Reference Guide, 268
Metalworking subcontractor, 233–243; *see also* Provan, case study of

Metrics
core set of, 153
lead time, 153–154
tracking and debugging, 153–155, 172–173
Minimal critical specification, as sociotechnical success factor, 358–360
The Monetary Value of Time (Warnacut), 274, 291, 298
Mortágua, João, 409
Motivation for POLCA
operational environment, 51–52
symptoms, 50–51
Multi-functionality, as sociotechnical success factor, 362

N

National Oilwell Varco (NOV), 273
New Product Introduction (NPI), 297
Nicolet Plastics, 274
Non-synchronized arrivals of components, dealing with, 133–134
Norling, Kurt, 185
NOV, *see* National Oilwell Varco (NOV)
NPI, *see* New Product Introduction (NPI)

O

OEE, *see* Overall Equipment Effectiveness (OEE)
Office operations, QRM for, 296
Operational benefits of POLCA, 27–45
avoiding congestion and excessive build-up of WIP, 32–33
being an adaptive system, 33
building on human capabilities, 42–44
controlling inventory for low-volume products, 37–39
costs, 29
customer-supplier relationships, 40–41
decentralized decision-making, 6, 44
ease of implementation, 29–30
effective use of capacity, 31–32
ERP system, 30–31
flexibility in product routings, 39–40
framework for improvement activities, 36
HMLVC production, 27–28
holding off on jobs, 33–36
visual management, 42
Order release decision, 393–404
based on shop floor state, 395–396
quality of capacity signal, 396–398
readiness assurance, 398
timing of, 394–395
Organizational goals, for POLCA implementation, 52–53
Organizational structure, QRM for, 291–294
changes to, 293–294
Outside operations, POLCA Chains for, 92–94, 93f
Overall Equipment Effectiveness (OEE), 200–201
Overlapping loops, 21–22, 21f

P

Paired-cell loops of cards, 16–18; *see also* POLCA cards; POLCA loops
Paired-cell Overlapping Loops of Cards with Authorization (POLCA), *see* POLCA
Patheon, case study of, 199–210
company overview, 199–200
POLCA system
boards and dashboard, 206–209
kaizen workshop, 201–205
life with, 209–210
number of cards, 201–205
senior leadership, 205–206
training, 206
prior to POLCA, 200–201
Periodic audits, 175–177
PERT, *see* Program Evaluation and Review Technique (PERT)
Peters, Jaap, 215
Pharmaceutical environment, 199–210; *see also* Patheon, case study of
Phoenix Products Company, 274
Planned lead times, 399–400
Planning Cell, 73–74
lead time for, 85–86
POLCA Chains in, 83–85, 84f
tasks performed in, 85–86

Planning system
  effective, 57–59
  MRP, *see* Materials requirements planning (MRP)
  prerequisites related to, 58–59
POLCA; *see also* Case studies
  acronym, 4
  bridging gap between theory and practice, 354
  business implications and benefits, 3–4
  designing, *see* Designing POLCA System
  ERP system and, 30–31
  factors contributing to success of, 5–7
  features, 4–5
  implementation, *see* Implementation of POLCA
  launched in in three days, 157–167; *see also* Canadian facility, POLCA implementation in
  operational benefits, 27–45
  order release in, 396
  recommended for situations, 97–98
  trends addressed by, 11
POLCA Board, 17, 17f
POLCA cards, 99–109
  calculating number of cards in each loop, 106–109, 108t
  color-coded, 17, 17f
  description and illustration, 16–18, 17f
  designing, 99–100, 99f
  determining process for returning, 103–106
  formula for number of, 375–379
  hedging with, 113
  paired-cell loops of, 16–18
  quantum, 100–103
  safety through, 114–117
POLCA Chains, 83–98
  defined, 75
  ending points for, 89–92
  for outside operations, 92–94, 93f
  starting point for, 83–85
    in a subset of shop floor, 87–88
POLCA Champion, 72, 144
POLCA-enabling cells, 55–57
POLCA Lite, 251–252, 252f
POLCA loops; *see also* POLCA cards; POLCA Chains
  convention for drawing, 74–75, 75f
  description and illustration, 16, 16f
  determining, 72–83
  overview, 72–73
  stocking points not to be included in, 77–78
  in subset of shop floor, 78–81
Post-implementation activities, 169–180
  audits, 175–178
    periodic, 175–177
    random, 178
  continuous improvement activities, 179–180
  Steering Committee, 169–172
  surveys and feedback, 173–174
  tracking metrics, 172–173
Power and authority, as sociotechnical success factor, 361–362
Preparation, POLCA implementation, 61–69
  commitment of management, 68–69
  ensuring availability of Authorization Dates and List, 61–67
  manufacturing metrics do not conflict with POLCA rules, 67–68
Prerequisites, for POLCA implementation, 53–61
  effective planning before control, 57–59
  POLCA-enabling cells, 55–57
  resolving issue of missing material or part shortages, 59–61
  shop floor organizational structure, 54–55
Preter, case study, 255–261
  cell formation and launch, 258–260
  implementation result, 260–261
  opportunities and challenges, 257–258
  overview, 255
  reviewing situation of, 256–257
  shop floor layout, 257, 257f
  timeaxx software, 255, 258–261
Priority rules, POLCA and, 409
Process-oriented sociotechnical success factors, 355–358, 356f
  compatibility, 356–357
  incompletion, 358
  transitional organization, 357–358

Production control systems; *see also* Order release decision
academic contributions, 381–382
application of POLCA, 384–393
literature, 382
pull, 382, 383
push, 382–383
reasons why companies need, 57–58
Product routings, flexibility in, 39–40
Proesmans, Ben, 233
Program Evaluation and Review Technique (PERT), 271
PROPOS, 223–230, 240, 241
Protzman, Charles W., 315, 407
Provan, case study of, 233–243
color-coded visual work management, 235–237
company overview, 233
POLCA system, 238–243
commercial success, 242–243
customized for use of scarce resources, 240–241
electronic solution, 238–239
personnel development, 241–242
shop floor system, 233–234
stove cell, 234

Q

QRM Number, 280–282; *see also* Manufacturing Critical-path Time (MCT)
advantages, 280–281
calculation, 281t
defined, 280
graph, 282f
Q-ROC, *see* Quick Response Office Cell (Q-ROC)
Quality Deviation Cards, 204
Quality problem, dealing with, 127–128
Quality Tracker, 127–128
Quantum, 408
Capacity Clusters, 338–342
POLCA cards, 100–103
Quick Response Manufacturing (QRM), 285–298, 361
challenges to reducing, 286–287
development of, 285
organizational structure, 291–294
realizing power of time, 287–291
securing future with, 298
system dynamics, 294–296
time-based decisions, 298
unified strategy for whole enterprise, 296–297
variability
dysfunctional, 286
strategic, 286
Quick Response Manufacturing (Suri), 353
Quick Response Office Cell (Q-ROC), 296

R

Random audits, 178
Release-and-Flow (RFPOLCA), 94–96
Authorization Dates, 95–96, 95f
Authorization List, 96
Decision Time rule, 96, 97f
Release list
frequency of being updated, 403
generation of, 399
as order backlog or order book, 401
parameters to be decided upon for, 402–403
time bucket granularity, 403
time horizon, 403
RenewAire, 274
Rhineland model, 215–216
The Rhineland Way (Peters and Weggeman), 215
Riezebos, Jan, 216, 397, 401, 405, 406, 407
Roadmap, for POLCA launch, 143–155
designing checks ensuring rules are being followed, 152–153
planing roll out adjustments and corrections, 151–152
scheduling review sessions and management updates, 151
standard process for secondary activities, 150–151
team and, 143–154
training, 145–149
for core group, 145–148, 147f
for rest of organization, 148–149

S

Safety
  lead time, 113–114
  POLCA cards, 114–117
Safety Card, 103
  aim of, 123
  concept, 122–123
  deployment of, 124–125
  example, 123f
  Shortage Tracker, 125–126, 126f
  unavailability of, 126
Schabel, Tom, 185, 275
Schedule changes
  circumstances under, 131–133
  seamless, dealing with, 131–133
Scheduling algorithms, 354
Scheduling software packages, 5–6
  cost of purchasing/leasing, 29
  drawbacks, 6
  POLCA *vs.* other over implementation, 29
Self-steering teams, POLCA for, 42–44
Shop floor
  implementing POLCA
    in a portion of operation, 78–81
    throughout operations, 73–76
  metrics, 67–68
  release decisions based on state of, 395–396
Shortage; *see also* Safety Card
  anticipated, 121
  components, 121
  dealing with, 120–126
  missing raw material, 120–121
  unanticipated, 121
Shortage Tracker, 125–126, 126f
Simulation game, 365–372
Six Sigma, 285, 286, 287f
Small companies, POLCA in, *see* Preter
Sociotechnical success factors, 353–363
  design-oriented factors, 358–362, 359t
    boundary control, 361
    information flow, 361
    minimal critical specification, 358–360
    multi-functionality, 362
    power and authority, 361–362
    support congruence, 362
    variance control, 360–361
  process-oriented factors, 355–358
    compatibility, 356–357
    incompletion, 358
    transitional organization, 357–358
Sourcing decisions, MCT in, 279
Spaghetti flow, 54
Spearman, Mark L., 382, 395, 396, 397
Steering Committee, 169–172
Stevenson, Mark, 315, 407, 409
Stocking points, in POLCA loops, 77–78
Stove cell, 234
  color coding, 235–236
  production control system in, 237
  success of, 237
Supply management
  MCT in, 279–280
  modern approaches to, 279
  QRM and, 297
Support congruence, as sociotechnical success factor, 362
Suri, Rajan, 365
Surveys, for qualitative benefits assessment, 173–174
System dynamics, QRM and, 294–296
Szklo, case study, 245–254
  cutting operations, 247–248
  factory, 246–247
  finishing processes, 249
  machining department, 248
  manufacturing processes, 246
  overview, 245–246
  POLCA system
    Authorization portion of, 251
    initial design, 249–251
    moving to POLCA Lite, 251–252, 252f
    results of, 253–254

T

Team, 143–144
  decision-making process, 17–21
  self-steering, 42–44
  training, 145–149

Thürer, Matthias, 315, 407, 409
Time, *see* Lead time
Timeaxx software, 255, 258–261, 260f
Time-based decisions, QRM for, 298
Time-phased release decisions, 394–395
effectiveness of, 395
negative effect, 395
positive effect, 395
Time phasing, MRP and, 399
Tracking metrics, 153–155, 172–173
Training, 145–149
for core group, 145–148, 147f
for rest of organization, 148–149
Transitional organization, as sociotechnical success factor, 357–358

U

Unanticipated shortages, 121
Unexpected shortage of component, 120–126

V

Value Stream Mapping (VSM), 277–279, 278f
Vanhees, Luc, 233
Variance control, as sociotechnical success factor, 360–361
Variass Electronics BV, 355
Visual management, 6, 42
color coded, 235–237
VSM, *see* Value Stream Mapping (VSM)

W

Warnacut, Joyce, 274
Wastes, *see* Enterprise-wide waste
Weggeman, Mathieu, 215
White Space, 270
Workcenter, 54–55
Work-in-process (WIP), 4
avoiding excessive build-up of, 32–33
Work order, 53

About the Author

Rajan Suri is Emeritus Professor of Industrial Engineering at the University of Wisconsin–Madison. He received his bachelor's degree from Cambridge University (England) and his MS and PhD from Harvard University.

Dr. Suri is the inventor of the POLCA system and published the first description of POLCA in his book *Quick Response Manufacturing: A Companywide Approach to Reducing Lead Times* (Productivity Press, 1998). Dr. Suri is also the author of over 100 technical publications and several other books including *It's About Time: The Competitive Advantage of Quick Response Manufacturing* (Productivity Press, 2010).

Professor Suri is the Founding Director of the Center for Quick Response Manufacturing (QRM) at the University of Wisconsin–Madison, a consortium in which around 300 companies have worked with the University on understanding and implementing QRM strategies. He is internationally regarded as an expert on the analysis of manufacturing systems, and served as Editor-in-Chief of the *Journal of Manufacturing Systems.*

Dr. Suri has consulted for leading firms including 3M, Alcoa, AT&T, Boeing, Danfoss, Ford, Harley-Davidson, Hewlett Packard, IBM, John Deere, Mitsubishi, National Oilwell Varco, Pratt & Whitney, Rockwell Automation, and TREK Bicycle. Consulting assignments in Europe and the Far East, along with projects for the World Bank, have given him an international perspective on manufacturing competitiveness.

Professor Suri has received awards from the American Automatic Control Council, The Institute of Management Sciences, and the IEEE.

In 1999, Dr. Suri was made a Fellow of the Society of Manufacturing Engineers (SME), and in 2006 he received SME's Albert M. Sargent Progress Award for the creation and implementation of the QRM strategy.

In 2010, Dr. Suri was one of only 10 people to be inducted into the Industry Week 2010 Manufacturing Hall of Fame for the development of QRM and for his long-standing efforts in helping U.S. manufacturers stay competitive in the global marketplace.

You can keep up to date on Professor Suri's activities and publications through his website: rajansuri.com.

For Product Safety Concerns and Information please contact our EU representative GPSR@taylorandfrancis.com Taylor & Francis Verlag GmbH, Kaufingerstraße 24, 80331 München, Germany